Methods in Molecular Biology™

Series Editor
John M. Walker
School of Life Sciences
University of Hertfordshire
Hatfield, Hertfordshire, AL10 9AB, UK

For other titles published in this series, go to
www.springer.com/series/7651

2 4 FEB 2011
DRILL HALL LIBRARY

RT-PCR Protocols

Second Edition

Edited by

Nicola King

University of New England, School of Science and Technology, Armidale, Australia

Editor
Nicola King, Ph.D.
University of New England
School of Science and Technology
Armidale NSW 2351
Australia
nking20@une.edu.au

ISSN 1064-3745 e-ISSN 1940-6029
ISBN 978-1-60761-628-3 e-ISBN 978-1-60761-629-0
DOI 10.1007/978-1-60761-629-0
Springer New York Dordrecht Heidelberg London

Library of Congress Control Number: 2010923597

Printed on acid-free paper

Humana Press is a part of Springer Science+Business Media (www.springer.com)

Preface

Thirty years ago, the investigation of gene sequences and, in particular, disease-causing mutations was a tedious, highly skilled operation; however, this field was revolutionized in the mid-1980s with the introduction and subsequent development of a technique which enabled the reliable amplification of minute quantities of starting DNA into readily detectable levels. Nowadays, this technique, otherwise known as reverse-transcription polymerase chain reaction (RT-PCR), has become routine in most laboratories. Indeed, such is the popularity of RT-PCR that a protocol has been published describing how this could be carried out in a kitchen using common household equipment. The aim of this volume is to translate RT-PCR theory into practice. To achieve this, a comprehensive guide to currently available RT-PCR techniques is given in the form of user-friendly protocols. These protocols contain precise information about all the necessary chemical, consumable, and equipment resources and detailed instructions about how to perform each stage of the different methods. Furthermore, each protocol concludes with a comprehensive notes section, where authors provide helpful hints, trouble-shooting tips, and other must-know information in a format which is accessible to the beginner.

RT-PCR protocols were the subject of an earlier edition in the *Methods in Molecular Biology*™ series, which was published in August 2002. In this second edition, some of the contents of the previous edition are revisited bringing these technologies up to date, for example, competitive RT-PCR, nested RT-PCR, RT-PCR from single cells, and RT-PCR for cloning. In addition, the second volume also describes the newer technologies that have been developed and applied in the last 7 years including multiplex RT-PCR and RT-LATE-PCR. This growth and development is reflected in the wide selection of basic RT-PCR techniques which are presented in the first section entitled "The RT-PCR Detective: Hunting Down the Right Method". Arguably, however, the greatest advances in RT-PCR have come in the field of real-time quantitative RT-PCR, and this, along with all the other means of quantifying PCR products, is explained in the second section, "The RT-PCR Mathematician: Assessing Gene and RNA Expression". Finally, since designing RT-PCR experiments requires both the correct recipe and the best ingredients, the last section, "The RT-PCR Master Chef: Finding the Best Ingredients," is devoted to recent advances in some of the individual elements that go together to make the optimum RT-PCR reaction, e.g. RNA extraction, primer design, and reverse transcription.

This volume is not intended to be a hard-core technical manual that is accessible to a few Nobel Laureate molecular biologists. Rather, the goal is that it should act as a handy companion to anyone who wants to explore the marvels of gene expression. This includes students and their tutors, researchers, laboratory managers, and technologists from many diverse disciplines ranging from biochemistry to zoology and forensics to physiology.

Armidale, NSW ***Nicola King***

Contents

Contributors

ALEJANDRA CASTILLO ALVAREZ • *Biochip Innovations Pty Ltd., 8 Mile Plains, Brisbane, QLD, Australia*
KENNETH M. BALDWIN • *Physiology and Biophysics Department, University of California Irvine, Irvine, CA, USA*
ROSS BARNARD • *School of Chemistry & Molecular Biosciences, The University of Queensland, St. Lucia, QLD, Australia*
VICTORIA BOYD • *CSIRO Livestock Industries, Australian Animal Health Laboratory (AAHL), Geelong, VIC, Australia*
ANUSHKA BROWNLEY • *Complete Genomics, Inc., Mountain View, CA, USA*
RICHARD J. BYERS • *Histopathology Department, Manchester Royal Infirmary, Manchester, UK*
MARTHA Z. CARLETTI • *Department of Molecular and Integrative Physiology, University of Kansas Medical Center, Kansas City, KS, USA*
VINCENT T.K. CHOW • *Human Genome Laboratory, Department of Microbiology, Infectious Diseases Program, Yong Loo Lin School of Medicine, National University Health System, National University of Singapore, Kent Ridge, Singapore*
LANE K. CHRISTENSON • *Department of Molecular and Integrative Physiology, University of Kansas Medical Center, Kansas City, KS, USA*
KRIS DE CLERCQ • *Department of Virology, Veterinary Agrochemical Research Center, Ukkel, Belgium*
ADRIAN M. DI BISCEGLIE • *Division of Gastroenterology and Hepatology, Department of Internal Medicine, Saint Louis University Liver Center, Saint Louis University School of Medicine, St. Louis, MO, USA*
MICHAEL E. DONALDSON • *Environmental & Life Sciences Graduate Program, Trent University, Peterborough, ON, Canada*
HONGXIN FAN • *Molecular Diagnostics Laboratory, Department of Pathology, The University of Texas Health Science Center at San Antonio, San Antonio, TX, USA*
XIAOFENG FAN • *Division of Gastroenterology and Hepatology, Department of Internal Medicine, Saint Louis University Liver Center, Saint Louis University School of Medicine, St. Louis, MO, USA*
STEPHANIE D. FIEDLER • *Department of Molecular and Integrative Physiology, University of Kansas Medical Center, Kansas City, KS, USA*
MAURO GIACCA • *Molecular Medicine Laboratory, International Centre for Genetic Engineering and Biotechnology (ICGEB), Trieste, Italy*
ANA SILVIA GONZÁLEZ-REICHE • *Centro de Estudios en Salud, Centers for Disease Control and Prevention, Universidad del Valle de Guatemala, Regional Office for Central America and Panama, Guatemala City, Guatemala*

THOMAS HAAF • *Institute of Human Genetics, Johannes Gutenberg University, Mainz, Germany*
FADIA HADDAD • *Physiology and Biophysics Department, University of California Irvine, Irvine, CA, USA*
CRISTINA HARTSHORN • *Department of Biology, Brandeis University, Waltham, MA, USA*
HANS G. HEINE • *CSIRO Livestock Industries, Australian Animal Health Laboratory (AAHL), Geelong, VIC, Australia*
ERIC C.H. HO • *Department of Medical Biophysics, Sunnybrook Health Sciences Centre, University of Toronto, Toronto, ON, Canada*
BARBARA W. JOHNSON • *Diagnostic & Reference Laboratory, Arbovirus Diseases Branch, Division of Vector-Borne Infectious Diseases (DVBID), Centers for Disease Control and Prevention (CDC), Fort Collins, CO, USA*
TAMÁS KAROSI • *Department of Otolaryngology Head and Neck Surgery, Medical and Health Science Center, University of Debrecen, Debrecen, Hungary*
NICOLA KING • *School of Science and Technology, University of New England, Armidale, NSW, Australia*
JÓZSEF KÓNYA • *Department of Medical Microbiology, Medical and Health Science Center, University of Debrecen, Debrecen, Hungary*
RICHARD LAI • *Biochip Innovations Pty Ltd., 8 Mile Plains, Brisbane, QLD, Australia; School of Chemistry & Molecular Biosciences, The University of Queensland, St. Lucia, QLD, Australia*
TONY LE • *Department of Research and Development, TriLink BioTechnologies, Inc., San Diego, CA, USA*
LEONG WAI FOOK • *Institute of Molecular and Cell Biology, Proteos, Singapore*
KELVIN LI • *J. Craig Venter Institute, Rockville, MD, USA*
WOLFGANG MANN • *Olympus Life Science Research Europe, München, Germany*
JANELLE P. MOLLICA • *Department of Zoology, LaTrobe University, Bundoora, VIC, Australia*
MARÍA DE LOURDES MONZÓN-PINEDA • *Centro de Estudios en Salud, Centers for Disease Control and Prevention, Universidad del Valle de Guatemala, Regional Office for Central America and Panama, Guatemala City, Guatemala*
MARÍA EUGENIA MORALES-BETOULLE • *Centro de Estudios en Salud, Centers for Disease Control and Prevention, Universidad del Valle de Guatemala, Regional Office for Central America and Panama, Guatemala City, Guatemala*
JENS MÜLLER • *Institute for Experimental Haematology and Transfusion Medicine, University of Bonn, Bonn, Germany*
MELISSA OLIVEIRA-CUNHA • *Hepatobiliary Surgery Unit, Manchester Royal Infirmary, Manchester, UK*
NATASHA PAUL • *Department of Research and Development, TriLink BioTechnologies, Inc., San Diego, CA, USA*
SANDY PINEDA • *Biochip Innovations Pty Ltd., 8 Mile Plains, Brisbane, QLD, Australia*
FIZA RASHID-DOUBELL • *School of Applied Sciences, University of Northumbria, Newcastle upon Tyne, UK*

RYAN S. ROBETORYE • *Molecular Diagnostics Laboratory, Department of Pathology, The University of Texas Health Science Center at San Antonio, San Antonio, TX, USA*
BARRY J. SAVILLE • *Forensic Science Program and Environmental & Life Sciences Graduate Program, Trent University, Peterborough, ON, Canada*
JONATHAN SHUM • *Department of Research and Development, TriLink BioTechnologies, Inc., San Diego, CA, USA*
AJITH K. SIRIWARDENA • *Hepatobiliary Surgery Unit, Manchester Royal Infirmary, Manchester, UK*
ANITA SZALMÁS • *Department of Medical Microbiology, Medical and Health Science Center, University of Debrecen, Debrecen, Hungary*
ISTVÁN SZIKLAI • *Department of Otolaryngology Head and Neck Surgery, Medical and Health Science Center, University of Debrecen, Debrecen, Hungary*
ELISE VANDEMEULEBROUCKE • *Department of Virology, Veterinary Agrochemical Research Center, Ukkel, Belgium*
FRANK VANDENBUSSCHE • *Department of Virology, Veterinary Agrochemical Research Center, Ukkel, Belgium*
LAWRENCE J. WANGH • *Departments of Biology and Molecular Diagnostics, Brandeis University, Waltham, MA, USA*
LORENA ZENTILIN • *Molecular Medicine Laboratory, International Centre for Genetic Engineering and Biotechnology (ICGEB), Trieste, Italy*

Part I

The RT-PCR Detective: Hunting Down the Best Method

Chapter 1

Single Cell RT-PCR on Mouse Embryos: A General Approach for Developmental Biology

Wolfgang Mann and Thomas Haaf

Abstract

Preimplantation development is a complicated process, which involves many genes. We have investigated the expression patterns of 17 developmentally important genes and isoforms in early mouse embryos as well as in single cells of the mouse embryo. The comparison is an excellent example for showing the importance of studying heterogeneity among cell populations on the RNA level, which is being increasingly addressed in basic research and medical sciences, particularly with a link to diagnostics (e.g. the analysis of circulating tumor cells and their progenitors). The ubiquitously expressed histone variant *H3f3a* and the transcription factor *Pou5f1* generated mRNA-derived products in all analyzed preimplantation embryos (up to the morula stage) and in all analyzed blastomeres from 16-cell embryos, indicating a rather uniform reactivation of pluripotency gene expression during mouse preimplantation development. In contrast, genes that have been implicated in epigenetic genome reprogramming, such as DNA methyltransferases, methylcytosine-binding proteins, or base excision repair genes revealed considerable variation between individual cells from the same embryo and even higher variability between cells from different embryos. We conclude that at a given point of time, the transcriptome encoding the reprogramming machinery and, by extrapolation, genome reprogramming differs between blastomeres. It is tempting to speculate that cells expressing the reprogramming machinery have a higher developmental potential.

Key words: Blastomere, Genome reprogramming, Mouse preimplantation embryo, Single cell analysis, RT PCR, Cell heterogeneity

1. Introduction

The genetic analysis of single cells to investigate heterogeneity of cell populations at the DNA and RNA level is a challenging field in biology and medicine. Single cell expression profiling is attracting an increasing number of researchers, and a plethora of cell types have been investigated (1, 2). Based on a new technology called AmpliGrid that amplifies the genetic content of very small amounts of biological material (including single cells), we describe

Nicola King (ed.), *RT-PCR Protocols: Second Edition*, Methods in Molecular Biology, vol. 630,
DOI 10.1007/978-1-60761-629-0_1,

a routine workflow for studying single cells by Reverse Transcription Polymerase Chain Reaction (RT PCR) in an easy and efficient way. The workflow consists of deposition of biological material onto a glass slide (the AmpliGrid), cell lysis by heating, specific RT of the mRNA species of interest, distribution of aliquots of the reaction volume onto other reaction spots, and finally end point PCR. The standard reaction volume for all reactions is one microliter enabling a very effective and reproducible workflow. Furthermore, the biological sample on the glass slide can be monitored by microscopy to confirm the investigation of a single cell and to provide a "What you see is what you amplify" process.

The set of genes for multiplex RT PCR has been designed based on the fact that genome reprogramming during preimplantation development must be regulated by maternal and/or embryonic expression of genes, controlling DNA demethylation and remethylation, chromatin modification, and pluripotency. DNA cytosine-5-methyltransferases (DNMTs) are the enzymes that are responsible for the establishment and maintenance of genomic methylation patterns (3). Inactivation of *Dnmt1*, *Dnmt3a*, and *Dnmt3b* caused hypomethylation of the genome and embryonic lethality, indicating that these genes are essential for early mammalian development (4, 5). DNMT1 has a high affinity for hemimethylated sites that are generated transiently during DNA replication and is largely responsible for maintaining methylation patterns throughout cell divisions. DNMT3A and DNMT3B are thought to function in de novo methylation (6, 7). DNMT3L lacks the motifs for transmethylation 3 processes but stimulates de novo methylation through direct interactions with DNMT3A and DNMT3B (8). MECP2, MBD1, MBD2, MBD3, and MBD4 comprise a family of nuclear proteins, which include a 5-methyl-CpG binding domain (MBD), and with the exception of MBD3, all bind specifically to methylated DNA sequences (9). They recruit repressor and chromatin remodeling complexes to methylated DNA (i.e., promoters), causing gene silencing (10). Interestingly, MBD2 has been reported to function as a demethylase (11), but this could not be confirmed. Instead, active demethylation of nonreplicating DNA may be achieved by base excision repair, a highly conserved mechanism for the repair of damaged DNA bases (12). Type 1 DNA glycosylases remove modified (i.e. methylated) bases and leave an apurinic/apyrimidinic (AP) site which is then cleaved by the AP endonuclease APEX1 (13). The DNA backbone at the AP site is subsequently repaired by DNA polymerase(s) and ligase(s). It is noteworthy that MBD4 also functions as a mismatch glycosylase (14). Maternal cytoplasmic factors control active demethylation in the fertilized egg and can even reprogram a somatic cell nucleus that has been transferred into an oocyte (15). The mouse embryonic genome is already activated at the two-cell stage (16). One key factor for the control of early embryonic development is the transcription factor POU5F1 (OCT4). The fertilized mouse oocyte and early cleavage stages contain residual maternal *Pou5f1* mRNA. Embryonic *Pou5f1* expression is initiated at the four to eight-cell stage (17). *Pou5f1* appears to be required in blastomeres throughout all preimplantation stages to maintain pluripotency (18). The histone variant H3F3A can replace histone H3 and thus epigenetically mark chromatin. It shows an asymmetric distribution between male and female pronucleus in the mouse zygote and is present in the nuclei of all preimplantation stages up to the blastocyst (19).

2. Materials

2.1. Mouse Embryos and Single Embryonic Cells

1. Six to eight week old female C57BL6/J mice
2. Male C57BL6/J mice
3. Hormones: 7.5 IU of eCG(equine chorionic gonadotropin) and hCG (human chorionic gonadotropin)
4. Acid Tyrode's solution (Sigma-Aldrich)
5. PBS containing PVP (1 L): 8 g of NaCl, 0.2 g of KCl, 1.44 g of Na_2HPO_4, 0.24 g of KH_2PO_4, adjust pH to 7.4; Polyvinylpyrrolidone is added to a final concentration of 1 mg/ml; aliquots are prepared on the day of use
6. Glass micropipettes
7. Bunsen burner, flame polished glass pipettes

2.2. Reverse Transcription of RNA, Amplification (RT PCR) and Primers

1. OneStep RT PCR Kit (Qiagen)
2. Multiplex PCR kit (Qiagen)
3. Primers, oligonucleotides
4. Staining solution: 1% $AgNO_3$ in double distilled H_2O
   (a) Developer: 0.1% formaldehyde solution, alkaline, containing NaOH (3 pellets/100 ml of solution)
   (b) AmpliGrid 480F (Beckman Coulter)
5. AmpliSpeed Cycler, for running PCR in the 1 µl regime on AmpliGrid (Beckman Coulter)
6. Covering solution for running PCR (Beckman Coulter)
7. Double distilled H_2O

2.3. Gel Electrophoresis of Amplicons

1. 8% polyacrylamide/bis solution (37:1), containing TEMED (N,N,N,N′-Tetramethylethylenediamine, toxic)

2. 50×TBE: 540 g of Tris(hydroxymethyl)methylamine, 275 g of Orthoboric acid, 100 ml of 0.5 M ethylenediaminetetraacetic acid (EDTA) pH 8.0, combined in 1 L of double distilled H_2O
3. Loading dye: 0.25% bromphenolblue, 30% glycerol in H_2O

3. Methods

In principle, the single cell platform (AmpliGrid glass slide) allows for the optical inspection of the sample (cell or cells, sperm, etc.) that is to be amplified. Quality and integrity of the biological material can be monitored during the preparation process under the microscope even after staining procedures or other kinds of manipulation (see Note 1).

3.1. Preparation of Single Cells and Embryos

1. Superovulate 6–8 week old female C57BL6/J mice by intraperitoneal injection of 7.5 IU eCG and 44–48 h later by injection of 7.5 IU hCG.
2. Mate the hormone treated females with C57BL6/J males.
3. Flush one-cell, two-cell, four-cell, and 16-cell stage embryos from the oviducts at 10–12 h, 33 h, 41 h, and 60 h after fertilization, respectively (see Note 2).
4. Isolate individual blastomeres by washing the 16-cell embryos in calcium- and magnesium-free PBS containing 1% PVP followed by incubation in acid Tyrode's solution (pH 2.3) for about 10 s at 37°C.
5. When thinning of the zona pellucida is observed, transfer the embryos into a drop of fresh PBS/PVP. Remove the residual zona by gentle pipetting with a glass micropipette (see Note 3).
6. Flush zona-free embryos with PBS/PVP and then incubate in a new drop of PBS/PVP for 5 min at 37°C.
7. Disaggregate blastomeres by pipetting up and down with a flame-polished glass micropipette (see Note 4).
8. Collect blastomeres in a 15-μl drop of PBS medium at 37°C (Fig. 1a, b).

3.2. RT PCR on AmpliGrid Chips

1. Transfer either individual preimplantation embryos or individual blastomeres in approximately 0.3 μl of PBS each onto an AmpliGrid slide (Fig. 1c) (see Note 5).
2. Carry out reverse transcription with RT primer mixture (Table 1) using a final primer concentration of 0.3 μM each, using buffer and polymerases of the OneStep RT PCR Kit according to the kit's instructions (see Note 6).

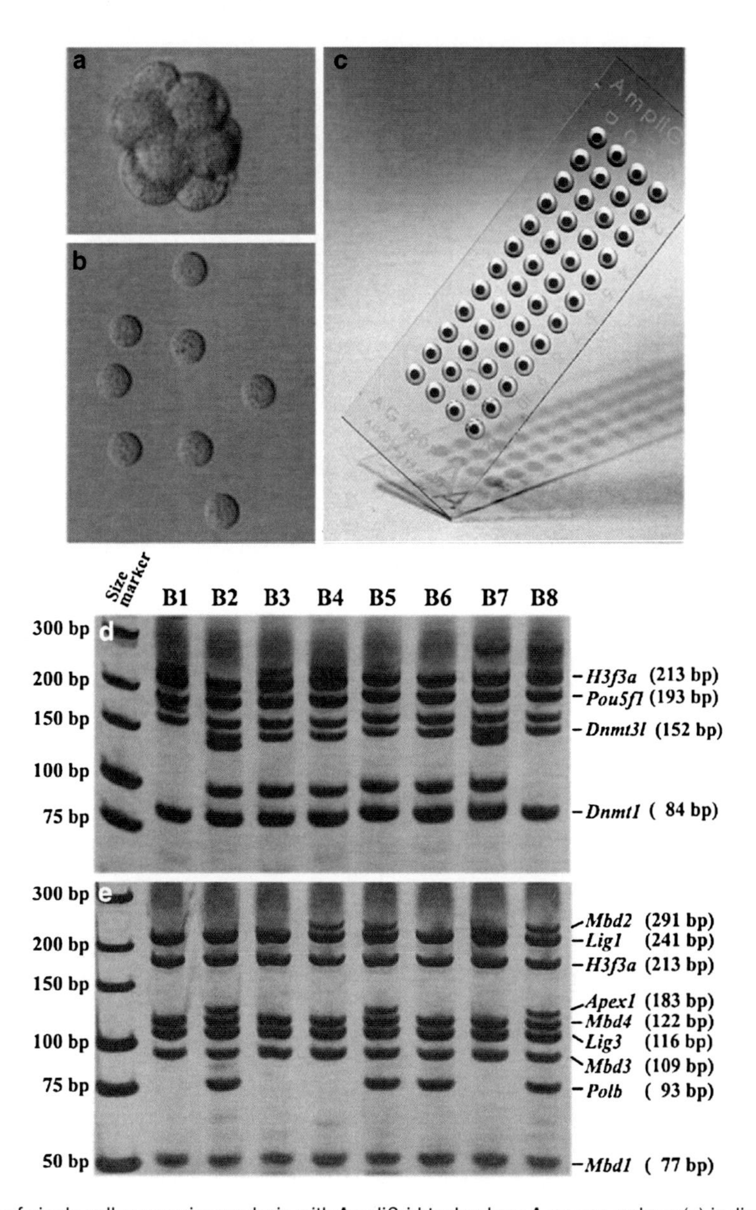

Fig. 1. Overview of single cell expression analysis with AmpliGrid technology. A mouse embryo (**a**) is disaggregated into individual blastomeres (**b**) which are placed onto discrete reaction sites of the AmpliGrid (**c**) for multiplex RT PCR. The PCR products from individual blastomeres (B1 to B8) are then resolved by gel electrophoresis (**d**, **e**). To facilitate the resolution of gene products on minigels, PCR was performed in two different reactions, each using one quarter of the RT product as starting material and amplifying products of the reference gene *H3f3a* and 8 study genes/isoforms. In the top gel (**d**), the bands representing *H3f3a*, *Pou5f1*, *Dnmt3l,* and *Dnmt1* are indicated on the right side. The unlabeled bands represent different amplicons of *Dnmt3a* and *Dnmt3b*. In the bottom gel, each band represents an individual gene, as indicated on the right. Please note that *H3f3a* gene products are detected in each cell, whereas most other gene products, i.e. *Polb* are only present in a subset of cells

Table 1
Primer mixtures for RT and PCR reactions. This primer set facilitates investigation of the expression of the DNA methyltansferase genes: *Dnmt1*, *Dnmt3a* (two isoforms), *Dnmt3b* (four isoforms), and *Dnmt3l*, the methycytosine-binding protein genes *Mbd1*, *Mbd2*, *Mbd3*, and *Mbd4*, the base excision repair genes *Apex1*, *Lig1*, *Lig3*, and *Polb*, the pluripotency marker *Pouf51*, and the histone variant *H3f3a*. The oligonucleotide sequences of the primers have been published elsewhere (20)

RT primer mixture	PCR primer mixture 1		PCR primer mixture 2	
Dnmt3a 1er	*Dnmt3a* 1el+r	270 bp	*Lig1* 1cl+r	241 bp
Dnmt3a 2gr	*Dnmt3a* 2gl+r	165 bp	*Lig3* 1fl+r	116 bp
Dnmt3b 1br	*Dnmt3b* 1bl+r	102 bp	*Mbd2* 1el+r	291 bp
Dnmt3b 2br	*Dnmt3b* 2bl+r	138 bp	*Mbd1* 1dl+r	77 bp
Dnmt3b 3pr	*Dnmt3b* 3pl+r	145 bp	*Mbd4* 1dl+r	122 bp
Dnmt1 1gr	*Dnmt1* 1gl+r	84 bp	*Mbd3* 1dl+r	109 bp
Dnmt3l 1fr	*Dnmt3l* 1fl+r	152 bp	*Apex1* 1bl+r	183 bp
Pou5f1 1cr	*Pou5f1* 1cl+r	193 bp	*Polb* 1bl+r	93 bp
Lig1 1cr	*H3f3a* l+r	213 bp	*H3f3a* l+r	213 bp
Lig3 1fr				
Mbd2 1er				
Mbd1 1dr				
Mbd4 1dr				
Mbd3 1dr				
Apex1 1br				
Polb 1br				
H3f3a r				

3. Prepare a master mix as described above calculating 1.05 × number of 1 μl reactions that are to be run on the AmpliGrid (see Note 7).
4. Add 1 μl of master mix to each reaction spot (negative control, positive control, respective biological material such as single cell).
5. Cover each sample with 5 μl of covering solution to prevent evaporation.
6. Place the prepared slide on the AmpliSpeed Cycler.

7. Run the appropriate thermal profile for reverse transcription on a preheated cycler, e.g., 60°C for 30 min.
8. Remove the 1 µl reaction volume (containing cDNA) from the chip, dilute 1:4 in double distilled H_2O and pipette 1 µl onto empty spots of an AmpliGrid.
9. Evaporate template cDNA to dryness on two different reaction sites of a second AmpliGrid (see Note 8).
10. Carry out two PCR reactions with the multiplex PCR kit and primer mixture 1 and 2 respectively (Table 1). Prepare a master mix with all buffer substances, Taq polymerase and either PCR primer mix 1 or 2 at a final primer concentration of 0.3 µM each according to the kit's manual. Use the Q solution at 0.3-fold concentration.
11. Apply 1 µl each of master mix 1 and 2 to the two different reaction sites with the template and cover with 5 µl of sealing solution.
12. Carry out PCR on a slide cycler with an initial denaturation step of 95°C for 10 min and 40 cycles of 94°C for 30 s, 60°C for 60 s, and 72°C for 60 s, and a final 10 min extension step at 72°C (see Note 9).

3.3. Gel Electrophoresis of Amplicons and Gel Staining

1. Add 4 µl of loading dye onto each reaction site by carefully pipetting on top of the mineral oil.
2. Let the aqueous phase of 1 µl combine with the loading dye.
3. Pipette the resulting 5 µl from underneath the mineral oil and transfer to one slot of the polyacrylamide gel.
4. Load the remaining samples to be analyzed in this way.
5. Run electrophoresis at 10 V/cm for 30 min.
6. Remove gel from gel electrophoresis chamber.
7. Stain the gel by soaking in staining solution for 5 min.
8. Wash the gel twice in ddH_2O.
9. Visualize the Amplicons by shaking the gel in developer for 3–5 min.
10. Stop developing by washing the gel two times in ddH_2O.

3.4. Documentation

1. Transfer results in the presence or absence of transcripts onto the result table (see Note 10).

4. Notes

1. This feature is unique to the AmpliGrid technology and is the key for reliable results for single cell RT PCR. Whereas in a conventional tube, an optical control of the target cell to be

amplified is impossible by means of a standard microscope; this is possible on the AmpliGrid platform. Depending on the specific application, it will be extremely useful to check for the integrity of the nucleus, the total morphology of the cell, stained parts of the cell, etc.

2. Only "high quality" embryos (by morphological criteria) were used for further analysis. Embryos with developmental delay or signs of fragmentation were discarded.
3. For its optical properties, glass pipettes are used for micromanipulation rather than plasticware.
4. It is important to work with micropipettes that are free of any nucleic acid or protein (like RNAses). In order to avoid contamination in one of the subsequent reaction steps, it is recommended to flame-polish the micropipette for a few seconds using a Bunsen burner.
5. This is a glass chip with a surface structure for the specific positioning of one single microliter on each of 48 discrete reaction sites. For a detailed description of the structuring, see product description on www.advalytix.com.
6. The Q solution was used at onefold concentration.
7. As in conventional tube-based reactions, the total volume of mastermix should be 5% more compared to the exact calculated volume to ensure that the number of reactions can definitely be processed. This is especially true for very low volumes like 1 μl PCR, and we recommend to take into account a factor of 1.05 × number of 1 μl reactions.
8. The end concentration of the reaction mix must not be diluted by additional liquid that might have stayed on the reaction sites. Carefully monitor the drying process of cDNA that will take a few minutes at room temperature and normal humidity.
9. Success rates of single cell PCR will strongly depend on the set up of the total workflow for single cell manipulation as well as on the assay optimization. The total workflow for analyzing single cells under optical control is a new technology that enables routine investigation. For an illustration of the workflow, see Fig. 1. Secondly, PCR conditions (time and temperatures during PCR) will influence the outcome dramatically. For a technical evaluation of sensitivity, some of the genes of interest should be cloned into expression vectors and the amount of synthetic gene sequences that can be detected should be known (see Fig. 2).
10. In principle, the analysis of gene expression in single cells results in the presence/absence of data transcripts in a respective single cell when carrying out end-point-PCR. Analysis of this kind of data has been shown to result in meaningful findings

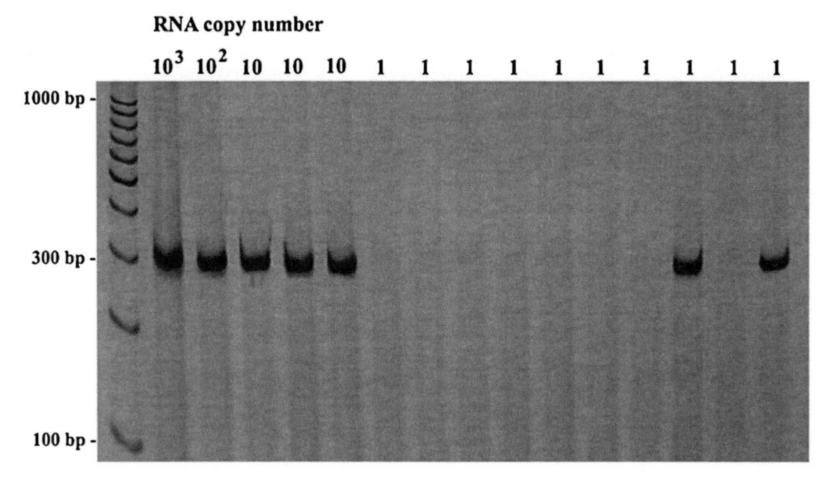

Fig. 2. RNA dilution series of a test RNA, showing the sensitivity of AmpliGrid technology. On this particular gel, the expected amplification product was detected in all trials with 1,000, 100, and 10 RNA copies and in two of ten trials with one RNA copy. When we generated RNAs of four of our study genes by in vitro transcription of *Apex1*, *Lig1*, *Lig3,* and *Mbd2* cDNA clones, a gene-specific product was detected in all trials with 1,000 or 100 RNA copies, in 68–89% of trials with ten copies, and in 4–24% of trials with one copy. The water controls were always negative. When working with limiting dilution (on average one copy per reaction), approximately 30% of the reactions may not contain any RNA molecule, whereas another 30% may contain two or more RNA molecules. Thus, the detection rate with one to two RNA copies is around 20%, with ten copies around 80%, and with 100 or more copies 100%

after Principal Components Analysis (20). Of course we are aware of the fact that real-time PCR can be regarded as state-of-the-art technology and can also be applied on single cells. Both methods are comparable in sensitivity (20). However, routine analysis demands technologies that are easy to handle and reveal a high level of accuracy in quality control. The optical properties of the AmpliGrid make this an excellent tool for achieving this purpose. In addition, one could speculate that gene expression in the early stages of the developing embryo might be characterized by large differences in mRNA level of the respective transcripts that are relevant for cell programming. Binary signatures of these cells are easy to interpret and can be generated in high throughput.

References

1. Levsky JM, Singer RH (2003) Gene expression and the myth of the average cell. Trends Cell Biol 13:4–6
2. Raj A, Peskin CS, Tranchina D, Vargas DY, Tyagi S (2006) Stochastic mRNA synthesis in mammalian cells. PLOS Biol 4:e309
3. Bestor TH (2000) The DNA methyltransferases of mammals. Hum Mol Genet 9:2395–2402
4. Li E, Bestor TH, Jaenisch R (1992) Targeted mutation of the DNA methyltransferase gene results in embryonic lethality. Cell 69:915–926
5. Okano M, Bell DW, Haber DA, Li E (1999) DNA methyltransferases Dnmt3a and Dnmt3b are essential for de novo methylation and mammalian development. Cell 99:247–257
6. Aoki A, Suetake I, Miyagawa J, Fujio T, Chijiwa T, Sasaki H, Tajima S (2001) Enzymatic properties of de novo-type mouse DNA (cytosine-5) methyltransferases. Nucl Acids Res 29:506–512

7. Chen T, Ueda Y, Dodge JE, Wang Z, Li E (2003) Establishment and maintenance of genomic methylation patterns in mouse embryonic stem cells by Dnmt3a and Dnmt3b. Mol Cell Biol 23:5594–5605
8. Suetake I, Shinozaki F, Miyagawa J, Takeshima H, Tajima S (2004) DNMT3L stimulates the DNA methylation activity of Dnmt3a and Dnmt3b through a direct interaction. J Biol Chem 279:27816–27823
9. Ballestar E, Wolffe AP (2001) Methyl-CpG-binding proteins. Targeting specific gene repression. Eur J Biochem 268:1–6
10. Jaenisch R, Bird A (2003) Epigenetic regulation of gene expression: how the genome integrates intrinsic and environmental signals. Nat Genet 33:245–254
11. Bhattacharya SK, Ramchandani S, Cervoni N, Szyf M (1999) A mammalian protein with specific demethylase activity for mCpG DNA. Nature 397:579–583
12. Jost JP (1993) Nuclear extracts of chicken embryos promote an active demethylation of DNA by excision repair of 5-methyldeoxycytidine. Proc Natl Acad Sci USA 90:4684–4688
13. Evans AR, Limp-Foster M, Kelley MR (2000) Going APE over ref-1. Mutat Res 461:83–108
14. Sancar A, Lindsey-Boltz LA, Unsal-Kacmaz K, Linn S (2004) Molecular mechanisms of mammalian DNA repair and the DNA damage checkpoints. Annu Rev Biochem 73:39–85
15. Wilmut I, Beaujean N, de Sousa PA, Dinnyes A, King TJ, Paterson LA, Wells DN, Young LE (2002) Somatic cell nuclear transfer. Nature 419:583–586
16. Schulz RM (1993) Regulation of zygotic gene activation in the mouse. BioEssays 8:531–538
17. Palmieri SL, Peter W, Hess H, Schöler HR (1994) OCT-4 transcription factor is differentially expressed in the mouse embryo during establishment of the first two extraembryonic cell lineages involved in implantation. Dev Biol 166:259–267
18. Nichols J, Zevnik B, Anastassiadis K, Niwa H, Klewe-Nebenius D, Chambers I, Schöler HR, Smith A (1998) Formation of pluripotent stem cells in the mammalian embryo depends on the POU transcription factor Oct4. Cell 95:379–391
19. Torres-Padilla ME, Bannister AJ, Hurd PJ, Kouzarides T, Zernicka-Goetz M (2006) Dynamic distribution of the replacement histone variant H3.3 in the mouse oocyte and preimplantation embryos. Int J Dev Biol 50:455–461
20. May A, Kirchner R, Müller H, Hartmann P, El Hajj N, Tresch A, Zechner U, Mann W, Haaf T (2009) Multiplex RT-PCR expression analysis of developmentally important genes in individual mouse preimplantation embryos and blastomeres. Biol Reprod 80:194–202. doi:10.1095/biolreprod.107.064691

Chapter 2

Poly(A) cDNA Real-Time PCR for Indicator Gene Measurement in Cancer

Melissa Oliveira-Cunha, Ajith K. Siriwardena, and Richard J. Byers

Abstract

Microarray gene expression profiling has identified gene signatures or "*Indicator*" genes predictive of outcome in many cancer types including lymphoma, and more recently pancreatic cancer. This has identified novel and powerful diagnostic and prognostic and generically applicable markers, promising more specific diagnosis and treatment, together with improved understanding of pathobiology. There is now an urgent need to translate these signatures to clinical use. However, gene microarrays rely on relatively large amounts of fresh starting tissue obviating measurement of *Indicator* genes in routine practice, and there is a need for development of another, simple, robust, relatively inexpensive and sensitive method for their translation to clinical use. We have piloted the use of real-time PCR measurement of specific prognostic genes, so called "Indicator" genes, in globally amplified polyA cDNA for this purpose.

Poly(A) PCR coordinately amplifies cDNA copies of all polyadenylated mRNAs, thereby generating a PCR product (polyA cDNA) whose composition reflects the relative abundance of all expressed genes in the starting sample. Poly(A) PCR enables global mRNA amplification from picogram amounts of RNA and has been routinely used to analyse expression in small samples including single cells. The poly(A) cDNA pool generated is also indefinitely renewable and as such represents a "molecular block". Real-time PCR measurement, using gene-specific primers and probes, of the expression levels of specific *Indicator* genes then allows gene signatures to be detected within the poly(A) cDNA, thereby enabling expression profiling of very small amounts of starting material. This chapter details this method as applied to fresh and paraffin embedded tissue and to pancreatic juice. In this chapter, we have concentrated on application of the method to pancreatic cancer, but the generic nature of the method renders it applicable to any cancer type, thereby representing a novel platform for cancer diagnosis across all tumour types.

Key words: PolyA cDNA, RT-PCR, Quantitative RT-PCR, Pancreatic cancer and pancreatic cancer gene expression

1. Introduction

Early detection of pancreatic cancer, in high-risk populations with chronic pancreatitis, will enable earlier surgical intervention and improved outcome. However, the pancreas is difficult to biopsy,

Nicola King (ed.), *RT-PCR Protocols: Second Edition*, Methods in Molecular Biology, vol. 630,
DOI 10.1007/978-1-60761-629-0_2, © Springer Science+Business Media, LLC 2010

making such detection difficult. Pancreatic cells are shed into the pancreatic juice and may provide a surrogate marker for pancreatic disease. We have used real-time PCR to test the hypothesis that detection of gene signatures predictive of tumour behaviour, so called *Indicator* genes, measured by globally amplified Poly(A) cDNA in pancreatic ductal juice from patients with pancreatic cancer will provide a high degree of correlation with *Indicator* genes detected by the same method in the tumour samples of these patients. If this hypothesis is correct, the second, logical hypothesis to follow from this is that globally amplified Poly(A) cDNA from the pancreatic juice of patients in a pre-defined high-risk population – long-standing chronic pancreatitis – can be used to quantify the risk of development of malignancy in these populations.

1.1. Basic Considerations in Reverse Transcription Real-Time PCR

Reverse Transcription Real-Time PCR (RT-PCR) was first described by Powel and colleagues in 1987 (1) and is a powerful method for the detection of minimal quantities of any given RNA. It relies upon the successful application of two separate enzymatic activities (1, 2):

1. Reverse transcription of RNA
2. Amplification of cDNA resulting from step 1

1.1.1. Reverse Transcription

The reverse transcription step can be primed using specific primers, random hexamers or oligo-dT primers, and the choice of primers requires careful consideration. The use of specific mRNA-specific primers decreases background priming, whereas the use of random and oligo-DT primers maximises the number of mRNA molecules that can be analysed from a small sample of RNA (2). The efficiency of transcription of a given template differs not only with the source of reverse transcriptase but also with the divalent cation used in the reaction. All reverse transcriptases require divalent cations in order to function, and their degree of functionality is influenced by the species of cation. The metals Mg^{2+} and Mn^{2+} are the divalent cations most often determined to be optimum with RT of avian virus origin preferring Mg^{2+} and those of mammalian virus origin preferring Mn^{2+} (3).

1.1.2. Polymerase Chain Reaction

The theoretical concept of amplifying DNA by a cycling process using DNA polymerase and oligonucleotide primers was first described by Kleppe and colleagues (4). There is a wide choice of enzymes that differ in their processivity, fidelity, thermal stability and ability to read modified triphosphates. The most commonly used enzyme Taq DNA polymerase has 5′–3′ nuclease activity but lacks a 3′–5′ proof reading exonuclease activity. This makes its use less advisable when fidelity is the main consideration (2). In addition, Mg^{2+} and dNTP concentrations require strict control, because Mg^{2+} affects enzyme activity and imbalanced dNTP mixtures will reduce polymerase fidelity. In addition, Mg^{2+} increases the melting temperature™ of double-stranded DNA, and forms soluble complexes with dNTP to produce the actual substrate that the polymerase recognises. Therefore, high concentrations of dNTPs interfere with polymerase activity and affect primer annealing by reducing free Mg^{2+} (2).

1.1.3. RT-PCR

Real-time PCR instruments measure the kinetics of product accumulation in each PCR tube. Generally, no product is detected during the first few cycles as the fluorescent signal is below the detection threshold of the instrument. However, most of the combinations of machine, fluorescence and reporter are capable of detecting the accumulation of amplicons before the end of the exponential amplification phase (5).

Conceptually, there are two fundamentally different means of quantification – absolute and relative. The absolute quantification determines tile input copy number, usually by relating the PCR signal to a standard curve and relative quantification. For relative quantification, the accurate determination of a gene's expression levels requires tile use of a ubiquitously expressed internal control, such as a housekeeping gene (6). An ideal housekeeping gene should maintain a constant RNA transcription level in all cell types and tissues, it should be resistant to regulative factors and its expression should not vary. However, differences in the expression levels of endogenous reference genes have been reported between different tissues and pathological states (6).

RT-PCR is based on the use of the 5′ nuclease activity of *Taq* polymerase to cleave a non-extendible hybridization probe during the extension phase of PCR. The approach uses dual-labelled fluorogenic hybridization probes; one fluorescent dye serves as a reporter (FAM (i.e., 6-carboxyfluorescein)) and its emission spectra is quenched by the second fluorescent dye, TAMRA (i.e., 6-carboxy-tetramethylrhodamine). The nuclease degradation of the hybridization probe releases the quenching of the FAM fluorescent emission, resulting in an increase in peak fluorescent emission at 518 nm. The use of a sequence detector (ABI Prism) allows measurement of fluorescent spectra of all 96 wells of the thermal cycler continuously during the PCR amplification. Therefore, the reactions are monitored in real time (7).

1.2. Global Gene Amplification: Poly(A) RT-PCR Method

A number of techniques were reported in the late 1980s for the analysis of DNA and RNA transcripts from samples as small as one cell (8, 9). These techniques relied on the use of specific primers and were only able to amplify a single known sequence at a time. Furthermore, the procedures destroyed the starting sample, precluding further analysis. Poly(A) reverse transcription polymerase chain reaction was developed by Brady et al. (10, 11) to overcome this.

Global cDNA amplification allows marker genes to be detected at very low levels, improving sensitivity, whilst also providing an improved level of specificity through the ability to analyse multiple genes. Critically, global cDNA allows analysis of very small numbers of cells. This is of particular relevance since it will enable measurement of gene expression profiles from cells isolated from pancreatic juice, providing a relatively non-invasive method for diagnosis.

The technique, illustrated in Fig. 1, has three steps namely Poly(A) primed reverse transcription to produce a first strand cDNA for each mRNA, polyadenylation of the first strand cDNA,

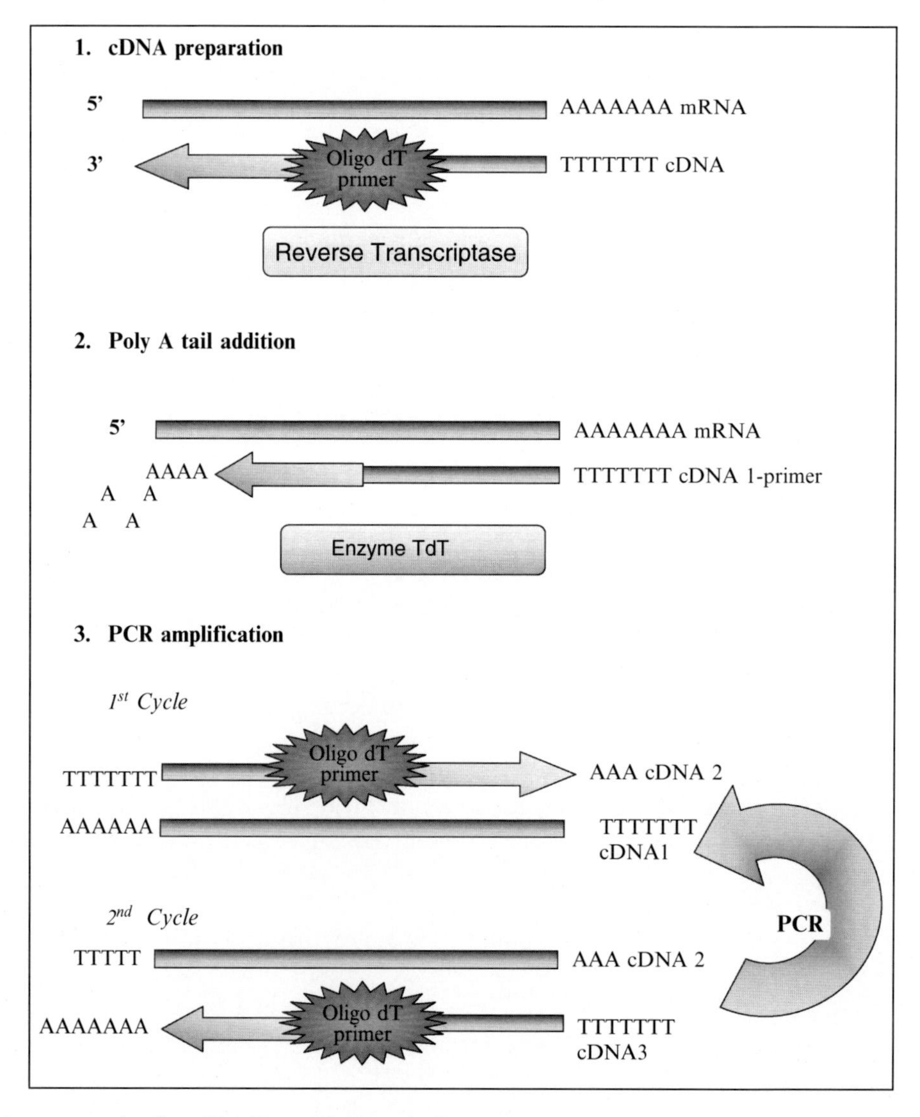

Fig. 1. Summary and outline of Poly(A) amplification of mRNA (Adapted from Brady and Iscove 1993). (11)

rendering it defined at both ends, and polymerase chain reaction using a Poly(dT) primer, which anneals to the Poly(A) site on the first strand cDNA (11).

The initial transcription can be performed from a defined quantity of purified RNA, or from fresh cells, from which the RNA is released by lysis in the presence of RNase inhibitors. Importantly, the first reverse transcription step is time limited, to 15 min, termination being by heat denaturation of the reverse transcriptase enzyme. Since complementary bases are added to the first strand cDNA at the rate of approximately one every 3 s, this limits the size of the cDNA to approximately 300 bp, although in practice, the cDNA produced lies between 100 and 500 bp, yielding a smear when visualized by gel electrophoresis. As a result, each of the cDNA molecules is amplified to the same efficiency in the polymerase chain reaction, while if the cDNAs were of different lengths, some would be preferentially represented in the PCR product (11, 12). Consequently, the relative abundance of the cDNAs in the PCR reflects that of the mRNA from which they were generated rather than the kinetics of PCR of cDNAs of differing size. Formal tests have shown that mRNA representing as little as 0.025% of the total is represented with fidelity (10). Furthermore, since the entire mRNA repertoire of the cell is represented by cDNA defined at one end with a Poly(A) sequence, and at the other by a Poly(dT) sequence, it is indefinitely renewable by Poly(dT) primed PCR, so-called Poly(A) PCR, as such represents a "molecular block", whilst multiple rounds of PCR can generate microgram quantities of cDNA from the single cells used. These properties make the technique extremely powerful for gene expression profiling (12).

The technique has been most extensively used in the analysis of both normal (13–15) and leukaemic haematopoiesis (16). The PolyA cDNA can then be assayed for the expression of particular genes, either by hybridization with cDNA microarrays or by real-time PCR. Real-time PCR measurement, however, enables more precise quantitation of the expression levels of specific *Indicator* genes. As such it is better suited for measurement in clinical practice, whilst focusing on diagnostically relevant *Indicator* genes. This approach thereby enables gene signatures to be detected within very small amounts of pathological material (17). As such it can be applied to the investigation of virtually any problem (12).

2. Materials

2.1. RNA Extraction from Pancreatic Juice

1. Isogen-LS
2. RNeasy Mini Kit (Qiagen)
3. RNase-Free DNase set (Qiagen)

4. 100% Chloroform
5. 100% Ethanol, ACS grade or higher quality
6. Sterile RNase-free set of pipette tips
7. Eppendorf tubes
8. Microcentrifuge

2.2. RNA Extraction from Pancreatic Tissue

1. RNeasy Mini Kit (Qiagen)
2. RNase-Free DNase set (Qiagen)
3. RNA later (Qiagen)
4. 14.3 M β-mercaptoethanol (β-ME)
5. 100% Ethanol, ACS grade or higher quality
6. Sterile RNase-free set of pipette tips
7. Microcentrifuge
8. Materials for disruption of tissue
9. Mixer Mill MM 300 homogenizer
   (a) Stainless bead (5 mm)
   (b) Blunt needle and scalpel

2.3. RNA Extraction from Formalin-Fixed Paraffin Embedded Tissue

1. RecoverAll™ Total nucleic acid isolation kit (Ambion)
2. 100% Xylene, ACS grade or higher quality
3. 100% Ethanol, ACS grade or higher quality
4. 2-ml microcentrifuge tubes
5. Sterile RNase-free set of pipette tips
6. Heat blocks set at 37°C, 50°C, and 95°C
7. Microtomes
8. Microcentrifuges

2.4. RNA and DNA Quantification

1. Spectrophotometer – Nanodrop® 1,000 A (Thermo Scientific)
2. DEPC water
3. Sterile RNase-free set of pipette tips

2.5. Reverse Transcription

2.5.1. First Strand cDNA Synthesis

1. cDNA/Lysis buffer
   (a) 1 ml of 5× reverse transcriptase buffer (250 mM Tris–HCl, 40 mM $MgCl_2$, 150 mM KCl, 5 mM dithiothreithol, pH 8.5 (20°C)
   (b) 10 μl of Bovine Serum Albumin (BSA), molecular biology grade
   (c) 250 μl of 10% Nonidet P-40
   (d) 3.55 ml of UltraPure™ DEPC water
   (e) Stored in 200-μl aliquots

2. Primer mix (PM):
   (a) 800 μl of 2.5 mM dNTP mix (mixture of dATP, dCTP, dGTP, dTTP – each 2.5 mM)
   (b) 24 μl of NOT I-24 (200 μM)
   (c) Stored in 50-μl aliquots
   NOT I-24 sequence: 5′ – CAT CTC GAG CGG CCG C $(T)_{24}$ – 3′
3. RNase inhibitor (40U/μl)
4. AMV reverse transcriptase (20U/μl)
5. Sterile RNase-free set of pipette tips
6. 0.2-ml thin walled PCR tubes
7. Thermal cycler

2.5.2. Poly(A) Tail Addition

1. 2× Tailing buffer
   (a) 1 ml of 5× tailing buffer (1M potassium cacodylate, 125 mM Tris–HCl, BSA, 1.25 mg/ml, pH 6.6 at 25°C)
   (b) 25 μl of 100 mM dATP (final concentration 1 mM)
   (c) 1.475 μl of water UltraPure™ DEPC water
   (d) Stored in 200-μl aliquots
2. Terminal deoxynucleotidyl transferase (TdT) (400 U/μl)
3. 0.2-ml thin walled PCR tubes
4. Sterile RNase-free set of pipette tips
5. Thermal cycler

2.5.3. Poly(A) PCR

1. MMM buffer
   (a) 1 ml of 10× Ex-Taq buffer (20 mM Tris–HCl (pH8.0), 100 mM KCl, 0.1 mM EDTA, 1 mM DTT, 0.5% Tween 20, 0.5% Nonidet P-40, 50% Glycerol)
   (b) 375 μl of 25 mM dNTP (from 100 mM stock dNTP)
   (c) 10 μl of Bovine serum albumin
   (d) 100 μl of 10% Triton X-100
   (e) 20 μl of 1 M $MgCl_2$
   (f) 2.35 ml of UltraPure™ DEPC water
2. NOT-I 24 oligonucleotide
   NOT I-24 sequence: 5′-CAT CTC GAG CGG CCG C $(T)_{24}$-3′
3. Ex-Taq DNA Polymerase – 5 U/μl (TaKaRa Bio Inc)
4. 0.2-ml thin walled PCR tubes
5. Sterile RNase-free set of pipette tips
6. Thermal cycler

2.6. GAPDH Specific Gene PCR

1. GAPDH primer design

The gene sequence can be accessed using BLAST (National Centre of Biotechnology Information, 2003) and the primers designed towards the 3′ end of each gene using a web design programme (Stanford Genomic Resources, 2003). PCR is performed using these primers using standard PCR conditions.

2.7. Reamplification of PolyA PCR

1. UltraPure™ DEPC water;
2. 10× Ex-Taq Buffer (20 mM Tris–HCl (pH 8.0), 100 mM KCl, 0.1 mM EDTA, 1 mM DTT, 0.5% Tween 20, 0.5% Nonidet P-40, 50% Glycerol)
3. dNTP mix (mixture of dATP, dCTP, dGTP, dTTP – each 2.5 mM)
4. Ex-Taq DNA polymerase – 5 U/µl (TaKaRa Bio Inc)
5. NOT I-16 (200 µM);
   Sequence of NOT I-16: 5′-GCG GCC GC $(T)_{16}$-3′
6. Sterile RNase-free set of pipette tips
7. Thermal cycler

2.8. Gel Electrophoresis

1. 5× TBE buffer (RNA use)
   (a) Tris(hydroxymethyl)methylamine – 54 g
   (b) Orthoboric acid – 27.5 g
   (c) 0.5 M ethylenediamintetraacetic acid (EDTA) pH 8.0 – 20 ml
   (d) DEPC water – 800 ml
2. DNA loading buffer
   (a) Glycerol (50%)– 50 ml
   (b) 0.2M EDTA, pH 8 – 50 ml
   (c) Bromophenol blue (0.05%) – 0.05 g
3. 100 bp ladder

2.9. Real-Time PCR

1. TaqMan® Universal PCR Master Mix (Applied Biosystems)
2. Custom TaqMan® Gene Expression Assays Primers and probes (Applied Biosystems)
3. UltraPure™ DEPC water
4. Sterile RNase-free set of pipette tips
5. 96-well optical plates
6. Adhesive
7. ABI Prism 7700 sequence detection system (Applied Biosystems).

3. Methods

3.1. RNA Extraction from Pancreatic Juice

1. Before starting the protocol in the laboratory under laminar flow hood, pre-cool the centrifuge to 4°C for 20 min. (see Note 1)
2. Place pancreatic juice in an Eppendorf tube and add Isogen-LS in the following proportions: 750 μl of Isogen-LS to 250 μl of liquid sample. Incubate the samples for 5 min at room temperature. (see Note 2)
3. Chloroform is then added (200 μl), and the sample is shaken vigorously (by hand) for 15 s. Then, samples are incubated for 2 min at room temperature. Centrifuge samples at 12,000 × *g* for 15 min at 4°C.
4. After centrifugation, three phases can be identified in the tube: clear first phase (RNA) and second and third phases corresponding to DNA and to proteins. The aqueous phase should be transferred to a new tube and add an equal volume of 70% of ethanol. Mix the sample by pipetting it up and down. (see Note 3)
5. Transfer the mixture to an RNeasy mini column placed in a 2-ml collection tube; a maximum 700 μl of the sample can be added at each spin. Close the tube and centrifuge for 15 s at 8,000 × *g*. Discard the flow through. Then, add 350 μl of buffer RW1, as supplied by RNeasy Mini Kit, to the RNeasy column. Close the tube and centrifuge for 15 s at 8,000 × *g* to wash the spin column membrane. Discard the flow through. (see Note 4)
6. The next step is DNase treatment. Mix 10-μl DNase I stock solution to 70-μl Buffer RDD, as supplied by RNeasy Mini Kit, gently inverting the tube, and centrifuge briefly to collect residual liquid from the sides of the tube. Mixing should only be carried out by gently inverting the tube, not by vortexing. Add DNase I incubation mix (80 μl) directly to the RNeasy spin column membrane, and place on the bench top (20–30°C) for 15 min. Then, add 350-μl Buffer RW1, as supplied by RNeasy Mini Kit, to the RNeasy spin column, close the lid gently and centrifuge for 15 s at 8,000 × *g*. Discard the flow through.
7. Transfer RNeasy column to a new 2-ml collection tube, and add 500 μl of RPE buffer, as supplied by RNeasy Mini Kit. Close the tube and centrifuge for 15 s at 8,000 × *g* to wash the column. Discard the flow through. Note that RPE buffer is supplied as a concentrate and ensure that four volumes of 100% of ethanol are added before using it. The above step is repeated using 500 μl of RPE buffer, centrifuge for 15 s and then discard the flow through.

8. Centrifuge the RNeasy column for 1 min at 8,000 ×*g* to dry the silica membrane. For RNA elution, the RNeasy column is placed to a new collection tube. Pipette 30 μl of RNase-free water directly onto the RNeasy silica gel membrane and centrifuge for 8,000 ×*g* for 1 min. The RNA extracted should then be analysed by Nanodrop and stored at –80°C.

3.2. RNA Extraction from Fresh Pancreatic Tissue

1. In a sterile collection round bottom tube pipette 600 μl of RNA lysis buffer – RLT (highly denaturating guanidine-thiocynate-containing buffer, RNeasy Mini Kit), add one stainless bead (5 mm) and 30 mg of pancreatic tissue that was stored in RNAlater.
2. Homogenize the mixture using the Mixer Mill MM 300 for 4 min at 20 Hz. (see Note 5). Subsequently, the sample is centrifuged (including the bead) for 3 min at maximum speed. Transfer the supernatant to a new sterile tube by pipetting and immediately adding 600 μl of 75% ethanol to the cleared lysate.
3. Transfer the lysate including any precipitate to an RNeasy spin column, centrifuge for 30 s at 8,000 ×*g* and discard the flow-through. Into the same column, add 350 μl of Buffer RW1 and centrifuge for 30 s at 8,000 ×*g*. Subsequently, the sample is treated with DNase. Add DNase I incubation mix (80 μl) directly to the RNeasy spin column membrane, and place on the bench top (20–30°C) for 15 min. Then, add 350 μl of Buffer RW1 to the RNeasy spin column, close the lid gently and centrifuge for 15 s at 8,000 ×*g*. Discard the flow through.
4. Transfer RNeasy column to a new 2-ml collection tube and add 500 μl of RPE buffer. Close the tube and centrifuge for 15 s at 8,000 ×*g* to wash the column. Discard the flow through. The above step is repeated using 500 μl of RPE buffer, centrifuge for 15 s, after which the flow through is discarded.
5. Place the spin column in a new collection tube and add 30 μl of RNase-free water directly to the spin column membrane and centrifuge for 1 min at 8,000 ×*g*. The RNA extracted is then analysed by Nanodrop and stored at –80°C.

3.3. RNA Extraction from Formalin-Fixed Paraffin Embedded Tissue

1. Cut five tissue sections at 20 μm thickness each using a microtome. Place samples in a sterile 2-ml microcentrifuge tube. Add, 1 ml of 100% xylene and vortex briefly to mix. Place samples in a heating block for 3 min at 50°C to melt the paraffin. After heating, centrifuge the samples for 2 min at room temperature and maximum speed to pellet the tissue. Remove xylene without disturbing the pellet. (see Note 6)

2. The next step consists of washing the pellet twice with 1 ml of 100% ethanol followed by air drying the pellet for 15 min. The ethanol washes remove xylene from the sample and accelerate drying of the tissue. For RNA isolation only, pipette 400 μl of digestion buffer (supplied by RecoverAll™ Total Nucleic Acid Isolation kit) and 4 μl of protease in each sample. The tube is swirled to mix and immerse the tissue. Incubate the sample in a heat block for 3 h at 50°C.
3. After the incubation step, add 480 μl of isolation additive to each sample and vortex to mix. Then, pipette 1.1 ml of 100% ethanol onto each sample. The solution should become clear at this point. For each sample, place a filter cartridge in one of the collection tubes and add 700 μl of the sample/ethanol mixture onto the filter cartridge. Centrifuge the filter cartridges at 8,000 × *g* for 60 s to pass the mixture through the filter. Repeat this step until all the sample mixture has passed through the filter.
4. Wash the samples by adding 700 μl of wash 1 solution onto the filter cartridge and then centrifuge the filter cartridges at 8,000 × *g* for 30 s. Discard the flow through and pipette 500 μl of wash 2/3 solution, then centrifuge the filter cartridges at 8,000 × *g* for 60 s.
5. The last step includes nuclease digestion and final nucleic acid purification. The following solutions are combined to make the DNase mix: 6 μl of 10× DNase buffer, 4 μl of DNase and 50 μl of nuclease-free water. Pipette the DNase mix to the centre of each filter cartridge. Incubate the samples for 30 min at room temperature.
6. After nuclease digestion, wash the samples with Wash 1 solution and then Wash 2/3 as described above. After centrifugation, transfer the filter cartridge to a fresh collection tube and add 30 μl of elution solution, or nuclease-free water, heated to 95°C, to the centre of the filter. Sit the sample at room temperature for 1 min followed by centrifugation at 8,000 × *g* for 1 min. The volume of collected eluate is then analysed by Nanodrop and stored at –80°C.

3.4. RNA Quantification

1. The Nanodrop machine is set up and calibrated according to the manufacturer's instructions. DEPC water is used as blank reference.
2. Pipette 1.5 μl sample of RNA onto the end of a fibre optic cable (the receiving fibre). A second fibre optic cable (the source fibres) is then brought into contact with the liquid sample causing the liquid to bridge the gap between the fibre optic ends. The instrument is controlled by special software run from a personal computer (PC), and the data is logged in an archive file on the PC. (see Note 7)

3. The ratio 260/230 is used as a measure of nucleic acid purity. The 260/230 values for "pure" nucleic acid are often higher than the respective 260/280 values. The final preparation of total RNA should be free of DNA and proteins, for which a 260/280 ratio of 1.6–1.9 should be expected. (see Note 8)

3.5. Reverse Transcription Poly(A) PCR

3.5.1. First Strand cDNA Synthesis

1. Fresh first lysis buffer (FFL) is prepared by adding 4 μl of RNase inhibitor and 4 μl of fresh dilute primer mix (PM) to 192 μl of cDNA/Lysis buffer.
2. In 0.2-ml thin walled PCR tubes, pipette 10 μl of FFL buffer. Add 1 μl of DNase treated RNA isolated from pancreatic juice or tissue or 2.5 μl of DNase treated RNA isolated from formalin-fixed paraffin embedded tissue (FFPET) samples. Place the tubes in a thermal cycler. (see Table 1).
3. Following this, add 0.5 μl of AMV reverse transcriptase into each sample tube. For detection of genomic DNA contamination, prepare an RT negative control by substitution of the AMV reverse transcriptase enzyme with 0.5 μl of DEPC water. Place the mixtures in a thermal cycler. (see Table 2).

3.5.2. Poly(A) Tail Addition

1. Prepare a master mixture of 0.5 μl of TdT per sample and 11.5 μl of tailing buffer. An equal volume of this master mix is added to each tube, and then heated in a thermal cycler as before. (see Table 3).

Table 1
Thermal cycling conditions for denaturing mRNA for Poly(A) RT-PCR

Step	Denaturation	Incubation	Cooling
Temperature	65°C	25°C	4°C
Time	1 min	3 min	15 min

Table 2
Thermal cycling conditions for first strand synthesis in Poly(A) RT-PCR

Step	Activate enzyme	Denature enzyme	Cooling
Temperature	37°C	65°C	4°C
Time	15 min	10 min	min

Table 3
Thermal cycling conditions for Poly(A) tail addition

Step	Activate enzyme	Denature enzyme	Cooling
Temperature	37°C	65°C	4°C
Time	15 min	10 min	min

Table 4
Thermal cycling conditions for Poly(A) PCR

Step	Taq polymerase activation	Denaturation	Annealing	Extension
Number of cycles	X1	X25		
Temperature	94°C	94°C	42°C	72°C
Time	3 min	1 min	2 min	6 min
Linked to				
Number of cycles		×25		
Temperature		94°C	42°C	72°C
Time		1 min	1 min	2 min

3.5.3. Poly(A) PCR

1. Prepare a Poly(A) PCR mix in the following ratios: 40 μl of MMM buffer mixed with 2.7 μl of NOT-I 24 oligonucleotide and 1.5 μl of Ex-Taq DNA Polymerase.
2. For each 7.5 μl aliquot of tailed cDNA, add 15 μl of the poly(A) PCR mix into the tube, mix thoroughly and then pulse centrifuge to collect all reagents at the bottom.
3. Place the tubes in a thermal cycler. (see Table 4).
4. The Poly(A) cDNA products might be either visualised on a 1.5% agarose gel, used to amplify specific genes using PCR, reamplified or stored at – 20°C.
5. Figure 2 illustrates a gel with Poly (A) cDNA smears.

3.6. GAPDH Specific Gene PCR

Specific PCR for GAPDH is performed using 3′ end directed primers and standard PCR conditions. Input cDNA is prepared for GAPDH PCR by dilution of 2 μl of the polyA cDNA in 198 μl of DEPC water, followed by brief vortexing to give a 1 in 100 dilution to use in the PCR reaction. For confirmation that the method was efficient, GAP PCR of pancreatic juice and tissue samples was performed and positive samples had a clear band (Fig. 3).

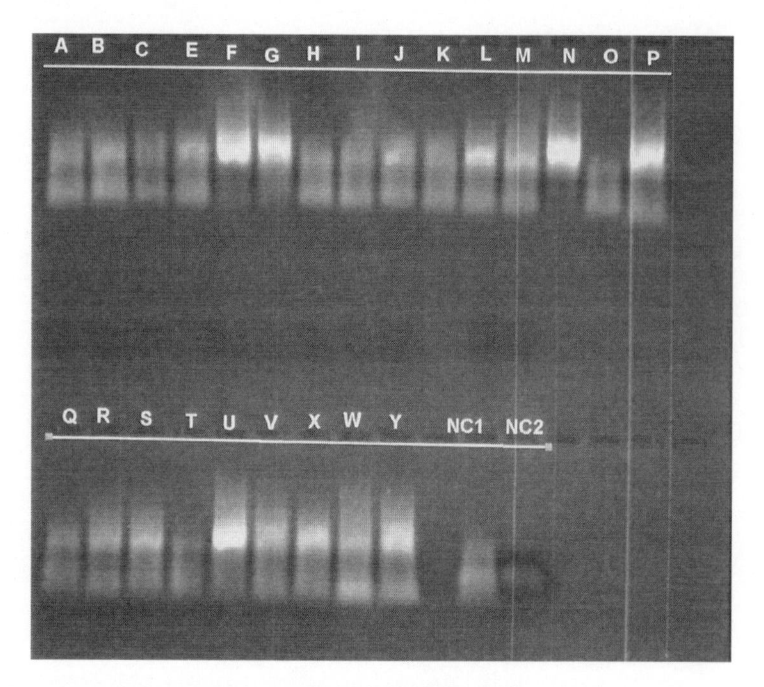

Fig. 2. Electrophoresis gel of PolyA PCR – Poly A cDNA smears

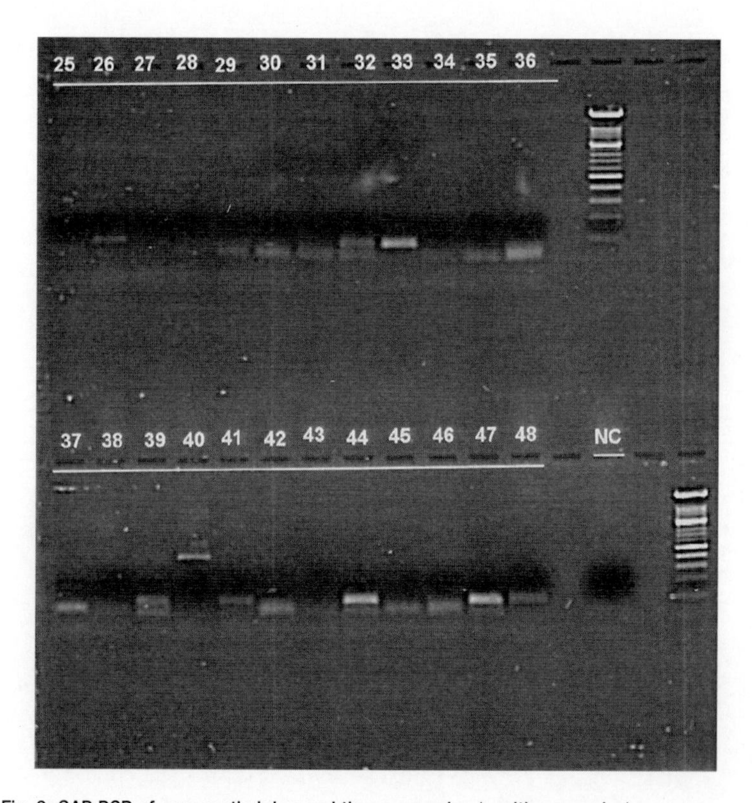

Fig. 3. GAP PCR of pancreatic juice and tissue samples (positive samples)

3.7. Reamplification of PolyA cDNA

1. Prepare dilutions of PolyA cDNA by adding 2 µl of sample to 198 µl of distilled H_2O.
2. Prepare a PCR mixture using the following reagents per sample:
   (a) 17.5 µl of DEPC water;
   (b) 2.5 µl of 10× Ex-Taq Buffer;
   (c) 2.5 µl of dNTP mix;
   (d) 0.25 µl of NOT I-16 (200 µM);
   (e) 0.25 µl of Ex-Taq DNA polymerase.
3. Add in each 0.2-ml thin walled tube 23 µl of the PCR mixture and 2 µl of each sample of 1:100 polyA cDNA.
4. Place samples in thermal cycler (see Table 5).
5. The reamplified polyA cDNA can be run on 1.5% gel, used to amplify specific genes or stored for later use at –20°C.

3.8. Gel Electrophoresis of PCR Products

3.8.1. Agarose Gel Preparation – 1.5% gel

1. Set up a gel tray in the electrophoresis room.
2. Measure 100 ml of 0.5× TBE into a cylinder and pour into a conical flask.
3. Add 1.5 g of agarose powder into the flask and swirl to mix.
4. Cover the flask with cling film and pierce small holes in the top. The mixture is microwaved at full power for 3 min with intermittent swirling every 30 s.
5. Pipette 2 µl of ethidium bromide into the flask.
6. The flask is cooled down under running water and swirling to prevent the gel from setting.
7. Pour the gel slowly into the tray to avoid air bubbles and leave it to cool and set for 30 min.

3.8.2. Gel Electrophoresis

1. Label the required number of 0.6-ml thick walled tubes and add 3 µl of loading buffer into each tube.
2. Pipette 10 µl of PCR product into the corresponding labelled tube with loading buffer.

Table 5
Thermal cycling conditions for Reamplification of Poly A cDNA

Step	Denaturation	Annealing	Extension
Number of cycles	×25		
Temperature	94°C	42°C	72°C
Time	1 min	1 min	2 min

3. Place the gel into the tank and submerge.
4. Pipette samples into the wells and in the final well, pipette 4 μl of 100 bp ladder with its loading buffer.
5. Set to 100 V and run for 90 min; when the gel has run far enough, take it to UV transilluminator.

3.9. Real-Time PCR

3.9.1. Primers and Probes

1. Taqman™ PCR primers and probes were designed for three indicator genes (CPB1, TFF2, and KLK3) and three housekeeping genes (GAPDH, B2M, and PSMB6) using Custom TaqMan® Gene Expression Assays. (see Note 9)
2. The reference and Indicator genes are demonstrated in Table 6. (see Note 10)

3.9.2. Real-Time PCR

1. Set up a 96 well plate.
2. Prepare a PCR mixture using the following proportions per reaction:

Table 6
Reference (RG) and Indicator genes (IG)

Gene	Accession number	Primers
GAPDH (RG)	NM_002046	1. ACCACTTTGTCAAGCTCATTTCCT 2. GAGGTCCACCACCCTGTTG 3. CTGTAGCCAAATTCG
B2M (RG)	NM_004048	1. CCGCATTTGGATTGGATGAATTCC 2. ACATTTTGTGCATAAAGTGTAAGT-GTATAAGC 3. CAAGCAAGCAGAATTT
PSMB6 (RG)	NM_002798	1. CAAGGAAGAGTGTCTGCAATTCAC 2. CTGCCAGGCGGATCACT 3. CCCGCTCCATGGCCAA
CPB1 (IG)	NM_001871	1. TCGGTGAGAACAATGCTGAGTT 2. CGAAGTTCAAAGGTGAAGGAA-TATCTGA 3. CAGTCGTCAGAGCCC
TFF2 (IG)	NM_005423	1. GGTCCCCTGGTGTTTCCA 2. ACCTCCATGACGCACTGATC 3. CCCCTCCCAAAGCAA
KLK3 (IG)	NM_001648 variant 1	1. GCGGCGGTGTTCTGG 2. GTCCACACACTGAAGTTTCTTTGG 3. CAGCTGCCCACTGCAT

(a) Taq Man®Universal PCR Mastermix 12.5 μl

(b) Primer and probes – 1.25 μl

(c) DEPC water – 8.75 μl

(d) cDNA – 2.5 μl

3. Inspect all the wells for uniformity of volume, and note which wells do not appear to contain the proper volume.
4. Seal the plate with the appropriate cover.
5. Centrifuge the tube briefly to spin down the contents and to eliminate any air bubbles from the reaction mix.
6. Place samples in thermal cycler ABI Prism 7500 Real-Time PCR System sequence detection system for a standard run, as specified below:

   Select Standard Run for 96-Well Standard Plates (Protocol of Applied Biosystems). (see Table 7)

3.9.3. Calculating Fold Change

Comparative quantification according to $2^{\Delta\Delta Ct}$ method

$$\text{Ratio}: \frac{2^{Ct_{BGOI} - Ct_{AGOI}}}{2^{Ct_{BREF} - Ct_{AREF}}} = \frac{2^{\Delta Ct_{GOI}}}{2^{\Delta Ct_{REF}}} = 2^{\Delta\Delta Ct}$$

Ct: number of cycles required to obtain threshold level
B: sample B
A: sample A
GOI: gene of interest
REF: reference or housekeeping gene

3.9.4. Normalisation

1. The expression levels for three housekeeping genes (GAPDH, B2M, and PSMB6) are measured by RT-PCR in each sample.
2. Copy numbers obtained for the mean of the three housekeeping genes in each sample is divided by the highest in all samples resulting in a normalisation correction factor (18). (see Note 11)

Table 7
Thermal cycling conditions for RT-PCR

Step	Ampli Taq gold activation	Denaturation	Annealing/extension
	×1	×40	
Temperature	95°C	95°C	60°C
Time	10 min	15 s	60 s

3. Following real time PCR amplification and quantification of the selected genes, this factor is then used for normalisation of expression levels of each gene measured.

4. Notes

1. Unlike DNA, RNA is extremely delicate. For that reason, RNA extraction must be of the highest quality and should be free of DNA and nucleases. RNases are stable and active. Therefore, when working with RNA, designate an area of the laboratory only for RNA , keep the aliquots on ice and utilize aseptic techniques. The use of RNase-free pipettes, tubes and water is an essential practice.
2. Isogen-LS (Wako) combined with RNeasy (Qiagen) were chosen as kits for total RNA purification in pancreatic juice. This combination provides a high relative yield and purity of intact RNA.

   A number of the equipment and reagents used for the RNA purification are very hazardous. This includes phenol that is a component of Isogen-LS, chloroform, ethanol and ethidium bromide and all must be handled with caution.

   Phenol is absorbed through the skin, and it can affect the central nervous system and cause damage to liver and kidneys. It is also a mutagen, and there is evidence that phenol may be a reproductive hazard. Phenol should always be handled in a fume cupboard to ensure that the short-term exposure limit is not exceeded and no naked flames are in use nearby. Always wash your hands after handling phenol, even if gloves are used (19).
3. Isogen-LS is a homogenous solution, mainly consisting of phenol and guanidine thiocynate. The extraction steps which are employed induce a liquid phase separation, resulting in a conveniently tinted aqueous RNA layer. Following the protocol, RNA, DNA and protein are separated from one another in one series of steps. A good pipetting technique is essential for a good RNA purification and large tips should be used initially before finishing with 20 μl tips.
4. The RNeasy kit (Qiagen) combines the selective binding properties of a silica-based membrane with the speed of microspinning technology. A specialized high-salt buffer system allows RNA to bind to a silica membrane.
5. This step inactivates RNases and ensures purification of intact RNA.

6. For RNA purification from paraffin-embedded tissue, the samples were cut from the interior of the paraffin block to minimize nucleic acid damage by exposure to the atmosphere during storage.
7. The usual method for assessing purity of RNA is absorbance spectrometry. In this study, the isolated RNA was quantified using NanoDrop® ND-1000 spectrophotometer (NanoDrop Technologies, Delaware, USA). The NanoDrop® ND-1000 is a full-spectrum spectrophotometer that measures 1 μl samples with high accuracy and reproducibility. It utilizes a patented sample retention technology that employs surface tension alone to hold the sample in place. In addition, the ND-1000 has the capability to measure highly concentrated samples without dilution.
8. The development of the Agilent Bionalyzer and Agilent RNA 6000 Nano Kit are now unquestionably the method of choice for analyzing RNA preparations destined to quantitative RT-PCR assays. Another quality assessment technique is preparing a formaldehyde agarose gel. The assessment of RNA quality by gel electrophoresis and analysis of 28S and 18S rRNA bands requires a large volume of RNA when preparing it. The protocol can be found in Bustin & Nolan (20).
9. Custom TaqMan Gene Expression Assays permit good performances in both absolute quantification and relative quantification of gene expression profiling on any gene. The assays use the TaqMan® (5′ nuclease) chemistry for amplifying and detecting the *target* in cDNA samples. All assays are developed using proprietary assay-design software. Custom TaqMan Gene Expression Assay contains:
   (a) Two target-specific primers
   (b) One TaqMan® MGB FAM™ dye-labelled probe
10. When using comparative quantification, there is an assumption that the reference gene expression is constant, and that the efficiency of amplification for each gene of interest tested is the same. Therefore, the reference gene selection is essential. GeNorm and Normfinder are software that can aid in the choice of an appropriate reference gene. Also, a guideline to reference gene selection can be found in Radonic et al. (21)
11. The normalisation technique can be done according to:
   (a) Known amount of extracted RNA
   (b) Mass or volume of extracted tissue
   (c) One known and not regulated reference gene
   (d) According to a reference gene index containing more than three reference genes.

References

1. Powell L, Wallis S, Pease R, Edwards Y, Knott T, Scott J (1987) A novel form of tissue-specific of RNA processing produces apolipoprotein B-48 in intestine. Cell 50:831–40
2. Bustin SA (2000) Absolute quantification of mRNA using real-time reverse transcription polymerase chain reaction assays. J Mol Endocrinol 25:169–93
3. Wu A, Gallo R (1975) Reverse Transcriptase. CRC Crit Rev Biochem 3:289–347
4. Kleppe K, Ohtsuka E, Kleppe R, Molineux I, Khorana HG (1971) Studies on polynucleotides. XCVI. Repair replications of short synthetic DNA's as catalyzed by DNA polymerases. J Mol Biol 56:341–61
5. Logan J, Edwards K (2004) An overview of real-time PCR platforms. In: Edwards K, Logan J, Saunders N (eds) Real-time PCR - an essential guide. Horizon Bioscience, Wymondham, pp 13–29
6. Rubie C, Kempf K, Hans J, Su T, Tilton B, Georg T, Brittner B, Ludwig B, Schilling M (2005) Housekeeping gene variability in normal and cancerous colorectal, pancreatic, esophageal, gastric and hepatic tissues. Mol Cell Probes 19:101–9
7. Schena M, Shalon D, Davis RW, Brown PO (1995) Quantitative monitoring of gene expression patterns with a complementary DNA microarray. Science 270:467–70
8. Gyllensten UB, Erlich HA (1988) Generation of single stranded DNA by the polymerase chain reaction and its application to direct sequencing of the HLA-DQA locus. Proc Natl Acad Sci USA 85:7652–6
9. Rappolee DA, Wang A, Mark D, Werb Z (1989) Novel method for studying mRNA phenotypes in single or small number of cells. J Cell Biochem 39:1–11
10. Brady G, Billia F, Knox J, Hoang T, Kirsch IR, Voura EB, Hawley RG, Cumming R, Buchwald M, Siminovitch K (1995) Analysis of gene expression in a complex differentiation hierarchy by global amplification of cDNA from single cells. Curr Biol 5:909–22
11. Brady G, Iscove NN (1993) Construction of cDNA libraries from single cells. Methods Enzymol 225:611–23
12. Byers RJ, Hoyland JA, Dixon J, Freemont AJ (2000) Subtractive hybridization – genetic takeaways and the search for meanings. Int J Exp Pathol 81:391–404
13. Cumano A, Paige CJ, Iscove NN, Brady G (1992) Bipotential precursors of B cells and macrophages in murine fetal liver. Nature 356:612–5
14. Trevisan M, Iscove NN (1995) Phenotypic analysis of murine long-term hemopoietic reconstituting cells quantitated competitively in vivo and comparison with more advanced colony-forming progeny. J Exp Med 181:93–103
15. Quesenberry PJ, Iscove NN, Cooper C, Brady G, Newburger PE, Stein GS, Stein JS, Reddy GP, Pearson-White S (1996) Expression of basic helix-loop-helix transcription factors in explant hematopoietic progenitors. J Cell Biochem 61:478–88
16. Hoang T, Paradis E, Brady G, Billia F, Nakahara K, Iscove NN, Kirsch IR (1996) Opposing effects of the basic helix-loop-helix transcription factor SCL on erythroid and monocytic differentiation. Blood 87:102–11
17. Byers RJ, Sakhinia E, Joseph P, Glennie C, Hoyland JA, Menasce LP, Radford JA, Illidge T (2008) Clinical quantitation of immune signature in follicular lymphoma by RT-PCR based gene expression profiling. Blood 111:4764–70
18. Sakhinia E, Faranghpour M, Liu Yin JA, Brady G, Hoyland JA, Byers RJ (2005) Routine expression profiling of microarray gene signatures in acute leukaemia by real-time PCR of human bone marrow. Br J Haematol 130:233–48
19. Bustin SA, Nolan T (2004) Good laboratory practice. In: Bustin SA (ed) A–Z of quantitative PCR. International University Line, La Jolla, pp 126–39
20. Bustin SA, Nolan T (2004) Template handling, preparation and quantification. In: Bustin SA (ed) A–Z of Quantitative PCR. International University Line, La Jolla, pp 198–99
21. Radonic A, Thulke S, Mackay IM, Landt O, Siegert W, Nitsche A (2004) Guideline to reference gene selection for quantitative real-time PCR. Biochem Biophys Res Commun 313:856–62

Chapter 3

Transcriptome Profiling of Host–Microbe Interactions by Differential Display RT-PCR

Leong Wai Fook and Vincent T.K. Chow

Abstract

In recent years, DNA microarray has become increasingly popular as a tool to investigate global expression patterns compared to differential display RT-PCR. Although differential display RT-PCR can be labour-intensive, it has its own merits over those of DNA microarray. While the latter usually consists of a well-defined set of species-specific genes, differential display RT-PCR allows the investigation of host–microbe interactions without bias towards any mRNA transcripts. This means that the regulated transcript expression of both host and pathogen can be analysed simultaneously. In addition, novel transcripts and alternate splicing variants pertaining to the infection can also be discovered. We have investigated the response of rhabdomyosarcoma cells to infection with a neurovirulent strain of enterovirus 71 (EV71) at different time-points during the infection process compared with uninfected cells. Using differential display RT-PCR, we identified mRNAs that were up- or down-regulated. Less than half of the clones match known genes including those involved in mediating the cytoskeleton, cell cycle, cell death, protein translational machinery and cellular transport. The rest of the clones do not match any known genes, of which several are novel genes. Noteworthy is the discovery of an alternate splicing form of TRIP7, which is down-regulated during EV71 infection. The differential display technique has potentially wide applicability to elucidate the gene expression or transcriptomic profiles of host–microbe interactions, which can provide a better understanding of microbial pathogenesis.

Key words: Differential display, Enterovirus 71, Host–microbe interactions, Reverse transcription polymerase chain reaction, Transcriptome profiling

1. Introduction

Ever since its introduction in 1992 by Liang and Pardee (1), differential display has been widely used by numerous researchers to identify genes that are selectively expressed in microbial infections (2–5), developmental processes (6, 7), cancers (8, 9) and other diseases (10, 11) and during chemical or drug treatment (12, 13). The power of differential display lies in its simplicity, harnessing a combination of three frequently employed molecular

Nicola King (ed.), *RT-PCR Protocols: Second Edition*, Methods in Molecular Biology, vol. 630, DOI 10.1007/978-1-60761-629-0_3, © Springer Science+Business Media, LLC 2010

biology techniques. First, high-quality total RNA is reverse-transcribed using reverse transcriptase, and one of three anchor primers is designed to anneal to the 3′ poly(A) tail of mRNA. The resulting cDNA species are subsequently included as templates in a polymerase chain reaction (PCR), utilizing the same anchor primer from the reverse transcription (RT) step in combination with an arbitrary primer. Second, an expression fingerprint of several cDNA species can be visualised after subjecting the differential display PCR products to electrophoresis in a denaturing polyacrylamide gel. If fingerprints from two or more samples are visualised side-by-side, then these samples can be easily compared, and reveal differences in gene expression. Third, the genes showing such differences can subsequently be isolated, cloned and characterised.

The first optimized protocol for differential display RT-PCR (DD RT-PCR) utilized radioactive labelling of differential display amplified products (14). Several years on, fluorescent differential display (15, 16) was developed that allowed better safety, convenient data acquisition and higher sample processing per setting than the traditional method. Nevertheless, radioactive labelling is still preferred by many researchers as it does not require expensive sophisticated equipment. In recent years, DNA microarray has become a more popular choice for differential transcription analysis due to the fact that it can interrogate global transcription patterns rapidly since as many as tens of thousands of genes can be analysed simultaneously (17, 18). This strategy is especially feasible since most of the human genes have been annotated. However, DNA microarray is relatively expensive and may be technically challenging. The advantage of differential display RT-PCR is that novel genes or alternate transcripts can be isolated, whereas DNA microarray is limited by the oligonucleotides printed onto the microarray slides, which represent known cDNAs.

The global analysis of gene expression can provide a better understanding of the temporal and spatial changes in host RNA transcription that occur upon interaction with a virus. This host–virus interaction can lead to numerous changes to host cell pathways and networks including modulation of RNA expression, target receptor induction, signal transduction, cytoskeletal rearrangements and vacuolar trafficking (19–22). Transcriptional profiling utilizing differential display RT-PCR offers an efficient strategy for analyzing these global alterations, and it paves the way for identifying specific pathways and molecules as potential drug targets. Thus, differential display can play a complementary role rather than competing with microarray technology for transcriptional profiling analysis (23).

Here, we exemplify the detailed methodology of differential display RT-PCR to investigate the transcriptome of cells infected with enterovirus 71 (EV71), a pathogen of increasing public health importance (24). Figure 1 illustrates the overall experimental strategy. Insertion point for Fig. 1.

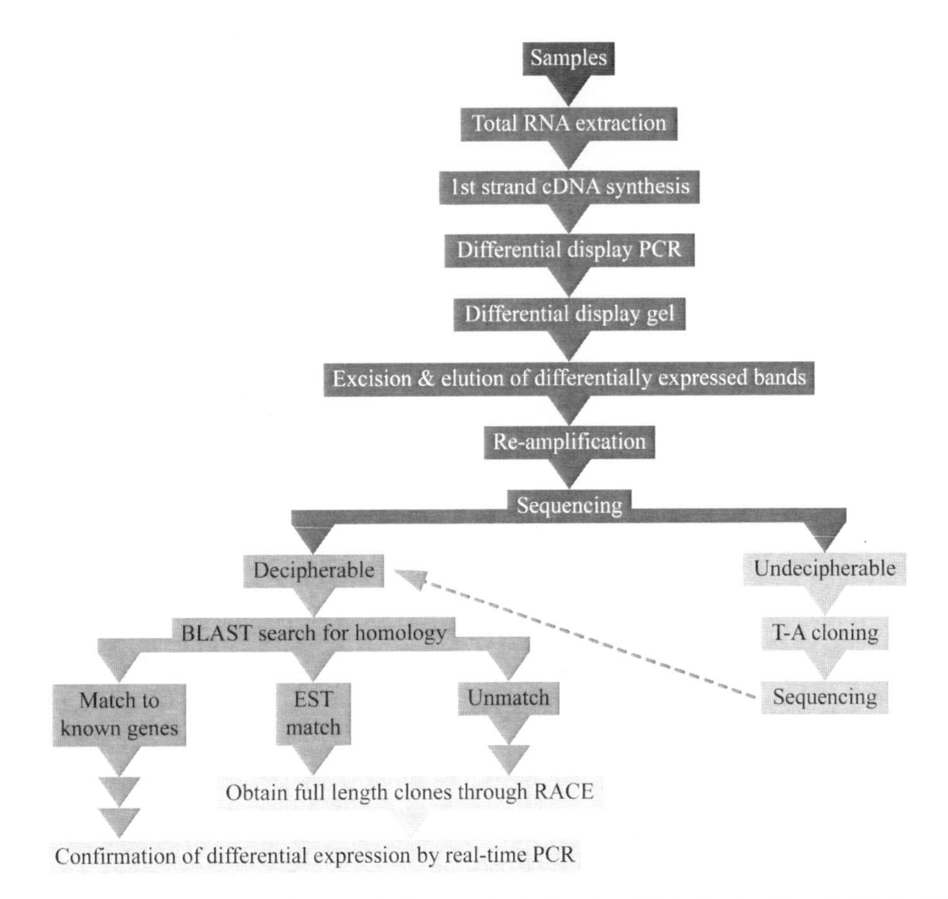

Fig. 1. Flowchart illustrating the overall strategy for the analysis of enterovirus 71 infection by differential display RT-PCR, and the confirmation of differential expression

2. Materials

2.1. Cell Culture, Virus Inoculation and Cell Harvesting

1. Growth medium: Minimum Essential Medium supplemented with 5% fetal calf serum (FCS), 1.05 g/L sodium bicarbonate, 20 mM HEPES, 0.1 mM non-essential amino acids, 1% penicillin and streptomycin. Store at 4°C.
2. Maintenance medium: As above except 2% FCS is used.
3. Solution of trypsin (0.25%) and ethylenediamine tetraacetic acid (EDTA) (1 mM). Store at 4°C.
4. Trypan blue solution (0.4%).
5. Phosphate-buffered saline (PBS): 0.2 g/L KCl, 0.24 g/L KH_2PO_4, 8 g/L NaCl, 1.44 g/L Na_2HPO_4 (1× solution). Adjust pH to 7.4 and store at 4°C.
6. Neubauer hemocytometer.

7. Sterile cell scrapers.
8. RNase-free certified or diethyl pyrocarbonate (DEPC)-treated pipette tips and microtubes.
9. RNA lysis buffer from SV Total RNA Isolation System (Promega, Madison, WI) (see Note 1).

2.2. Total RNA Extraction, Quantification and RNA Gel Electrophoresis

1. SV Total RNA Isolation System.
2. 20-gauge needle with 1 mL syringe.
3. 70°C heat block.
4. 100% absolute ethanol.
5. DEPC-treated deionised water: Stir 1 mL DEPC in 1 L deionised water for 4 h and autoclave (see Note 2). Store at room temperature.
6. RNA markers.
7. FA gel buffer (10× solution): 10 mM EDTA, 200 mM MOPS, 50 mM sodium acetate (NaOAc). Top up with DEPC-treated deionised water, adjust pH to 7.0 and store in the dark. Before using as running buffer, dilute to 1× with DEPC-treated deionised water, to a final 2% formaldehyde solution.
8. RNA loading buffer (5× stock): 4 mM EDTA, 7.2% formaldehyde solution, 30.84% formamide, 40% 10× FA gel buffer, 20% glycerol, 0.16% saturated bromophenol blue solution. Store at 4°C (shelf-life of 3 months).
9. 1.2% denaturing agarose gel: Add 1.2% agarose (w/v) in an appropriate volume of 1× FA gel buffer. Melt the agarose completely in a microwave oven, and cool to about 65°C before obtaining final concentrations of 1.8% formaldehyde solution and 50 ng/mL ethidium bromide. Mix and pour onto gel support to solidify. Equilibrate the gel in 1× running buffer for at least 30 min.

2.3. cDNA Synthesis and Differential Display PCR

1. Verso™ cDNA Kit (Abgene, Epsom, UK). Store at –20°C.
2. RNAimage Kits (Genhunter, Nashville, TN). Store at –20°C.
3. GoTaq® DNA Polymerase (Promega). Store at –20°C.
4. 5× Colorless GoTaq® Reaction Buffer (Promega). Store at –20°C.
5. α-(^{33}P) dATP, 3,000 Ci/mmol. Special precautions to be observed for handling radioactive material. Store at –20°C.
6. GeneAmp® PCR System 9700 (Applied Biosystems).

2.4. Gel Electrophoresis and Isolation of Differentially Amplified Bands

1. Sequi-Gen GT sequencing cell (38 cm H × 30 cm W) (Bio-Rad).
2. Tris-borate-EDTA (TBE) running buffer (10× stock): 108 g/L Tris, 55 g/L boric acid, 20 mM EDTA (pH 8.0).

3. Acrylamide stock solution: Mix 100 mL 10× TBE, 57 g acrylamide, 3 g N,N′-methylenebisacrylamide, 420 g urea, and top up with deionised H_2O to 1 L. Filter through 0.45 μm membrane and store at 4°C.
4. 10% ammonium persulfate: 0.1 g ammonium persulfate in 1 mL deionised H_2O.
5. N,N,N′,N′-Tetramethylethylenediamine (TEMED).
6. Stop solution: 95% formamide, 20 mM EDTA, 0.05% bromophenol blue, 0.05% xylene cyanol FF. Store at 4°C.
7. 3MM Chr paper (35 × 43 cm).
8. Gel dryer.
9. X-ray film.
10. Fine needle.
11. Surgical blade size 11 with Bard-Parker handle.
12. 10% sodium hypochlorite solution (bleach).
13. 1.6% TBE agarose gel.
14. 10 mg/mL ethidium bromide (carcinogenic) or GelRed stain (Biotium, Hayward, CA).

2.5. Identification of Differentially Expressed Genes

1. Nucleospin Extract II kit (Macherey-Nagel, Düren, Germany).
2. BigDye® Terminator v3.1 Cycle Sequencing Kit (Applied Biosystems).
3. 3 M sodium acetate (NaOAc).
4. 125 mM EDTA.
5. 70%, 100% ethanol.
6. pGEM® – T Easy Vector System.
7. Chemically competent bacterial cells (e.g. *Escherichia coli* DH5α).
8. 42°C water-bath.
9. Luria-Bertani (LB) broth.
10. Ampicillin (50 mg/mL)
11. X-gal LB agar plates.
12. NucleoSpin® Plasmid kit (Macherey-Nagel, Düren, Germany).
13. *Eco*RI restriction enzyme (10,000 U/mL).
14. 1.4% agarose gel.

2.6. Confirmation of Differentially Expressed Genes by Real-Time RT-PCR

1. LightCycler® 480 SYBR Green I Master.
2. LightCycler® 480 Real-Time PCR system.

3. Methods

3.1. Preparation of EV71-Infected Samples for DD RT-PCR Analysis

1. Using trypsin/EDTA, passage the rhabdomyosarcoma (RD) cells when they reach 85–90% confluency, at a split ratio of 1:5 in 75 cm^2 flasks with growth medium.
2. A 25 cm^2 flask of cells is required for a viable cell count using trypan blue staining in addition to each experimental time-point (see Note 3). Once the cells have reached about 90% confluency, trypsinise them for cell counting. Infect the cells with a neurovirulent EV71 strain at a multiplicity of infection (MOI) of 10^{-3} and with inactivated virus (control) (see Note 4), in a virus inoculum volume of 4 mL diluted with maintenance medium.
3. After virus adsorption for 2 h at 37°C, remove the inoculum, and add 8 mL of maintenance medium. Perform further incubations of the infected cells for various experimental periods before harvesting.
4. During harvesting, remove the medium and rinse the cells twice with cold 1× PBS. Remove excess PBS prior to the next step.
5. Lyse the cells by scraping directly in the flask with 200 μL of RNA lysis buffer. Carefully transfer the lysate into a 1.5 mL microtube and proceed directly to total RNA isolation, or store at –70°C to be processed later.

3.2. Total RNA Extraction, Quantitation and RNA Gel Electrophoresis

Below is a modified protocol of the SV Total RNA Isolation System. All buffers and columns mentioned below can be found in the kit. Centrifuge all samples at room temperature unless otherwise stated.

1. If the RNA samples are frozen, thaw them at room temperature. Homogenize the samples carefully using a 20-gauge needle with a 1 mL syringe for 5–8 times in order to shear the genomic DNA and reduce the viscosity. Always change to a new needle/syringe for each sample (see Note 5).
2. Dilute the lysate with 350 μL of RNA dilution buffer and mix by inverting 5–6 times. If the lysate is a little viscous, mix by pipetting up and down a few times, and if necessary, pass the diluted lysate through a 20-gauge needle (see Note 6).
3. Place the diluted lysates in a 70°C water-bath or heat block for a maximum of 3 min before centrifuging at 15,000×*g* for 10 min.
4. Carefully remove the supernatant to a new 1.5 mL microtube. Avoid the pellet debris, and if necessary, leave some cleared lysate behind.
5. Add 200 μL of 95% ethanol and mix by pipetting up and down 5–6 times. Transfer the mixture to the spin column assembly, and centrifuge at 15,000×*g* for 1 min.

6. Empty the collection tube and place the spin column back. Add 600 μL of RNA wash buffer and centrifuge at 15,000×*g* for 1 min.
7. Prepare the DNase incubation mix fresh by combining 40 μL of yellow core buffer, 5 μl of 0.09 M $MnCl_2$ and 5 μL of DNase I enzyme per sample in a clean microtube. Add 50 μL of DNase mix directly on the silica membrane, and incubate at room temperature for 20 min (see Note 7).
8. Add 200 μL of DNase stop solution, and centrifuge at 15,000×*g* for 1 min.
9. Wash all impurities with 600 μL of RNA wash buffer and centrifuge at 15,000×*g* for 1 min. Discard the flow-through and add another 250 μL of wash buffer. Centrifuge at maximum speed for 2 min to remove excess wash buffer.
10. Remove the collection tube and place the spin column into a fresh 1.5 mL microtube. Add 50 μL of nuclease-free water directly on the silica membrane, and leave for 1 min before centrifugation at 15,000×*g* for 1 min. Add another 30 μL of nuclease-free water to the column and centrifuge once more. Discard the spin column and cap the microtube. The RNA sample can be quick frozen in liquid nitrogen and store at –70°C (see Note 8).
11. Measure the absorbance reading at 260 nm, and determine the RNA concentration by multiplying the OD_{260} by the RNA extinction coefficient of 40 μg/mL.
12. In order to check the integrity of the total RNA, mix about 1 μg RNA with 5× RNA loading buffer, and electrophorese in a 1.2% denaturing agarose gel at 100 V. When viewed under the UV transilluminator, two distinctive bands of about 1.9 and 5.0 kb for human RNA should be observed (see Note 9).

3.3. cDNA Synthesis and Differential Display PCR

1. For each sample, set up three 20 μL reaction mixes (each with the respective single-base anchored oligo(dT) primer found in the RNAimage kit) using the Verso™ cDNA kit as follows: 250 ng of total RNA (see Note 10), 4 μL of 5× cDNA synthesis buffer, 2 μL of dNTP mix, 2 μL of single-base anchored primer (2 μM), 1 μL of RT enhancer, 1 μL of Verso enzyme mix, and top up to 20 μL with RNase-free water. Mix the reagents well, centrifuge briefly, and place on ice.
2. Set up the reaction mixes to run on a thermal cycler with a profile of 47°C for 1 h followed by inactivation at 95°C for 2 min before holding at 4°C. Store the cDNA samples at –20°C.
3. Differential display PCR is performed in duplicate using paired combinations of the three single-base anchored

primers and eight arbitrary primers from the RNAimage kit 1, giving a total of 24 primer pairs (see Note 11).

4. Mix 2 μL of cDNA template, 2 μL of single-base anchored primer (2 μM) (use the corresponding anchored primer that was used in the cDNA synthesis), 2 μL of arbitrary primer (2 μM) (RNAimage kit), 1.6 μL of dNTP (25 μM) (RNAimage kit), 4 μL of 5× Colorless GoTaq® Reaction Buffer, 0.2 μL of α-(^{33}P) dATP, 0.2 μL of GoTaq® DNA Polymerase, 8 μL of deionised H_2O in a 0.2 mL thin-wall PCR tube (see Note 12). Perform the following cycling profile in the GeneAmp® PCR System 9700: 94°C for 15 s (denaturation), 40°C for 2 min (annealing) and 72°C for 30 s (extension) each for 40 cycles, followed by 72°C for 5 min (final extension) and a 4°C hold. Store at –20°C if not processed immediately by gel electrophoresis.

3.4. Gel Electrophoresis and Isolation of Differentially Amplified Fragments

1. Set up the Sequi-Gen GT sequencing cell and prepare a 6% denaturing polyacrylamide gel. Degas 100 mL acrylamide stock solution, and add 500 μL of 10% ammonium persulfate and 50 μL of TEMED. Mix well while avoiding bubble formation.
2. Once fully polymerised (at least 2 h) (see Note 13), pre-run the gel for 30 min at 60 W.
3. Thaw the DD RT-PCR samples if stored frozen, mix and centrifuge briefly. Mix 4 μL of amplified product with 2 μL of stop solution and heat at 95°C for 2 min. Place the tubes immediately on ice.
4. Flush the urea from all wells with the running buffer using a micropipette or syringe with needle, and load the samples.
5. Electrophorese at 60 W constant power (keep voltage limit to 1,900 V) for about 3.5 h or until the xylene cyanol dye front reaches near the bottom. Stop the electrophoresis and pry open the gel plates. Blot the gel onto a piece of Whatman 3MM Chr paper. Cover the gel with a plastic wrap, and place it in the gel dryer to dry at 70°C for 1.5 h with vacuum.
6. Align the X-ray film to the top edge of the dried gel, and tape them together at the top edge with two small pieces of masking tape. Make a few V-shape cuts at various sides for easy alignment of film with gel at a later step. Expose the film at –80°C for 3 days.
7. After developing the film, align the autoradiogram to the gel based on the V-shape cuts made previously and tape them together.
8. Locate differentially expressed bands that are greater than 100 bp (Fig. 2) by punching through at the four corners of each band of interest on the autoradiograph with a needle (see Note 14).

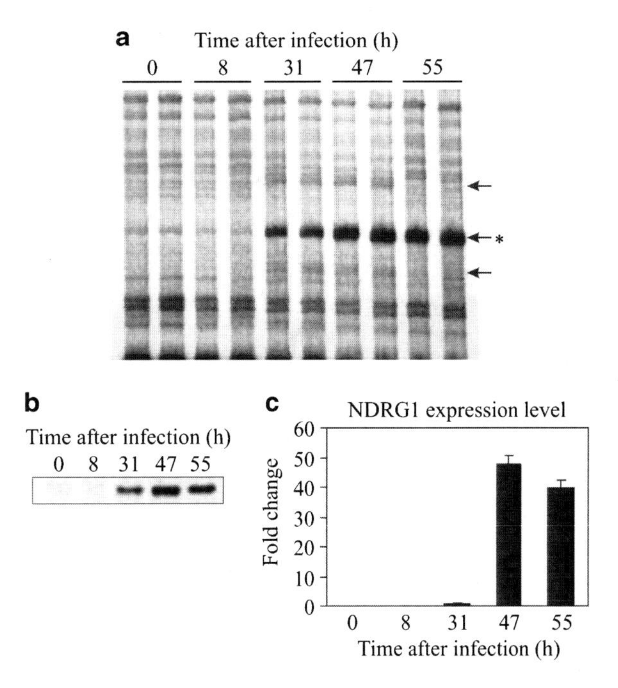

Fig. 2. Differential display RT-PCR analysis of enterovirus 71 infection in rhabdomyosarcoma cells. (**a**) A representative autoradiograph of polyacrylamide gel electrophoresis of differential display RT-PCR products (in duplicates) showing the upregulation of NDRG1 gene transcripts (*) during the process of EV71 infection. *Arrows* indicate potential differentially regulated transcripts. (**b**) Conventional RT-PCR confirming the upregulation of NDRG1 transcript. (**c**) Real-time PCR confirmation of NDRG1 upregulation of at least 48-fold at the later stages of infection

9. Excise the located band on the dried gel with a clean surgical blade (size 11) and place it in a clean 1.5 mL heat-proof microtube. Clean the blade with 10% bleach solution and rinse in deionised H_2O before use.
10. Add 100 μL of deionised H_2O to the microtube and boil for 15 min.
11. Centrifuge at maximum speed for 5 min to pellet the gel and paper debris, and transfer the supernatant to a clean 1.5-mL microtube.
12. Re-amplify the PCR fragment using 5 μL of the collected supernatant (see Note 15) with the same pair of primers used in the initial differential display PCR in a 40-μL mix consisting of 4 μL of single-base anchored primer (2 μM), 4 μL of arbitrary primer (2 μM), 3.2 μL of dNTP (250 μM) (RNAimage kit), 8 μL of 5× Colorless GoTaq® Reaction Buffer, 0.4 μL of GoTaq® DNA Polymerase, 15.4 μL of deionised H_2O in a 0.2 mL thin-walled PCR tube. Perform the cycling profile as before except changing the annealing step to 45°C for 45 s.

13. Resolve 35 µL of each PCR product in a 1.6% agarose gel stained with 0.8 µg/mL ethidium bromide or a safer alternative such as GelRed at 10,000× dilution. Check whether the re-amplified fragment sizes correspond to their sizes from the polyacrylamide gel.

3.5. Identification of Differentially Expressed Genes

1. Excise the re-amplified PCR fragment with a clean surgical blade, and extract the DNA from the gel using NucleoSpin Extract II kit. Elute the DNA in 25 µL of elution buffer found in the kit.
2. Perform direct DNA sequencing using 5–10 µL of purified DNA (see Note 16) in a 20 µL reaction mix with 4 µL of BigDye ready mix, 2 µL of BigDye sequencing buffer, 160 nM of corresponding arbitrary primer. A typical cycle sequencing profile is 96°C for 1 min (initial denaturation), 30 cycles each of 96°C for 10 s (denaturation), 50°C for 5 s (annealing), 60°C for 4 min (extension), and a final hold at 4°C.
3. Purify the sequencing product with 2 µL of EDTA (125 mM), 2 µL of NaOAc (3 M) and 50 µL of 100% ethanol. Mix and incubate at room temperature for 15 min.
4. Centrifuge at maximum speed for 20 min and decant the supernatant.
5. Wash the pellet with 200 µL of 70% ethanol, and centrifuge at maximum speed for 10 min.
6. Remove the supernatant and leave the pellet to dry.
7. Process the samples on a suitable automated DNA sequencer (e.g. ABI PRISM 377 DNA sequencer).
8. Once clean and readable sequences are obtained, compare it to sequences available in GenBank using the BLAST program. Also compare with sequences in the expressed sequence tags (EST) database.
9. For unreadable sequences, set up a ligation reaction with 3 µL of purified DNA (from step 1), 1 µL of pGEM®-T Easy vector, 5 µL of 2× rapid ligation buffer and 1 µL of T4 DNA ligase. Leave the ligation mix overnight at 4°C.
10. Transform 3 µL of the ligation mix into chemically competent *E. coli* cells (such as DH5α) at 42°C, recover in LB broth at 37°C for 1 h, and plate onto X-gal LB agar containing 50 µg/mL ampicillin for blue-white selection.
11. Select a few (about 5) white colonies, and incubate each overnight with shaking in 3 mL of ampicillin-LB broth. Extract the plasmid using the NucleoSpin® plasmid kit. Set up a restriction digest with 1 µL of the extracted plasmid, 0.5 µL of *Eco*RI, 1 µL of 10× reaction buffer and 7.5 µL of deionised H_2O. Incubate at 37°C for about 1.5 h and resolve the

```
  1 CAGCATCCAGCGGCGGCGCCAGCAGTTCCAGTCCGTTGCTTTACTTTTTGCTTCACCGACATAGTCATTATGCCGAAGAGAAAGTCTCCA
                                                                                M  P  K  R  K  S  P
 91 GAGAATACAGAGGGCAAAGATGGATCCAAAGTAACTAAACAGGAGCCCACAAGACGGTCTGCCAGATTGTCAGCGAAACCTGCTCCACCA
     E  N  T  E  G  K  D  G  S  K  V  T  K  Q  E  P  T  R  R  S  A  R  L  S  A  K  P  A  P  P
181 AAACCTGAACCCAAACCAAGAAAAACATCTGCTAAGAAAGAACCTGGAGCAAAGATTAGCAGAGGTGCTAAAGGGAAGAAGGAGGAAAAG
     K  P  E  P  K  P  R  K  T  S  A  K  K  E  P  G  A  K  I  S  R  G  A  K  G  K  K  E  E  K
271 CAGGAAGCTGGAAAGGAAGGCACAGAAAACTGAATCTGTAGATAACGAGGGAGAATGAATTGTCATGAAAAATTGGGGTTGATTTTATGT
     Q  E  A  G  K  E  G  T  E  N ***
361 ATCTCTTGGGACAACTTTTAAAAGCTATTTTTACCAAGTATTTTGTAAATGCTAATTTTTTAGGACTCTACTAGTTGGCATACGAAAATA
451 TATAAGGATGGACATTTTATCGTCTCATAGTCATGCTTTTTGGAAATTTACATCATCCTCAAGTAAAATAAATATCAGTTAAATATTGGA
541 AGCTGTGTGTAAGATTGATTCAGCATTCCATGCACTTGCTTTAAAATTTAGTCCTGTGCATACTGTGGTGTTTTTACTGTGCATATTTGA
631 ATTTTTCATGCAGTTTTTCTAGAGCAATAATCAGTGGTGCTTTTGTACCTAGGTTTTATGTGATTTTAATGAAACATGGATAGTTGTGGC
721 CACCTGCTGACTATTTGTGGTTTAAAATAAAAGGTTTACTTGTCTGCAAAAAAAAAAAAAAAAAAAAAAA
```

Fig. 3. The finalized 790-bp product of rapid amplification of cDNA ends (RACE) and predicted amino acid sequences of RDEV154, a novel alternatively spliced variant of the thyroid receptor interactor, TRIP7. RDEV154 was initially isolated as a ~450-bp fragment showing a decreasing expression during EV71 infection. Sequence alignment revealed that nucleotides (nt) 22–289 and 290–465 of RDEV154 cDNA were identical to nt 16–283 (*light grey shaded area*) and 325–500 (*dark grey shaded area*) of TRIP7 cDNA. From this, it was evident that a 41-bp deletion within RDEV154 corresponded to nt 284–324 of TRIP7, which generated a frame-shift mutation and encoded a C-terminal truncated protein. *Asterisks* (*) denote an in-frame stop codon

digested products in a 1.4% agarose gel to check for positive clones through the release of the DNA insert.

12. Sequence the clones (use about 500 ng of plasmid), and if several clones of the same purified DNA source harbour the same sequences, this would identify the putative differentially expressed gene. Sequence alignment against known genes may also reveal alternatively spliced transcript variants (Fig. 3) (see Note 17).

3.6. Confirmation of Differentially Expressed Transcripts

1. Using 500 ng total RNA extracted previously, perform an RT reaction as outlined in Subheading 3.3, with the exception of using 1 μL of a blend of random hexamers and anchored oligo-dT (at 3:1 ratio) provided in the kit instead of the anchored single-base primer. The incubation at 47°C can be shortened to 30 min. Store the cDNA in −20°C.
2. Design an inter-exon primer pair to amplify a target fragment of 100–250 bp with the T_m of the primers within 56–60°C (see Note 18). Confirm the differential expression of the gene of interest with real-time RT-PCR (see Note 19).
3. Prepare a master mix containing 1 μL of cDNA template, 5 μL of SYBR Green mix, 0.5 μL of each forward and reverse primer (10 μM), 3 μL of PCR-grade water for each sample.
4. Set up the reactions to run with a corresponding set for a housekeeping gene, e.g. β-actin, HPRT, GAPDH or HMBS (see Note 20) in the LightCycler® 480 real-time PCR system.
5. Using the software provided, calculate the relative fold change in relation to the designated control sample or in our study, the uninfected sample at time zero (2, 20, 22).

Correlate the data with those obtained from differential display PCR to authenticate whether the identified gene is truly differentially expressed (Fig. 2).

4. Notes

1. Besides Promega, other commercial total RNA isolation kits can be used as long as they incorporate a DNase treatment step. Trizol® can also be used, but DNase treatment is required after extracting the total RNA, and must be removed by phenol–chloroform method prior to cDNA synthesis. This is important to remove any genomic DNA contamination which may give rise to false positive bands, and in order to detect gene expression changes at transcriptional level.
2. DEPC is considered as a hazardous chemical and is stored at 4°C. Before use, leave it at room temperature for 30 min, and open it in a fume hood. It is an effective nuclease inhibitor, and will decompose slowly to ethanol and carbon dioxide upon contact with moisture. Thus, it is important to autoclave the DEPC-treated solution before use.
3. For virus or microbe-related work, 25 cm^2 screw cap flasks are preferred over 60 mm culture dishes as any spillage, which can be hazardous to the operator, can be prevented. This is particularly important when handling pathogenic strains.
4. MOI refers to the ratio of infectious virus to target cells. For example, MOI of 0.1 is defined as one virus particle to ten host cells.
5. Prepare all solutions used in RNA-related work using DEPC-treated water. When performing RNA work, ensure that a fresh pair of gloves and face mask are worn at all times. Ensure that all consumables are RNase-free and use disposable wares whenever possible.
6. The step of reducing the viscosity of the lysate is very important. This is to ensure that major particulates are pelleted and to prevent the clogging of the silica membrane on the column during the purification step.
7. Occasionally, the DNase mix is not absorbed into the silica matrix, but forms a globule on the top of the matrix instead. This may be prevented by adding the mix onto the matrix with the pipette tip barely touching the surface of the matrix. Alternatively, while capping the column, apply a little downward pressure on the cap to help push the solution into the matrix.
8. Quick freezing in liquid nitrogen before storing at −70°C is crucial for maintaining the integrity of RNA. We have tried

several freeze-thaw cycles without significant impact on the absorbance reading as well as the RNA integrity as visualised by denaturing agarose gel electrophoresis.

9. The 18S and 28S ribosomal RNA (rRNA) species are visualised as 1.9 and 5.0 kb bands, respectively. This is typical for mammalian cells except that the sizes may vary slightly. The 28S rRNA should also be approximately twice the intensity of 18S rRNA. If the ratio is about equal or less, and the bands do not appear sharp but with smearing of smaller sized RNAs, it is possible that RNA degradation may have occurred during sample preparation.
10. Before adding the RNA into the reaction mix, the RNA can be heated to 70°C for 5 min, and placed immediately on ice to remove secondary structure. Do not heat for more than 5 min as this may affect the integrity of the RNA sample.
11. Each RNAimage kit consists of three single-base anchored primers and eight arbitrary primers. In order to increase the coverage of the total mRNA transcript population, there are ten RNAimage kits, each with different sets of eight arbitrary primers. Duplication of reactions helps to check for reproducibility.
12. Always make a PCR master mix based on $(n+1)$ number of samples (n) for each primer combination minus the cDNA template. This will account for any pipetting errors that otherwise might cause the last sample to have a lower reaction volume. This will also ensure that all samples are subjected to the same reaction conditions.
13. The gel can be left to polymerise overnight. If so, plastic wrap should be used to prevent the gel from drying out.
14. It will be easier to do this by placing the dried gel/autoradiograph on a light box tilted at about 45–60°. For safety, this can be performed behind an acrylic radiation shield.
15. If desired, the supernatant can be concentrated by adding 0.1 vol. of NaOAc (3 M), 0.5 mg/mL of glycogen (10 mg/mL stock), and 2.5 vol. of 100% ethanol. Leave it to precipitate at −80°C or on dry ice for about 30 min before centrifuging at maximum speed at 4°C for 10 min. Drain away the supernatant without disturbing the pellet. Rinse the pellet with 250 μL of cold 80% ethanol, and centrifuge at full speed for another 5 min. Remove the supernatant and dry the pellet. Re-dissolve the pellet in 10–15 μL of deionised H_2O. Use about 3–4 μl for the re-amplification.
16. If possible, quantify the amount of purified DNA by measuring the absorbance at 260 nm. Typically for amplification of fragments of less than 500 bp, use about 10–30 ng cDNA. Alternatively, estimate the amount of DNA against a DNA

marker of known concentration for each fragment size by electrophoresis in a 1.6% agarose gel.

17. If the sequences do not match any known genes or match only to the EST database, rapid amplification of cDNA ends (RACE) can be performed in order to isolate the full-length cDNA clone (25, 26).
18. Preferably design an inter-exon primer pair with a long intervening intron. This helps to ensure that in a real-time RT-PCR cycling profile, only the target transcript will be amplified without any amplicons from genomic DNA. A smaller amplified product will not overstretch the reactants when compared to a larger product, thus facilitating a more efficient amplification and quantification process.
19. In our laboratory, we prefer to confirm the differential expression of transcripts by real-time RT-PCR which is rapid and quantitatively more accurate. With several housekeeping genes available, real-time RT-PCR data can be easily normalized to obtain an accurate interpretation of fold change in gene transcription.
20. Selecting a good housekeeping gene is very critical. This gene should exhibit as minimal changes as possible in the model under study. In addition, it should preferably have a threshold cycle value (C_T) similar to that of the target gene in the control sample (27).

Acknowledgments

The projects using differential display in our laboratory were supported by the Biomedical Research Council, Singapore and the Microbiology Vaccine Initiative, National University of Singapore.

References

1. Liang P, Pardee AB (1992) Differential display of eukaryotic messenger RNA by means of the polymerase chain reaction. Science 257: 967–971
2. Leong PW, Liew K, Lim W, Chow VT (2002) Differential display RT-PCR analysis of enterovirus-71-infected rhabdomyosarcoma cells reveals mRNA expression responses of multiple human genes with known and novel functions. Virology 295:147–159
3. Liew KJ, Chow VT (2004) Differential display RT-PCR analysis of ECV304 endothelial-like cells infected with dengue virus type 2 reveals messenger RNA expression profiles of multiple human genes involved in known and novel roles. J Med Virol 72:597–609
4. Lim WC, Chow VT (2006) Gene expression profiles of U937 human macrophages exposed to *Chlamydophila pneumoniae* and/or low density lipoprotein in five study models using differential display and real-time RT-PCR. Biochimie 88:367–377
5. Talà A, De Stefano M, Bucci C, Alifano P (2008) Reverse transcriptase-PCR differential display analysis of meningococcal transcripts during infection of human cells: up-regulation of priA and its role in intracellular replication. BMC Microbiol 8:131

6. Neelima PS, Rao AJ (2008) Gene expression profiling during Forskolin induced differentiation of BeWo cells by differential display RT-PCR. Mol Cell Endocrinol 281:37–46
7. Watahiki J, Yamaguchi T, Enomoto A, Irie T, Yoshie K, Tachikawa T, Maki K (2008) Identification of differentially expressed genes in mandibular condylar and tibial growth cartilages using laser microdissection and fluorescent differential display: chondromodulin-I (ChM-1) and tenomodulin (TeM) are differentially expressed in mandibular condylar and other growth cartilages. Bone 42:1053–1060
8. Carles A, Millon R, Cromer A, Ganguli G, Lemaire F, Young J, Wasylyk C, Muller D, Schultz I, Rabouel Y, Dembélé D, Zhao C, Marchal P, Ducray C, Bracco L, Abecassis J, Poch O, Wasylyk B (2006) Head and neck squamous cell carcinoma transcriptome analysis by comprehensive validated differential display. Oncogene 25:1821–1831
9. Li J, Olson LM, Zhang Z, Li L, Bidder M, Nguyen L, Pfeifer J, Rader JS (2008) Differential display identifies overexpression of the USP36 gene, encoding a deubiquitinating enzyme, in ovarian cancer. Int J Med Sci 5:133–142
10. Corominola H, Conner LJ, Beavers LS, Gadski RA, Johnson D, Caro JF, Rafaeloff-Phail R (2001) Identification of novel genes differentially expressed in omental fat of obese subjects and obese type 2 diabetic patients. Diabetes 50:2822–2830
11. de Yebra L, Adroer R, de Gregorio-Rocasolano N, Blesa R, Trullas R, Mahy N (2004) Reduced KIAA0471 mRNA expression in Alzheimer's patients: a new candidate gene product linked to the disease? Hum Mol Genet 13:2607–2612
12. Miura N, Matsumoto Y, Miyairi S, Nishiyama S, Naganuma A (1999) Protective effects of triterpene compounds against the cytotoxicity of cadmium in HepG2 cells. Mol Pharmacol 56:1324–1328
13. Meier V, Mihm S, Ramadori G (2008) Interferon-alpha therapy does not modulate hepatic expression of classical type I interferon inducible genes. J Med Virol 80:1912–1918
14. Liang P, Zhu W, Zhang X, Guo Z, O'Connell RP, Averboukh L, Wang F, Pardee AB (1994) Differential display using one-base anchored oligo-dT primers. Nucleic Acids Res 22:5763–5764
15. Cho YJ, Meade JD, Walden JC, Chen X, Guo Z, Liang P (2001) Multicolor fluorescent differential display. Biotechniques 30:562–572
16. Meade JD, Cho YJ, Fisher JS, Walden JC, Guo Z, Liang P (2005) Automation of fluorescent differential display with digital readout. In: Liang P, Meade JD, Pardee AB (eds) Differential display methods and protocols, vol 317, 2nd edn. Humana, Totowa, NJ, pp 23–57
17. Schena M, Shalon D, Davis RW, Brown PO (1995) Quantitative monitoring of gene expression patterns with a complementary DNA microarray. Science 270:467–470
18. Chee M, Yang R, Hubbell E, Berno A, Huang XC, Stern D, Winkler J, Lockhart DJ, Morris MS, Fodor SP (1996) Accessing genetic information with high-density DNA arrays. Science 274:610–614
19. Cummings CA, Relman DA (2000) Using DNA microarrays to study host-microbe interactions. Emerg Infect Dis 6:513–525
20. Leong WF, Tan HC, Ooi EE, Koh DR, Chow VT (2005) Microarray and real-time RT-PCR analyses of differential human gene expression patterns induced by severe acute respiratory syndrome (SARS) coronavirus infection of Vero cells. Microbes Infect 7:248–259
21. Liew KJ, Chow VT (2006) Microarray and real-time RT-PCR analyses of a novel set of differentially expressed human genes in ECV304 endothelial-like cells infected with dengue virus type 2. J Virol Methods 131: 47–57
22. Leong WF, Chow VT (2006) Transcriptomic and proteomic analyses of rhabdomyosarcoma cells reveal differential cellular gene expression in response to enterovirus 71 infection. Cell Microbiol 8:565–580
23. Chang KC, Komm B, Arnold NB, Korc M (2007) The application of differential display as a gene profiling tool. Methods Mol Biol 383:31–40
24. Bible JM, Pantelidis P, Chan PK, Tong CY (2007) Genetic evolution of enterovirus 71: epidemiological and pathological implications. Rev Med Virol 17:371–379
25. Quek HH, Chow VT (1997) Genomic organization and mapping of the human HEP-COP gene (COPA) to 1q. Cytogenet Cell Genet 76:139–143
26. Sim DL, Chow VT (1999) The novel human HUEL (C4orf1) gene maps to chromosome 4p12–p13 and encodes a nuclear protein containing the nuclear receptor interaction motif. Genomics 59:224–233
27. Roche Applied Science (2005) Selection of housekeeping genes. Technical note LC 13/2005

Chapter 4

Quantitative RT-PCR Methods for Mature microRNA Expression Analysis

Stephanie D. Fiedler, Martha Z. Carletti, and Lane K. Christenson

Abstract

This chapter describes two methods to measure expression of mature miRNA levels using qRT-PCR. The first method uses stem-loop RT primers to produce cDNA for specific miRNAs, a technique that our laboratory has modified to increase the number of miRNAs being reverse transcribed within a single RT reaction from one (as suggested by the manufacturer) to five. The second method uses a modified oligo(dT) technique to reverse transcribe all transcripts within an RNA sample; therefore, target miRNA and normalizing mRNA can be analyzed from the same RT reaction. We examined the level of miRNA-132, a miRNA known to be upregulated in granulosa cells following hCG treatment, using both of these methods. Data were normalized to GAPDH or snU6 and evaluated by ΔΔCt and standard curve analysis. There was no significant difference ($P > 0.05$) in miRNA-132 expression between the stem-loop and modified oligo(dT) RT methods indicating that both are statistically equivalent. However, from a technical point of view, the modified oligo(dT) method was less time consuming and required only a single RT reaction to reverse transcribe both miRNA and mRNA.

Key words: Quantitative RT-PCR, miRNA expression, Stem-loop and modified oligo(dT) primers

1. Introduction

MicroRNA (miRNA) are endogenous, noncoding single-stranded RNA molecules that are ~22 nucleotides in length (1). Discovered less than 20 years ago, 695 miRNA have now been identified in the human genome (miRBase Database Version 12.0), and estimates indicate that >1,000 exist (2–4). MicroRNA have been associated with many biological processes including tissue development, proliferation, and differentiation (1, 5, 6), and research links specific miRNA to heart disease and many types of cancer (7–10). The sequential processing of a primary miRNA transcript (pri-miRNA) to yield a precursor miRNA (pre-miRNA)

Nicola King (ed.), *RT-PCR Protocols: Second Edition*, Methods in Molecular Biology, vol. 630, DOI 10.1007/978-1-60761-629-0_4, © Springer Science+Business Media, LLC 2010

and ultimately a mature, single stranded miRNA that associates with the RNA-induced silencing complex is well described (1, 11, 12). Both microRNA microarrays and Northern blotting provide reliable detection of mature miRNA, but are time consuming, require more starting material, and do not provide the accuracy or precision offered with quantitative RT-PCR (qRT-PCR; (13)).

Analysis of mature miRNA by PCR presents a unique problem as the final length of a miRNA transcript is similar to the primers needed to amplify them. Therefore, expression analysis of mature miRNA has required the development of specialized primers for reverse transcription (RT) or extension of the transcript prior to qRT-PCR. A stem-loop reverse transcriptase primer (described in Fig. 1 and Subheading 3.2.1; (14)) or a long primer (used in the miR-Q method; (15)) specific to each miRNA is used to convert total RNA into cDNA. Amplification is then achieved using a primer set containing a primer specific to the mature miRNA of

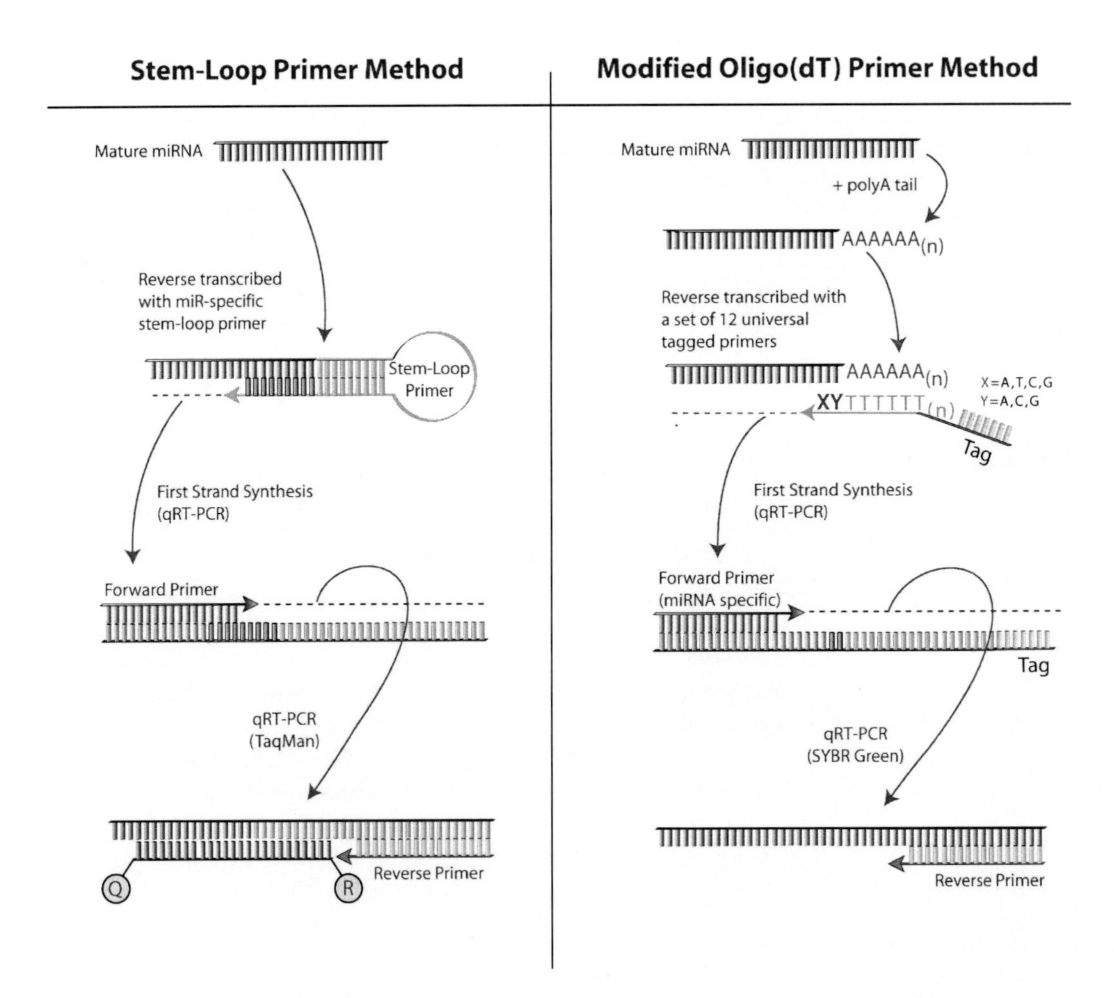

Fig. 1. Graphical representation of RT and PCR methods used to analyze the expression of mature miRNAs

interest and a primer specific to a region of the stem-loop sequence or miR-Q primer (14, 15). Another approach is to add polyA tails to all transcripts within a sample, then reverse transcribe using a set of 12 modified oligo(dT) primers containing a unique sequence tag at the 5′ end and two bases at the 3′ end (as described in Fig. 1 and Subheading 3.2.3; (16)). Amplification is then achieved using a PCR primer specific to the miRNA of interest and a primer specific to the tag. Unlike the stem-loop primer RT reaction and the miR-Q method, this RT reaction converts all miRNA and mRNA into cDNA within a single reaction. In this chapter, we discuss modifications that we have made to the stem-loop primer system that allow for simultaneous measurement of up to 5 miRNA. We also compare the stem-loop and polyA tail/modified oligo(dT) RT systems for miRNA-132 expression within murine granulosa using either a standard curve or ΔΔCt method of quantification.

2. Materials

2.1. RNA Extraction

2.1.1. Materials

1. 19 day old CF-1 mice (20–25 g)
2. Disposable latex gloves
3. 30-gauge needles
4. Dissecting forceps
5. Dissecting scissors
6. 35-mm sterile disposable plates
7. 53-μm Nitex filter fabric (cat. #24-C27, Wildlife Supply Company)
8. Pipets (P2, P20, P100, P1000)
9. Sterile, RNase/DNase free filter pipet tips (to dispense volumes between 0.2 and 1,000 μl)
10. Benchtop centrifuge (refrigerated if possible)
11. Sterile, RNase/DNase free 1.5-ml microcentrifuge tubes
12. –70°C freezer
13. NanoDrop-1000 Spectrophotometer

2.1.2. Solutions and Reagents

1. eCG (Calbiochem, cat. #367222)
2. hCG (Sigma, cat. # CG5-1VL)
3. 1× PBS
4. TRIZOL
5. Chloroform
6. Glycogen
7. Isopropanol

8. 75% Ethanol
9. Ice
10. DEPC water

2.2. Reverse Transcription

2.2.1. RT Materials

1. Vortex mixer
2. Benchtop microcentrifuge
3. Thermocycler
4. 0.5-ml sterile PCR tubes with caps
5. Heating block or water bath (capable of reaching 95°C)

2.2.2. Stem-Loop Primer Reagents

1. TaqMan MicroRNA Reverse Transcription Kit (Applied Biosystems, cat. #4366597)
2. hsa-miR-132 Human MicroRNA Assay Kit (Applied Biosystems, cat. #4373143) (see Note 1)
3. hsa-miR-212 Human MicroRNA Assay Kit (Applied Biosystems, cat. #4373087)
4. hsa-miR-30b Human MicroRNA Assay Kit (Applied Biosystems, cat. #4373290)
5. dNTPs
6. Random Primers
7. MMLV reverse transcriptase (with DTT and 5× RT Buffer)

2.2.3. Modified Oligo(dT) Primer Reagents

1. miScript Reverse Transcriptase Kit (Qiagen, cat #218061)

2.3. Quantitative RT-PCR

2.3.1. qRT-PCR Materials

1. Applied Biosystems 7900HT Sequence Detection System (may be substituted with other real-time machines)
2. Rainin 250-μl EDP-Plus electronic repeat pipetter
3. 384-well optical reaction plate
4. Optical adhesive covers
5. Optical cover applicator tool
6. Centrifuge with swinging bucket optical plate holders

2.3.2. Stem-Loop Primer PCR Reagents

1. hsa-miR-132 Human MicroRNA Assay Kit (also used in RT reaction, see Note 1)
2. TaqMan Rodent GAPDH Control Reagents (Applied Biosystems, cat. #4308313)
3. TaqMan 2× Universal PCR Master Mix, No AmpErase UNG (Applied Biosystems, cat. #4324018)
4. Small nuclear RNA U6 (snU6) primer set – f: ctcgcttcggcagcaca; r: aacgcttcacgaatttgcgt (17)
5. SYBR Green (Invitrogen, cat. #4309155)

2.3.3. Modified Oligo(dT) Primer RT Reagents

1. miScript SYBR Green PCR Kit (Qiagen, cat. #218073)
2. Mm_miR-132_1 miScript Primer Assay (Qiagen, cat. #MS00001561) (see Note 2)
3. TE Buffer, pH 8.0
4. TaqMan Rodent GAPDH Control Reagents (Applied Biosystems, cat. #4308313)
5. Small nuclear RNA U6 (snU6) primer set – f: ctcgcttcggcag-caca; r: aacgcttcacgaatttgcgt (17)

2.4. Data Analysis

1. Microsoft Office 2003
2. GraphPad Prism (version 4.0)

3. Methods

3.1. RNA Extraction and Dilution

1. Treat nine 19-day old CF-1 mice with 5 IU eCG by intraperitoneal injection to initiate follicular growth. After 44–48 h, sacrifice three of the mice by cervical dislocation for the 0 h samples, and inject the other six animals with 5 IU hCG to induce ovulation. Collect ovaries from these animals at 4 and 6 h ($n=3$ for each time point) following hCG treatment.
2. Collect ovaries into cold 1× PBS, then puncture the follicles with a 30-gauge needle. Separate the granulosa cells from other follicular cells by passing each sample through a 53-μm Nitex filter fabric. Place filtrates into 15-ml centrifuge tubes and centrifuge at $0.8 \times 1{,}000 \times g$ for 5 min at 4°C.
3. Remove supernatant and replace with 1-ml TRIZOL and lyse cells with gentle pipetting (see Note 3).
4. Isolate total RNA from each sample with a few modifications to the manufacturer's protocol as described below (see Note 4):
   a. Add chloroform (200 μl) to each tube. Vigorously shake each tube by hand for 15 s before allowing them to rest at room temperature for 5–10 min.
   b. Centrifuge tubes at $16{,}100 \times g$ for 10 min at 4°C.
   c. Carefully remove aqueous phase (top layer) from each tube, transfer into a new tube, and then add 2 μl of glycogen (20 μg/μl stock) to each tube (see Note 5). Gently mix the samples, then add 500 μl of isopropanol to each tube.
   d. Vortex samples, incubate at room temperature for 10 min, and then centrifuge at $16{,}100 \times g$ for 10 min at 4°C.
   e. Pour off supernatant (being careful not to lose the pellet), and then wash the sample with 75% ethanol.

f. Centrifuge the tubes at 16,100 × *g* for 10 min at 4°C, and then pour off the supernatant.

g. Air dry the RNA pellet until the edges of the pellet begin to look clear, then rehydrate the pellet with 20 µl of DEPC water. Gently pipette the samples several times to completely dissolve the pellet and then immediately place on ice until RNA quality and concentration determinations can be completed.

5. Determine the absorbance and concentration (ng/µl) for each RNA sample with a NanoDrop-1000 spectrophotometer. The ratio of the absorbance at 260 and 280 can be used to determine the purity of each sample and values greater than 1.8 are considered acceptable (see Note 6).

6. Dilute aliquots of all 9 RNA samples with DEPC water to yield 30–50 µl volumes of 50 ng/µl solutions of total RNA. Then, use these dilutions in all subsequent RT reactions. Store stock and diluted RNA at –70°C until RT reactions are performed.

3.2. RT Reactions

3.2.1. Stem-Loop Primer RT for Analysis of Multiple (3-5) MicroRNAs

1. Reverse transcribe the total RNA (250 ng) from the dilution prepared in Subheading 3.1.6 with a combination of three miRNA primers (miRNA-132, miRNA-212, and miRNA-30b; see Note 7; (18)). Simultaneous RT reactions for 4 or 5 miRNA are explained in Note 8.

2. The master mix for a 3 miRNA RT reaction contains 0.15 µl of dNTP mix, 1.0 µl of Multiscribe RT enzyme, 1.5 µl of 10× RT buffer, 0.19 µl of RNase Inhibitor, 1.16 µl of nuclease-free water, and 6 µl of RT primers (2 µl of each miRNA-132, miRNA-212 and miRNA-30b RT primers; see Note 9).

3. Pipette 10 µl of the master mix and 5 µl of total RNA (50 ng/µl; 250 ng) into PCR tubes, and load tubes into a thermocycler. Run reaction using the following settings: 16°C for 30 min, 42°C for 30 min, 85°C for 5 min, hold at 4°C.

4. After the reverse transcription reaction, dilute the samples to 1:15 with DEPC water. Store stocks and dilutions at –20°C.

3.2.2. Random Primer RT for Normalization of Stem-Loop qRT-PCR

1. Reverse transcribe total RNA (1 µg of the 50 ng/µl dilutions made in Subheading 3.1.4) for GAPDH and snU6 normalization using random primers (see Note 10).

2. Prepare a random primer master mix containing 2 µl of dNTPs, 0.5 µl of random primers (100 ng/µl), and 5.5 µl of DEPC water (see Note 9).

3. Place 8 µl of the random primer master mix and 20 µl of total RNA (50 ng/µl) into 1.5 ml microcentrifuge tubes then vortex, briefly spin down, and place in a 65°C water bath for 5 min.

4. Chill tubes on ice for 5 min immediately after removal from the water bath.

5. Prepare a second master mix containing 2 µl of DTT (0.1 M) and 8 µl of 5× RT Buffer (see Note 9). Add 10 µl of the second master mix and 2 µl of MMLV reverse transcriptase to each tube (see Note 11). Vortex each sample and spin down briefly, then place in a 37°C water bath for 2 h.
6. Inactivate the enzymatic reaction by placing tubes in a 72°C water bath for 15 min. Dilute reverse transcribed samples to 1:10 with DEPC water. Store stocks and 1:10 dilutions at –20°C.

3.2.3. Modified Oligo(dT) RT

1. Reverse transcribe total RNA (250 ng) from the sample dilutions prepared in Subheading 3.1.4 with the miScript Reverse Transcription Kit (Qiagen; see Note 12). Perform all RT reactions according to the manufacturer's instructions.
2. Prepare a master mix containing 4 µl of miScript RT Buffer (5×), 1 µl of miScript reverse transcriptase mix, and 10 µl of DEPC water (see Note 9).
3. Place 15 µl of master mix and 5 µl total RNA (50 ng/µl) into 1.5 ml microcentrifuge tubes then vortex, briefly spin down, and place in a 37°C water bath for 1 h.
4. Inactivate the enzyme reaction by incubating all samples on a 95°C heating block for 5 min. Dilute reverse transcribed samples to 1:5 with DEPC water. Store stocks and 1:5 dilutions at –20°C.

3.3. Quantitative Real Time PCR Reactions

3.3.1. Generation of Standards for qRT-PCR Assays (see Note 13)

1. Use the stock cDNA from the *stem-loop RT* reactions (Subheading 3.2.1, step 4) to generate a standard curve for miRNA-132 qRT-PCR reactions (see Note 14).
2. Use the stock cDNA from the *random primer RT* reactions (Subheading 3.2.2, step 6) to generate a standard curve for GAPDH and snU6 qRT-PCR reactions (see Note 14).
3. Use the stock cDNA from the *modified oligo(dT) RT* reactions (Subheading 3.2.3, step 4) to generate a standard curve for miRNA-132, GAPDH, and snU6 qRT-PCR reactions (see Note 14).
4. Generate standard curves by serial dilution of above stock cDNAs as outlined in Table 1 (see Notes 13 and 14).

3.3.2. qRT-PCR Using the Stem-Loop RT Product

1. For amplification and detection of mature miRNA-132, perform triplicate qRT-PCR reactions (10 µl reaction volumes) for standards, samples, and negative controls (see Note 15) using the miRNA-132 specific primers and probe provided in the hsa-miRNA-132 Human MicroRNA Assay Kit.
2. Prepare a master mix containing 0.35 µl of hsa-miR-132 primer set, 5 µl of TaqMan 2× Universal PCR Master Mix, and 3.32 µl of DEPC water (see Note 9) and aliquot 8.7 µl

Table 1
Serial dilutions for running a standard curve analysis.* The initial solution is composed of small amounts from several RT stocks. Each subsequent dilution is made from the previous dilution (i.e. dilution B made from dilution A, dilution C made from dilution B, etc). *It is crucial to vortex each tube well before making the next dilution in the series.

Dilution	cDNA (RT product)	Water	Total volume	Remaining volume	Dilution ratio	Arbitrary PCR value
A.	80	320	400	240	1:5	2,000
B.	160 (A)	160	320	240	1:10	1,000
C.	80 (B)	320	400	240	1:50	200
D.	160 (C)	160	320	240	1:100	100
E.	80 (D)	320	400	240	1:500	20
F.	F. 160 (E)	160	320	240	1:1000	10
G.	80 (F)	320	400	400	1:5000	2

into individual wells of a 384-well optical reaction plate using a repeat pipetter.

3. Then add 1.3 μl of sample, standard, or negative control to each well, seal the plate with an optical adhesive cover, then centrifuge at 1,000 × *g* for 2 min in a swinging bucket plate holder.
4. Load the plate into a sequence detector and run reaction with the following program:
   a. Stage 1 (TaqMan hot start): 95°C for 10 min.
   b. Stage 2 (amplification): 95°C for 15 s then 60°C for 1 min (40 cycles).

3.3.3. qRT-PCR for GAPDH and snU6 for Normalization of Stem-Loop qRT-PCR Results

1. For amplification and detection of GAPDH and snU6, perform triplicate qRT-PCR reactions (10 μl reaction volumes) for standards, samples, and negative controls (see Note 16) using TaqMan Rodent GAPDH Control Reagents or the snU6 primer set.
2. Prepare a GAPDH master mix containing 0.1 μl of each forward primer, reverse primer, and the probe for rodent GAPDH, 5.0 μl of TaqMan Universal PCR Master Mix, and 3.7 μl of DEPC water (see Note 9) and aliquot 9 μl into individual wells of a 384-well optical reaction plate using a repeat pipetter.
3. Prepare an snU6 master mix containing 90 nM snU6 forward and reverse primers (0.1 μl of 9 μM forward/reverse working

solution; see Note 16), 5.0 μl of SYBR green and 3.9 μl of DEPC water (see Note 9) and aliquot 9 μl into individual wells of a 384-well optical reaction plate using a repeat pipetter.

4. Then add 1 μl of sample, standard, or negative control to each well, seal the plate with an optical adhesive cover, then centrifuge at 1,000 × *g* for 2 min in a swinging bucket plate holder.
5. Load the plate into a sequence detector and run reaction with the following program:
   a. Stage 1: 50°C for 2 min
   b. Stage 2: 95°C for 10 min
   c. Stage 3: 95°C for 15 s then 60°C for 1 min (40 cycles)
   d. Stage 4: 95°C for 15 s, 60°C for 15 s, then 95°C for 15 s

3.3.4. Using Modified Oligo(dT) Primer RT Product

1. For amplification and detection of mature miRNA-132, GAPDH, and snU6, perform triplicate qRT-PCR reactions (10 μl reaction volumes) for standards, samples, and negative controls (see Note 15) using the miScript SYBR Green PCR Kit and the mmu_miR-132_1 miScript Primer Assay.
2. Prepare a miRNA-132 master mix consisting of 1 μl of 10× miScript Universal primer, 1 μl of 10× miScript miRNA-132 primer, 5 μl of 2× Quantitect SYBR Green PCR Master Mix (see Note 16), and 1 μl of DEPC water (see Note 9) and aliquot 9 μl into individual wells of a 384-well optical reaction plate using a repeat pipetter.
3. Prepare a GAPDH master mix consisting of 0.1 μl of each forward primer, reverse primer, and the probe for rodent GAPDH, 5.0 μl of TaqMan Universal PCR Master Mix, and 3.7 μl of DEPC water (see Note 9) and aliquot 9 μl into individual wells of a 384-well optical reaction plate using a repeat pipetter.
4. Prepare an snU6 master mix consisting of 90 nM snU6 forward and reverse primers (0.1 μl of 9 μM forward/reverse working solution; see Note 16), 5.0 μl of SYBR green and 3.9 μl of DEPC water (see Note 9) and aliquot 9 μl into individual wells of a 384-well optical reaction plate using a repeat pipetter.
5. Then add 1.0 μl of sample, standard, or negative control to each well, seal the plate with an optical adhesive cover, then centrifuge at 1,000 × *g* for 2 min in a swinging bucket plate holder.
6. Load the plate into a sequence detector and run reaction with the following program:
   a. Stage 1 (TaqMan hot start): 95°C for 15 min
   b. Stage 2 (40 cycles)

Denaturation:	94°C for 15 s
Annealing:	55°C for 30 s
Extension:	70°C for 34 s (see Note 17)

3.3.5. Normalization and Analysis Methods (see Note 18)

1. *ΔΔCt method.* This method bypasses the need for a standard curve by simply using the raw qRT-PCR Ct values. Comparisons are made between a target of interest and a normalizing mRNA as described in the formula below (see Note 19)

$$\text{Fold change} = 2^{-\left((\text{Average Ct of sample} - \text{Average Ct of GAPDH or snU6}) - \Delta\text{Ct}^{*}\text{ of control}\right)}$$

   *ΔCt = (Average Ct of sample – Average Ct of GAPDH or snU6), and the control sample is the control time point, treatment, etc.

2. *Standard curve method.* Arbitrary values for both miRNA and mRNA are determined by serial diluting cDNA samples produced under the same RT conditions to generate a standard curve. Serial dilutions are then run simultaneously with the samples, and arbitrary values (see formula below) are determined based on the standard curve (see Note 12).

$$\text{Arbitrary value} = \text{Average of sample/Average of GAPDH or snU6}$$

3. Analyze using two-way ANOVA comparisons of the stem-loop qRT-PCR (Fig. 2) or modified oligo dT (Fig. 3) results using the ΔΔCt and standard curve analyses.
   a. MicroRNA-132 levels measured by stem-loop qRT-PCR normalized with GAPDH (Fig. 2a) and snU6 (Fig. 2b) were different at 4 and 6 h post-hCG, but there was no significant difference between the ΔΔCt and standard curve analyses.

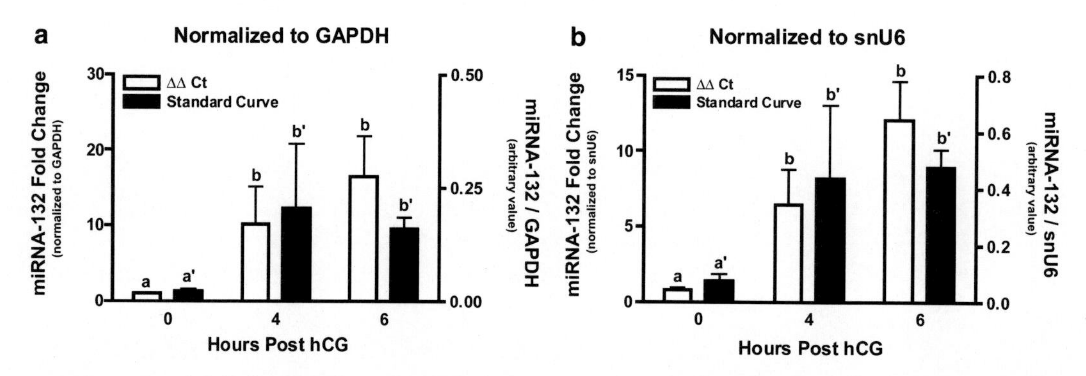

Fig. 2. Stem-loop qRT-PCR method – comparison of mural granulosa cell miRNA-132 levels before (0 h) and 4 and 6 h after hCG using the ΔΔCt and standard curve analysis for stem-loop qRT-PCR method. Mature miRNA-132 levels were normalized to GAPDH (**a**) or snU6 (**b**). [a,b] & [a′,b′] Means ± SEM within an analysis group (ΔΔCt or standard curve, respectively) with different superscripts are significantly ($P < 0.05$) different

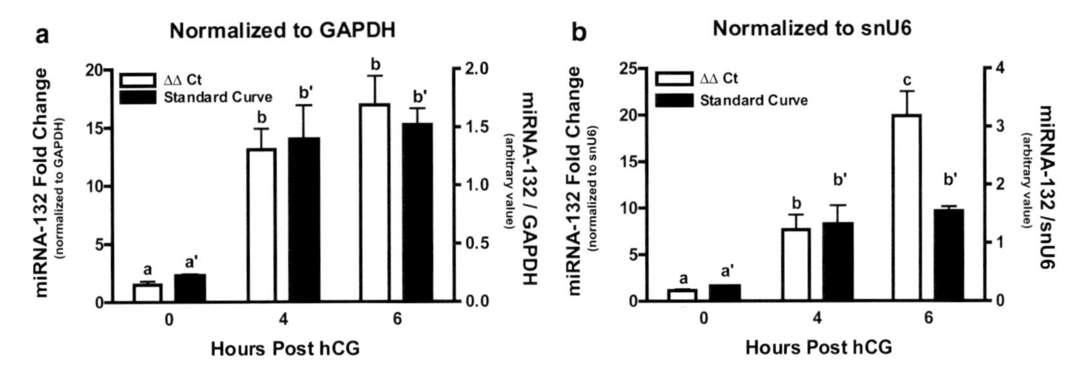

Fig. 3. Modified oligo(dT) qRT-PCR method – comparison of mural granulosa cell miRNA-132 levels before (0 h) and 4 and 6 h after hCG using the ΔΔCt and standard curve analysis for modified oligo(dT) qRT-PCR method. Mature miRNA-132 levels were normalized to GAPDH (**a**) or snU6 (**b**). [a,b,c] & [a′,b′] Means ± SEM within an analysis group (ΔΔCt or standard curve, respectively) with different superscripts are significantly ($P < 0.05$) different

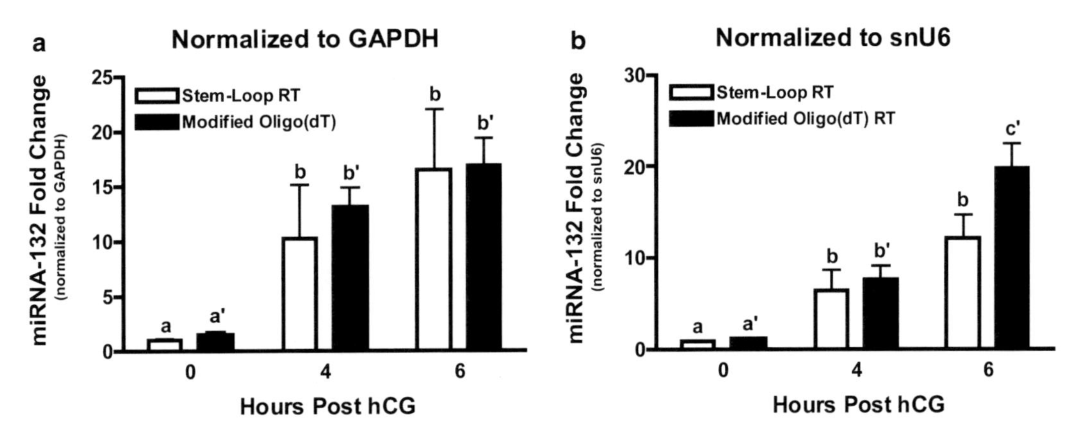

Fig. 4. Comparison of stem-loop and modified oligo(dT) qRT-PCR methods for quantification of mural granulosa cell miRNA-132 levels using ΔΔCt analysis. Mature miRNA-132 levels were normalized to GAPDH (**a**) or snU6 (**b**). Two-way ANOVA showed that method (stem loop versus modified oligo(dT)) and method by hour interactions were not significant, while day was significant ($P < 0.05$). [a,b] & [a′,b′,c′] Means ± SEM within an analysis group (ΔΔCt or standard curve, respectively) with different superscripts are significantly ($P < 0.05$) different. We observed a trend ($P = 0.07$) for a method difference in the GAPDH comparison

b. MicroRNA-132 levels measured by modified oligo(dT) qRT-PCR normalized with GAPDH (Fig. 3a) and snU6 (Fig. 3b) were different at 4 and 6 h post-hCG, but there was no significant difference between the ΔΔCt and standard curve analyses. In contrast, the snU6 normalization detected a difference ($P < 0.05$) between the 4 and 6 h time points using the ΔΔCt analysis that was not seen with the other methods/analyses.

4. Analyze using two-way ANOVA side-by-side comparisons of stem-loop versus modified oligo(dT) qRT-PCR ΔΔCt analyzed data normalized with either GAPDH (Fig. 4a) or snU6 (Fig. 4b; see Note 20).

a. Method (stem-loop versus modified oligo (dT)) of miRNA-132 quantification was not different ($P>0.05$) for either GAPDH or snU6 normalized data nor was there a method by hour interaction. We did observe a trend ($P=0.07$) for a method (stem-loop versus modified oligo (dT)) difference but only in the ΔΔCt analysis. Although no differences between the two methods were observed, the standard error values did appear to be decreased in samples reverse transcribed using the modified oligo(dT) RT system.

4. Notes

1. The mature forms of both human and murine miRNA-132, 212 and 30b are 100% homologous; therefore, the human primer sets can be used to analyze the expression of these mouse miRNA. Data analysis within this paper focused on miRNA-132 expression; therefore, qRT-PCR data for miRNA-212 and miRNA-30b are not shown.
2. Individual miRNA specific qRT-PCR primers can be designed and substituted for miScript primers within the PCR section of the modified oligo(dT) protocol. We have designed our own miRNA-132 specific primers and observed similar results as provided with primers from the kit. It is important to test designed primers for specificity and efficiency; in this case, this was done by comparing expression levels produced by the designed primer to those produced by the kit primer.
3. Cultured cells or tissues may be substituted in place of isolated granulosa cells. Follow manufacturer's recommendation for Trizol volumes.
4. Several kits and reagents are used within the methods described in this paper. These products are all provided with extensive protocols that include many tips and troubleshooting suggestions.
5. Glycogen is added to samples during the RNA isolation process to act as a carrier of nucleic acids and will not interfere with RT or PCR reactions when used at concentrations lower than 250 μg/ml. For our purposes, samples were isolated with 40 μg/ml glycogen, a concentration that provides higher RNA yields while remaining well below maximum levels.
6. RNA samples can be contaminated with proteins, salts, and DNA. To prevent these types of contamination, extra caution should be taken when transferring the aqueous layer during the isolation process. The 260/280 ratio should be at least 1.8–2.0 with lower values suggesting high levels of protein within a sample; while a 260/230 ratio below 1.8 can indicate the presence of organic contaminants including

Trizol. Although the NanoDrop spectrophotometer is a highly sensitive instrument capable of very accurately measuring these ratios as well as the RNA concentration using only 1 µl of the sample, any standard spectrophotometer can be used to determine absorbance and concentration values.

7. The instructions for the TaqMan MicroRNA Reverse Transcription kit suggest a starting amount between 1 and 10 ng of total RNA per 15 µl reaction. Following the PCR reaction, this amount of RNA consistently yielded high Ct values (i.e. low transcript levels) for miRNA within granulosa cells; therefore, we increased the starting amount of total RNA to improve the RT reaction and thus produce more robust qRT-PCR results.

8. The stem-loop RT method provides a quick, reliable way to measure the expression of mature miRNA using a relatively small starting amount of RNA. The manufacturer's protocol provides instructions to reverse transcribe one specific miRNA per RT reaction. Our laboratory has determined that it is possible to combine up to 5 RT primers in one reaction (18)*. To do this, adjustments have been made to the water volume, the primer amount and/or the total reaction volume. For the stem-loop RT reactions described in this chapter, three specific primers were combined (data for miR-132, one of the three miRNA is shown). The total volume of the reaction (15 µl reaction) follows the manufacturer's protocol, although the volume of water was adjusted to 1.16 µl and the amount of each primer was reduced to 2.0 µl of each (compared to 3 µl recommended by the manufacturer). A combination of either four or five RT primers requires the volume of the RT reaction to be increased from 15 µl to 20 µl. Volumes of all the reagents change accordingly (i.e., 0.2 µl of dNTP mix, 1.5 µl of Multiscribe, 2.0 µl of 10× RT buffer, 0.25 µl of RNase inhibitor, 3.05 µl (for 4 primers) or 1.05 µl (for 5 primers) of DEPC water, and 2 µl of each primer). Each RT reaction consists of 15 µl of master mix plus 5 µl of RNA (50 ng/µl). *It is important that expression levels of miRNA amplified in a combination RT reaction should first be compared to levels produced when each miRNA is reverse transcribed alone to make sure that certain primer combinations do not interfere with each other and affect the PCR reaction.*

9. All master mix reagent volumes are listed as per reaction. When making master mix solutions for reverse transcription and PCR reactions, increase the total master mix volume by 10% to prevent a shortage of master mix solution.

---

* Since submission Applied Biosystems has developed a set of two master RT-primer mixes that allows for the simultaneous RT of all miRNA for which they have developed qRT-PCR assays.

10. The TaqMan MicroRNA Reverse Transcription kit is designed to reverse transcribe only mature miRNA transcripts, therefore normalization to a nonchanging (i.e. constitutively expressed RNA) is done by performing two separate RT reactions, one for the miRNA and one for the normalizer (i.e. GAPDH or snU6) as described in this chapter. Of note, we have demonstrated that simultaneous reverse transcription of snU6 within the stem-loop primer RT reaction (data not shown) provides the same results as compared to two separate RT reactions (shown in Fig. 2).
11. Remove MMLV reverse transcriptase from freezer right before needed and return to the freezer immediately after using it.
12. One of the added benefits of the miScript Reverse Transcription Kit (shown in Fig. 1) is that both oligo(dT) primers and random primers are contained within the RT buffer. Therefore, miRNA and mRNA are reverse transcribed within the same tube allowing both miRNA and normalizing mRNA from a sample to be analyzed following a single RT reaction rather than two separate RT reactions as required in the stem-loop RT and miR-Q RT methods. The miScript Reverse Transcription Kit is optimized for starting amounts between 10 pg and 1 μg of total RNA.
13. Quantitative RT-PCR is only quantitative when the amount of cDNA is determined during the exponential doubling phase. Fluorescence is measured to establish threshold cycle (Ct) values which represent the amount of product amplified at a given point in the reaction. When more template is available at the beginning of the reaction, a fewer number of cycles is necessary for the fluorescent signal to measure above the background level. Therefore, diluting cDNA will cause an increase in the number of cycles needed to reach a given fluorescence value. This concept can be used to establish arbitrary fluorescence values by creating a standard curve from a serial dilution for a given sample set. To do this, an initial stock solution is made from small volumes of RT stocks *prepared under the same RT conditions*. Arbitrary values can be assigned to each dilution based on the fold differences between each of the dilutions in the series, and a relative value for an unknown can be extrapolated by comparing its signal to standard curve (13).
14. The standard curve should be made from samples that are known or predicted to have high transcript levels to ensure that the unknown sample values will fall within the range of arbitrary values generated by the curve.
15. Negative controls include samples that were not reverse transcribed and water blanks. The absence of fluorescence in nonreverse transcribed samples indicates the absence of DNA contamination. We typically run nonreverse transcribed negative controls for a single miRNA and GAPDH/actin for each sample dilution. If the results are negative for genomic contamination

for these genes, we make the assumption that the sample is negative for all DNA contamination. However, if the laboratory has used transfection agents, like pre-miRNA, siRNA, or expression plasmids for specific genes, a negative control should always be run for that specific transcript simultaneously with the reverse transcribed sample, as plasmid contamination is a major problem for qRT-PCR assays. Water blanks are included to confirm the absence of primer dimers (in the case of SYBR green qRT-PCR) and to confirm that contamination of the PCR product has not occurred within the space used to set up the qRT-PCR assays.

16. PCR primer sets designed without a probe (i.e. snU6 described above) rely on SYBR green detection. This dye binds nonspecifically to double stranded DNA causing the fluorescence signal to increase as the amount of double stranded DNA increases. This method of measuring fluorescence is vulnerable to primer dimers; therefore, a primer check should be performed to determine the highest concentration at which the primers can be used without forming dimers. The presence of primer dimers can be determined from dissociation curve analysis. The melting points for a given primer set should all be the same, while the shorter sequence formed by primer dimers will melt at a lower temperature which is indicated by a peak shift. It is not necessary to perform a primer check when primer sets provided within a kit are used (e.g. miScript Primer Assay), as the company will provide optimized conditions for each reagent.

17. The extension time recommended by the miScript System Handbook (30 s) is optimized for the ABI PRISM 7000, and an adjustment to 34 s is suggested when using Applied Biosystems 7300 and 7500 Systems. As an Applied Biosystems, 7900HT Sequence Detector was used for amplification of our samples, we chose a 34 s extension period.

18. It is not necessary to evaluate data with *both* ΔΔCt and standard curve analyses.

19. Normalization is performed to account for variation between unknown samples and can be achieved by comparing fluorescent values from a gene of interest to those from a gene expressed at a constant level within a tissue/cell. The ΔΔCt data analysis method can be used to determine a relative fold-change between samples, but the calculations used to determine these values depend on the assumption that the PCR reaction is equally efficient for both the transcript of interest and that of the normalizer (GAPDH/snU6; (13)).

20. As there was no significant difference between the ΔΔCt and the standard curve methods (Figs. 2 and 3), only the ΔΔCt analysis is shown in the side-by-side comparison of the two RT methods.

Acknowledgments

We would like to express our most sincere appreciation to Stan Fernald for assistance with graphic design as well as Lacey Luense and Caitlin Healy for their critiques of this work.

References

1. Bartel DP (2004) MicroRNAs: genomics, biogenesis, mechanism, and function. Cell 116:281–297
2. Griffiths-Jones S (2004) The microRNA registry. Nucleic Acids Res 32:D109–D111
3. Griffiths-Jones S, Grocock RJ, van Dongen S, Bateman A, Enright AJ (2006) miRBase: microRNA sequences, targets and gene nomenclature. Nucleic Acids Res 34:D140–D144
4. Griffiths-Jones S, Saini HK, van Dongen S, Enright AJ (2008) miRBase: tools for microRNA genomics. Nucleic Acids Res 36:D154–D158
5. Song L, Tuan RS (2006) MicroRNAs and cell differentiation in mammalian development. Birth Defects Res C Embryo Today 78:140–149
6. Du T, Zamore PD (2005) microPrimer: the biogenesis and function of microRNA. Development 132:4645–4652
7. van Rooij E, Sutherland LB, Thatcher JE, DiMaio JM, Naseem RH, Marshall WS, Hill JA, Olson EN (2008) Dysregulation of microRNAs after myocardial infarction reveals a role of miR-29 in cardiac fibrosis. Proc Natl Acad Sci U S A 105:13027–13032
8. Lu J, Getz G, Miska EA, Alvarez-Saavedra E, Lamb J, Peck D, Sweet-Cordero A, Ebert BL, Mak RH, Ferrando AA, Downing JR, Jacks T, Horvitz HR, Golub TR (2005) MicroRNA expression profiles classify human cancers. Nature 435:834–838
9. Chan JA, Krichevsky AM, Kosik KS (2005) MicroRNA-21 is an antiapoptotic factor in human glioblastoma cells. Cancer Res 65: 6029–6033
10. Lui WO, Pourmand N, Patterson BK, Fire A (2007) Patterns of known and novel small RNAs in human cervical cancer. Cancer Res 67:6031–6043
11. Kim VN (2004) MicroRNA precursors in motion: exportin-5 mediates their nuclear export. Trends Cell Biol 14:156–159
12. Lee Y, Jeon K, Lee JT, Kim S, Kim VN (2002) MicroRNA maturation: stepwise processing and subcellular localization. EMBO J 21:4663–4670
13. VanGuilder HD, Vrana KE, Freeman WM (2008) Twenty-five years of quantitative PCR for gene expression analysis. Biotechniques 44:619–626
14. Chen C, Ridzon DA, Broomer AJ, Zhou Z, Lee DH, Nguyen JT, Barbisin M, Xu NL, Mahuvakar VR, Andersen MR, Lao KQ, Livak KJ, Guegler KJ (2005) Real-time quantification of microRNAs by stem-loop RT-PCR. Nucleic Acids Res 33:e179
15. Sharbati-Tehrani S, Kutz-Lohroff B, Bergbauer R, Scholven J, Einspanier R (2008) miR-Q: a novel quantitative RT-PCR approach for the expression profiling of small RNA molecules such as miRNAs in a complex sample. BMC Mol Biol 9:34
16. Lee DY, Deng Z, Wang CH, Yang BB (2007) MicroRNA-378 promotes cell survival, tumor growth, and angiogenesis by targeting SuFu and Fus-1 expression. Proc Natl Acad Sci U S A 104:20350–20355
17. Kunkel GR, Maser RL, Calvet JP, Pederson T (1986) U6 small nuclear RNA is transcribed by RNA polymerase III. Proc Natl Acad Sci U S A 83:8575–8579
18. Fiedler SD, Carletti MZ, Hong X, Christenson LK (2008) Hormonal regulation of MicroRNA expression in periovulatory mouse mural granulosa cells. Biol Reprod 79:1030–1037

Chapter 5

Detection of Influenza A Virus Neuraminidase and PB2 Gene Segments by One Step Reverse Transcription Polymerase Chain Reaction

Alejandra Castillo Alvarez, Victoria Boyd, Richard Lai, Sandy Pineda, Cheryl Bletchly, Hans G. Heine, and Ross Barnard

Abstract

We describe a single step reverse transcription polymerase chain reaction protocol that can be used to amplify part of the neuraminidase gene segment (segment 6) from all nine subtypes of influenza A virus. The method has also been applied to amplify gene segment 1 of influenza A, which encodes the basic polymerase protein 2 (PB2). The method combines the use of mixed base primers with a "touchdown" thermal cycling program and is applicable to a wide range of nucleic acid targets in which there is genetic variability in the regions complementary to the PCR primers.

Key words: Reverse transcription, Polymerase chain reaction, Influenza A, PB2, Neuraminidase

1. Introduction

Apart from the challenges of sample quality and nucleic acid integrity, the design of nucleic acid-based assays faces two fundamental challenges. One challenge is to design the assays so that they will continue to detect the target organism despite the occurrence of genetic change. This is a particularly difficult challenge with RNA viruses because they exhibit a relatively high mutation rate and/or undergo reassortment of genetically divergent segments (1). The second, related challenge is that the design of primers generally requires a guess regarding the virus subtype that is present. Influenza A undergoes frequent reassortment, so it is not possible to predict which neuraminidase segment will be present in a clinical or wildlife sample. Incorrect guessing will result in the selection of incorrect primers and false negative results. We have addressed

Nicola King (ed.), *RT-PCR Protocols: Second Edition*, Methods in Molecular Biology, vol. 630,
DOI 10.1007/978-1-60761-629-0_5,

both of these challenges by using mixed-base primer designs, incorporating nonstandard nucleotides that are complementary to highly conserved regions but can tolerate a substantial sequence variation. These primer designs necessitate the use of a "touch-down" or "step-down" thermal cycling program (2) because the optimum annealing temperature for PCR will vary depending on the target that is present in the sample. In the touchdown protocol, the temperature of the reaction passes through successively lower annealing temperatures in the first few cycles of the PCR. The method has been applied to amplify part of the neuraminidase gene segment of influenza A, which contains molecular determinants of resistance to neuraminidase inhibitor drugs (3), and the PB2 segment, which is hypothesized to contain molecular determinants of virus pathogenicity (4). The neuraminidase reverse transcription polymerase chain reaction (RT-PCR) is a simple alternative to immunological methods for neuraminidase typing of Influenza A viruses.

2. Materials

2.1. Design of Oligonucleotides and Bioinformatic Analysis of Designed Oligonucleotides

1. Biological Sequence Alignment Editor Software (BioEdit, version 7.09 or higher, CA, US).
2. Sequence Database (e.g. National Center For Biotechnology Information. Influenza Virus Resource database (IVRD) (http://www.ncbi.nlm.nih.gov/genomes/FLU/)).
3. ClustalW alignment program (http://www.clustal.org/)

2.2. Obtaining the Sample

2.2.1. Growth of Influenza A Virus in Embryonic Eggs

1. PC3 containment facility.
2. Class 2 Laminar flow cabinet.
3. Specific-pathogen-free (SPF) fowl eggs or specific antibody negative (SAN) eggs.
4. Supernatant fluids of faeces or tissue suspensions.
5. Sterile filter tips.
6. Syringes.

2.3. Viral RNA Extraction

1. Bench area or laminar flow hood.
2. Calibrated sterile pipettes (P2, P10, P20, P200, P1,000).
3. RNase/DNase free sterile filter tips.
4. RNAase AWAY™ Reagent (Invitrogen).
5. Nuclease-free water.
6. Ethanol 70% (Industrial grade).

7. Gloves (nitrile or latex).
8. Clinical sharp bin disposal.
9. Sterile 1.5-, 0.6- and 0.2-mL microtubes.
10. Racks for 1.5-, 0.6- and 0.2-mL tubes.
11. Laboratory coat (preferably back tie, with stockinette cuffs and stockinette collar).
12. –80°C Freezer.
13. Small zip-lock plastic bags (to store RNA aliquots).
14. Plastic storage boxes.
15. 2-mercaptoethanol.
16. RNeasy extraction kit and QIAshredder columns.
17. MagNA Pure LC Total nucleic acid isolation kit (Roche) (see Note 1).
18. Ice bucket.

2.4. Materials for Conventional RT-PCR Assay

2.4.1. For Master Mix PCR and Primer Preparation Only

1. PCR cabinet or a laminar flow hood with UV light incorporated.
2. Calibrated sterile pipettes (P2, P10, P20, P200, P1,000).
3. RNase/DNase free sterile filter tips (e.g. 10 μL, 20 μL, 200 μL, 1,000 μL).
4. Kimberly-Clark Kimwipes® EXL Delicate Task wipes, 11.4 × 21.3 cm or lint-free paper.
5. Ethanol 70% (Industrial grade).
6. Gloves (nitrile or latex).
7. Sterile 1.5-, 0.6- and 0.2-mL Eppendorf tubes.
8. Racks for 1.5-, 0.6- and 0.2-mL tubes.
9. Lab coat (preferably with stockinette cuffs and stockinette collar).
10. Nuclease free water.
11. Vortex mixer.
12. Microcentrifuge.
13. Clinical sharps bin.
14. Ice bucket.

2.4.2. For Adding Template Only

1. PCR cabinet or a laminar flow hood with UV light incorporated.
2. Calibrated sterile pipette (P2, or P10).
3. RNase/DNase free sterile filter tips (e.g. 10 μL).
4. Kimberly-Clark Kimwipes® EXL Delicate Task wipes, 11.4 × 21.3 cm or lint-free paper.

5. Ethanol 70% (Industrial grade).
6. Gloves (nitrile or latex).
7. Minicentrifuge with 0.6- and 0.2-mL tubes adaptor.
8. PCR racks for 0.6- and 0.2-mL tubes.
9. Nuclease free water.
10. Lab coat (preferably a Lab coat with stockinette cuffs and stockinette collar).
11. Ice bucket.
12. Clinical sharp bin disposal.

2.5. Conventional RT-PCR Assay and DNA Clean Up

1. SuperScript™ III One-step RT-PCR System with Platinum® *Taq* DNA polymerase kit (Invitrogen, cat no. 12574-026).
2. Gene specific forward primer and gene specific reverse primer containing M13 "tag".
3. Molecular weight ladder.
4. 1× TAE buffer. (stock 50× TAE buffer is prepared by combining 242 g Tris, 100 mL of 0.5 M EDTA (pH 8.0), 57.1 mL of glacial acetic acid in a total volume of 1 L).
5. 1.5% Agarose gel made in 1× TAE buffer containing 0.5 μg/mL ethidium bromide.
6. QIAquick Gel Extraction Kit (QIAGEN).
7. Ethanol 100% (Molecular grade).
8. 3 M sodium acetate pH 5.
9. The Green Bag® Kit (Bio101).
10. 5× loading buffer.
11. BenchTop microcentrifuge.
12. 100 bp DNA Ladder.
13. Spectrophotometer capable of measuring absorbance at 260 and 280 nm.
14. Nuclease free water.

2.6. Equipment and Materials for Electrophoresis and Visualization of PCR Products in Agarose Gels

1. Calibrated pipettes (P20 and P200) for gel area *only* (see Note 2).
2. Sterile filter tips (e.g. 20 μL, 200 μL,) for gel area *only* (see Note 2).
3. Electrophoresis equipment (power pack and gel electrophoresis tanks).
4. UV transilluminator.
5. Gel documentation system (e.g. should include camera, software, and computer).
6. Printer.

7. Benchtop microcentrifuge for 1.5- and 0.6-mL tubes.
8. Thermocycler.
9. Sequencing facility or a reputable sequencing service provider.
10. Medical waste bins for incineration.
11. Beaker for ethidium bromide waste disposal.
12. Sterile blade or scalpel.
13. Gloves for gel electrophoresis area.

3. Methods

3.1. Design of Primers and Bioinformatic Analysis of Designed Primers for the Influenza A Segment 6 Encoding Neuraminidase

1. Design RT-PCR primers to amplify segment 6 based on sequence information obtained from the NCBI Influenza Virus Resource Database (IVRD) (5). To do the alignment, download all the full-length sequences to process and save them in FASTA format. Open Bioedit and click on "Add new". Open all the downloaded sequences in one file and make the alignment. To do this, go to the "Accessory application" tab and click in ClustalW Multiple alignment. Make sure that only full multiple alignment and bootstrap parameters are highlighted. The value for bootstrapping should be 1,000. Then, click on "Run ClustalW", and the program will start generating the alignment. A black screen will appear, and when the program finishes, the black screen closes and the alignment is done. Do not close the black screen and do not perform any other task on the computer while the alignment is being done as this will consume a large amount of memory from the computer, and it will generate the alignment very slowly. When the alignment is finished, save it as an alignment as it will be used to generate the tabular summary of the nucleotide composition. The tabular summary of alignments was performed using the alignment generated by the software Bioedit. Load the recently performed alignment into a new document of BioEdit. Go to the Tab "alignment" and then click "positional nucleotide numerical summary". A table will appear with the information needed: position number, number of sequences that have a determined nucleotide A/G/C/U in each position, and the GAP value (see Note 3).
2. The tabular summary of the nucleotide composition at each position in the alignment is used for the primer design and the strategy. All base positions in the selected primer regions have a GAP ≤ 5 (see Note 4).

3. Semi-conserved sequence regions of 18–25 nucleotides long with a multiplicity ≤195 should be sought (see Note 5). Multiplicity is minimized by inserting inosines at one or more positions (see Note 6).
4. Choose sequence regions that will produce a PCR product of the required length. Melting temperature analysis was performed on primers complementary to semi-conserved sequence regions (see Note 7).
5. In order to test the selected primer sequences against particular influenza A subtypes, an additional bioinformatic analysis of the designed primers is performed on the total number of 3,337 full-length influenza A neuraminidase gene sequences available from the IVRD (5). This analysis is performed by downloading, in FASTA format, each of the 9 NA subtypes sequences from the NCBI database (IVRD) and grouping the sequences in subtype specific folders. An alignment using potential primer sequences was carried out using BioEdit for each subtype. Each of the alignments was saved into the subtype specific folder (see Note 8).
6. If the alignment was considered to be of high quality (see Note 8), left click on the ruler which has the base pair number and select the sequence range to which the pair of primers is not going to anneal. Use the "delete to beginning" or "delete to the end" option in BioEdit. Once the aligned sequences have been "trimmed" to the area to which the primers will anneal, sequence logo software (6) is used to graphically present the data for each primer, and the percentage of sequences that matched identically at all five 3′ terminal bases is visually checked.

3.2. Design of Oligonucleotides and Bioinformatic Analysis of Designed Oligonucleotides for the Influenza A Segment 1 Encoding Polymerase Basic Protein 2 (PB2)

1. Mainly, full length segment 1 sequences from multiple subtypes from a range of host species and from different geographical locations are selected from the NCBI Influenza Virus Resource Database (IVRD) (5) and aligned using Biological Sequence Alignment editor software (7). The same informatics approach as described under Subheading 3.1, utilizing a table showing position number, number of sequences that have a particular nucleotide in each position and the GAP (see Note 4) is used to select primers.

3.3. Obtaining Viral RNA Cultures

3.3.1. Virus Culture

1. Culture virus by the inoculation of embryonated specific pathogen free (SPF) fowl eggs, or specific antibody negative (SAN) eggs, as follows (see Notes 10 and 11):
2. Clarify the supernatant fluids of faeces or tissue suspensions by centrifugation at 1000×*g*, and then inoculate into the allantoic sac of at least five embryonated SPF or SAN fowl eggs of 9–11 days' incubation.
3. Incubate the eggs at 35–37°C for 4–7 days.

4. Eggs containing dead or dying embryos as they occur, and all eggs remaining at the end of the incubation period, must be chilled to 4°C, and the allantoic fluids are tested for haemagglutination (HA) activity as follows (8):
5. Dispense 0.025 ml of PBS into each well of a plastic V-bottomed microtitre plate.
6. Place 0.025 ml of infected allantoic fluid in the first well. For accurate determination of the HA content, this should be done from a close range of an initial series of dilutions, i.e. 1/3, 1/4, 1/5, 1/6, etc.
7. Make twofold dilutions of 0.025 ml volumes of the virus suspension across the plate.
8. Dispense a further 0.025 ml of PBS into each well.
9. Dispense 0.025 ml of 1% (v/v) chicken RBCs (red blood cells) into each well.
10. Mix by tapping the plate gently and then allow the RBCs to settle for approximately 40 min at room temperature (approximately 20°C), or for 60 min at 4°C if ambient temperatures are high, by which time control RBCs should be settled to a distinct "button".
11. Determine haemagglutination, by tilting the plate and observing the presence or absence of tear-shaped streaming of the RBCs. The titration should be read to the highest dilution giving complete haemagglutination (no streaming); this represents 1 haemagglutination unit (HAU) and can be calculated accurately from the initial range of dilutions (see Note 11).

3.3.2. Viral RNA Extraction from Samples from Allantoic Fluids

1. Viruses from the reference collection at the Australian Animal Health Laboratory (AAHL) were grown in embryonated eggs. Samples included Influenza virus A from ducks ($n=9$), chickens ($n=6$), shearwater ($n=3$), gull ($n=3$), emu ($n=1$), other avian species ($n=7$), equine ($n=1$) and others ($n=2$).
2. RNA is extracted with the RNeasy mini kit as follows: Use gloves, eye protection, and appropriate level of physical containment (see Note 10). Viral RNA is extracted from 100 μL of amniotic fluid sample inactivated by addition of 600 μL of RLT buffer (guanidine thiocyanate denaturant) and 6 μL of 2-mercaptoethanol and thorough mixing by vortexing.
3. Lysis is completed by incubating for 5 min at room temperature.
4. Lysate was transferred by pipetting to a QIA shredder column sitting in a 2-ml microcentrifuge tube.
5. Centrifuge at approx 14,000 × g for 2 min in a microfuge.
6. Discard the QIAshredder column and retain the filtrate.

7. Add 600 μL of 70% ethanol to the filtrate and mix well by pipetting.
8. Place an RNeasy spin column in a 2 mL collection tube and carefully transfer 700 μL of sample from step 6 to the spin column. Do not moisten the rim of the column.
9. Close the cap. Centrifuge at 8,000 × *g* for 15 s.
10. Discard the filtrate and reuse the spincolumn. Carefully open the cap, add remaining sample from step 6 to the spin column. Do not moisten the rim of the column.
11. Close cap. Centrifuge at 8000 × *g* for 15 s.
12. Discard filtrate and reuse the spin column. Carefully open cap. Add 700 μL wash buffer RW1 from the RNeasy kit to the spin column. Close cap. Centrifuge at 8,000 × *g* for 15 s.
13. Transfer RNeasy spin column to clean 2 mL tube. Carefully open spin column, add 500 μL of wash buffer RPE. Centrifuge at 8000 × *g* for 15 s.
14. Discard flow through and reuse the collection tube. Carefully open the cap. Add 500 μL wash buffer RPE to spin column. Close cap. Centrifuge at 14,000 × *g* for 2 min.
15. Transfer the spin column to a new 1.5 mL tube. Carefully open the spin column and add 50 μL of RNase free water directly to the membrane in the column. Centrifuge at 8,000 × *g* for 1 min.
16. Store eluate at –80°C.

3.3.3. Viral RNA Extraction from Samples from Clinical Samples

Viral RNA was extracted from 200 μL of nasopharyngeal aspirates using MagNA Pure LC total nucleic acid isolation kit, using the external lysis protocol and the MagNA Pure LC instrument. (see Note 12) as follows:

1. Transfer 50–200 of serum/plasma sample or 50–100 μl whole blood sample into a suitable vial, e.g., the Sample Cartridge.
2. Add 300 μl Lysis/Binding Buffer.
3. Mix the samples thoroughly by pipetting.
4. Transfer the sample lysate (350–500 μl) into the Sample Cartridge.
5. Place the Sample Cartridge on the sample stage of the MagNA Pure LC instrument and start the "Total NA External_lysis" purification protocol (see Note 12) with final elution volume set to 100 μl of elution buffer provided in the kit.

3.4. One step Reverse Transcription PCR Assay for the Gene Encoding Influenza A Neuraminidase

1. Total RNA was extracted from animal samples and clinical samples as described in Subheading 3.3, items 2 and 3 using the RNeasy mini kit (from amniotic fluid samples) or MagNA Pure LC total nucleic acid isolation kit respectively.

2. One-step RT-PCR was performed in either 50 μL or 25 μL reaction volumes using SuperScript™ III One-Step RT-PCR System with Platinum® *Taq* DNA polymerase kit. 4 μmoles/L of each primer (or 1 μmoles/L of each primer for cultured samples) (see Note 13). 1–2 μL of RNA was added to each reaction tube, and the reaction is made up to volume with nuclease free water.
3. Primer sequences were:

   NA8F-M13 5′-GTA AAA CGA CGG CCA GT GRA CHC ARG ART CIK MRTG-3′ (multiplicity = 192)

   NA10R-M13 5′-CAG GAA ACA GCT ATG AC CCI IKC CAR TTR TCY CTR CA-3′ (multiplicity = 32)

   NA8F 5′-GRA CHC ARG ART CIK MRTG-3′ (multiplicity = 192)

   NA10R 5′-CCI IKC CAR TTR TCY CTR CA-3′ (multiplicity = 32) (see Notes 14–16)
4. Thermal cycling was performed with the following cycling conditions: 30 min at 46°C and 10 min at 60°C (reverse transcription), 3 min at 94°C (initial denaturation), 8 cycles of touch-down PCR consisting of 30 s at 94°C (denaturation), 30 s at 56°C (annealing) – decrease annealing temperature by 2°C each cycle until 42°C is reached; and 75 s at 68°C (extension) (see Note 17). Amplification of the final product was completed with 36 cycles of 30 s at 94°C, 30 s at 43°C, and 75 s at 68°C, with a final extension of 10 min at 68°C. For egg cultured and in vitro transcribed RNA samples, 36 cycles were used. For RNA extracted from clinical samples, 43 cycles were used.
5. In the negative control, water for injections BP or RNase free water was used instead of template RNA.
6. The positive control was influenza A subtype N2 (a clinical sample) or influenza A/H3N8/Avian/669/WA/78. PCR products (10 μL of sample plus 2 μL of bromophenol blue loading dye) were visualized by gel electrophoresis on 1.5% agarose containing 0.5 μg/ml ethidium bromide. The size of the products generated with M13 tags was approximately 253 basepairs (see Fig. 1), and the amplicons without the M13 tags were approximately 219 basepairs.

3.5. One Step Reverse Transcription PCR Assay for Gene Segment 1 Encoding Influenza A PB2

1. Total RNA is extracted from animal samples and clinical samples as described under Subheading 3.3, items 2 and 3.
2. One-step RT-PCR is performed in 50 μL reaction volume using SuperScript™ III One-Step RT-PCR System with Platinum® *Taq* DNA polymerase. Add 2 μL (320 picomole) of each primer (see Note 13). Add 1–2 μL of RNA to each reaction tube.

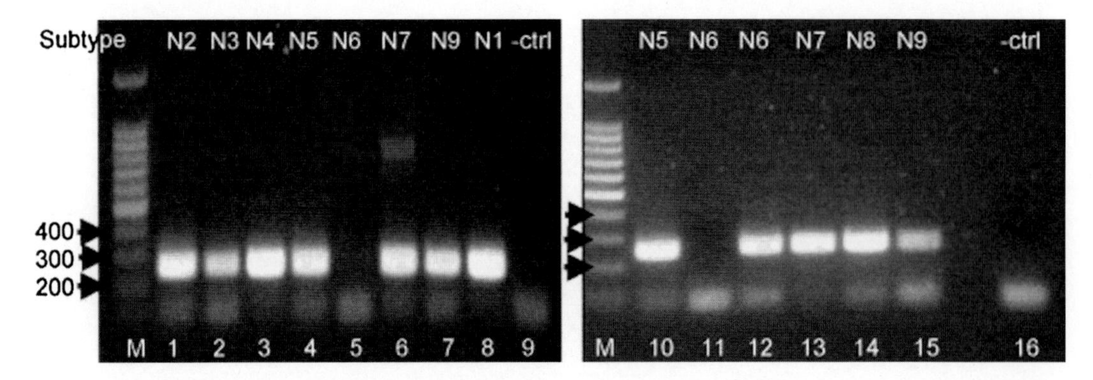

Fig. 1. One step RT-PCR amplification of the segment 6 from nine neuraminidase subtypes. A fragment of approximately 253 bp was amplified using the primers containing M13 "tags". M, 100 bp DNA Ladder (Promega); (1) H9N2, (2) H16N3, (3) H8N4, (4) H14N5, (5) H13N6, (6) H10N7, (7) H11N9, (8) H5N1, (9) negative control (water instead of template), (10) H6N5, (11) H13N6, (12) H14N6, (13) H7N7, (14) H3N8, (15) H11N9, (16) negative control (water) (reproduced from (13)). The H13N6 sample used in this experiment failed to amplify, but nine other N6 viruses were successfully amplified

3. Primer sequences are:

   PB2(1)-F AGYTCITCYTTYAGYTTYGG (multiplicity = 32)

   PB2(1)-R CIGGIGAYARKAGYAYRTTYC (multiplicity = 128)

   PB2(2)-F GAIGTIAGYGARACMCARGG (multiplicity = 16)

   PB2(2)-R AGTATYCTCATYCCWGANCC (multiplicity = 32) (see Note 9)

   (see Note 14 for the codes used to represent nonstandard bases and mixed bases).

   (see Note 15).

4. The reaction is set up as a single step procedure without the separate production of cDNA. This procedure is faster, requiring shorter cycle times and no manipulation between the RT and PCR steps.

   As follows:

25 μL of superscript III 2× reaction mix (contains 0.4 mM each dNTP, 3.2 mM $MgSO_4$)
2 μL of forward primer (PB2(1)F 160 μM solution)
2 μL of reverse primer (PB2(2)R 160 μM solution)
2 μL of Superscript III enzyme
17 μL of RNase free H_2O
2 μL of template RNA in RNase free water (added last)

"Touchdown" program for thermal cycler:

46°C	30 min
60°C	10 min

PCR cycle:

94°C	2 min
94°C	15 s
56°C	30 s

Decrease annealing temperature in 2°C increments for each cycle, until annealing temperature reaches 42°C, so that the last "touchdown" cycle is:

94°C	15 s
42°C	30 s
68°C	1 min 15 s
94°C	15 s
40°C	30 s
68°C	1 min 15 s
Repeat for 36 cycles	
68°C	5 min

(see Note 18).

5. In the negative control, water for injections BP or RNase free water was used instead of template RNA.
6. PCR products (10 μL of sample plus 2 μL of bromophenol blue loading dye) were visualized by gel electrophoresis on 1.5% agarose containing ethidium bromide (Fig. 2).

3.6. Sequencing

1. The 253 bp products with M13 tags, from the neuraminidase RT-PCR are directly sequenced as described below.
2. For RT-PCR products without M13 tags, indirect sequencing is performed after a second round of PCR (see Subheading 3.6, item 2).

3.6.1. Direct Sequencing

1. Whole reaction volumes of the 253 bp amplicon with M13 tags were loaded on 1.5% agarose containing ethidium bromide and gel purified using QIAquick Gel Extraction Kit (see Note 20).

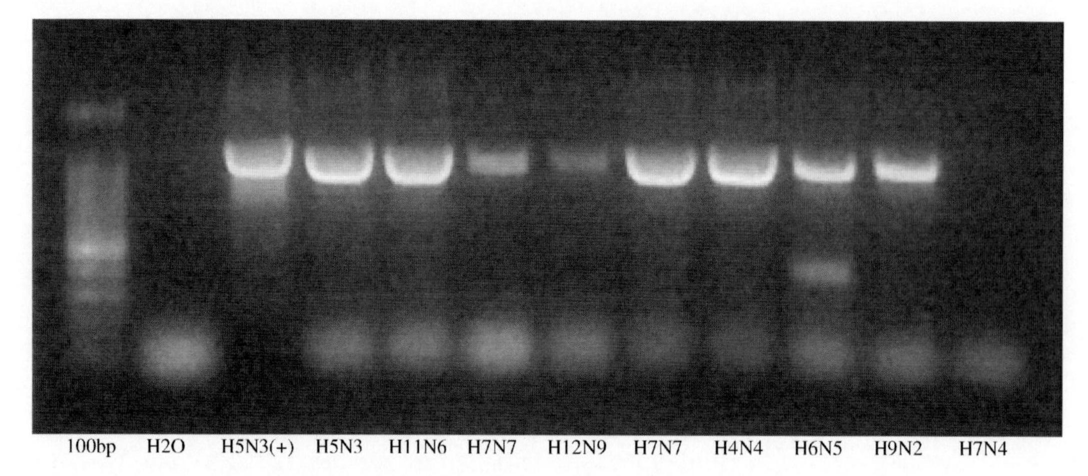

Fig. 2. PB2 gene segment of Australian Influenza A isolates (except H9N2). Product of approx. 986 base pairs generated using external PB2 primers (PB2(1)-F and PB2(2)-R) (see Note 19)

2. Sequencing reactions using M13 sequencing primers were completed at CSIRO-AAHL (see Note 21) or at the Australian Genome Research Facility (see Note 21), using an AB3730xl.
3. Results were analyzed and influenza virus subtypes determined by BLAST analysis of the full length sequence using the BLAST (9) alignment tool available at http://blast.ncbi.nlm.nih.gov/Blast.cgi.

3.6.2. Indirect Sequencing

1. PCR products (10 μL of reaction mix) of the 219 bp one-step RT-PCR product were initially visualized by gel electrophoresis on 1.5% agarose containing ethidium bromide.
2. The remaining reaction volumes were gel purified using QIAquick Gel Extraction Kit (see Note 20).
3. Six ng of gel-extracted cDNA (see step 2, above) was used as template for a PCR using M13-NA8F and M13-NA10R to produce the 253 bp fragment with M13 tags. The PCR was performed using *Taq* DNA polymerase, recombinant kit with the following modifications: 4 μmoles/L of each primer NA8F-M13 and NA10R-M13 were used.
4. Thermal cycling was performed with the following conditions: 3 min at 94°C (initial denaturation), 8 cycles of step-down PCR consisting of 30 s at 94°C (denaturation), 30 s at 56°C then decrease 2°C each cycle until 42°C; 75s at 68°C (extension), followed by 30 cycles of 30 s at 94°C, 30 s at 43°C, 75 s at 72°C, with a final extension of 10 min at 72°C. In the negative control, water for injection BP was used instead of template RNA. Positive controls included RNA from clinical samples of influenza A subtype N2.
5. PCR products (10 μL of sample) were visualized by gel electrophoresis on 1.5% agarose containing ethidium bromide.

Sequencing was performed as described for the direct sequencing method. Results were analyzed and influenza virus subtypes were determined using the BLAST (10) alignment tool available at http://blast.ncbi.nlm.nih.gov/Blast.cgi.

3.7. Sensitivity of the RT-PCR Assay

The analytical sensitivity of the one-step RT-PCR using NA8F-M13 and NA10R-M13 primers is determined as follows:

1. A plasmid (pGEM-T) containing the NA fragment amplified by NA8F and NA10R primers, from RNA from Influenza A/Chicken/Cambodia/1A/04/H5N1 was used to produce in vitro transcribed RNA as follows:
2. Plasmids containing the NA gene segment were linearized by digestion using the restriction enzyme Sty1 and were purified using the Charge Switch Pro PCR clean up kit, with elution of purified DNA into a 30 μL volume.
3. After purification, the DNA template was in vitro transcribed using the MAXI script Kit according to these specifications: 1 μL of 10 mM UTP, 1 μL of 10 mM GTP, 1 μL of 10 mM CTP, 1 μL of 10 mM ATP, 2 μL T7 polymerase enzyme mix, 2 μL of 10× transcription buffer, and 1 μg of DNA template (linearized plasmid) in a volume up to 6 μL; nuclease free water was added to make the volume to 20 μL.
4. All reagents were mixed thoroughly by spinning briefly in a bench top microcentrifuge and incubated for one hour at 37°C.
5. The reaction was DNase treated using 3 μL of TurboDNase for 15 min at 37°C.
6. 5 μL of RNA were aliquotted into 0.2 ml-plastic PCR tubes.
7. The expected size of RNA product was verified by electrophoresis in a 1.5% RNAse free agarose gel, stained with ethidium bromide (Fig. 2).
8. A PCR reaction using NA8F and NA10R primers was used to check for any DNA carryover. The protocol was as described under Subheading 3.4 except Taq polymerase was used alone, without reverse transcriptase.
9. The concentration of the transcribed RNA (ng/μL) was quantified using the Nanodrop® ND-1000 UV-Vis spectrophotometer. Conversion of ng/μL of single stranded RNA to pmol/μL was performed using the following formula: pmol/μL = ng/μL (of ssRNA) × (1 μg/1,000 ng) × (10^6 pg/1 μg) × (1 pmol/340 pg) × (1/N); where N = 324 bp, the number of bases of the RNA transcript, and 340 pg/pmol is the average molecular weight of a ribonucleotide.
10. The copy number/μL transcribed RNA was calculated as follows: copy number/μL RNA transcript = (RNA in mol/μL) × (Avogrado constant, 6.023×10^{23} molecules/mol).

11. Serial ten-fold dilutions of in vitro transcribed RNA were diluted down to 10 copies/μL in nuclease free water (see Note 22).
12. 2 μL of undiluted RNA stock was used as a positive control, and 2 μL of each serial dilution was used for the one-step RT-PCR.
13. Products (10 μL/sample) were visualized by gel electrophoresis on 1.5% agarose containing ethidium bromide (see Note 23).

4. Notes

1. The RNeasy kit is used to extract RNA from cultured virus, whereas the MagNA Pure LC total nucleic acid isolation kit (Roche cat. No. 03038505) is used for clinical samples.
2. It is paramount to work in three separate areas. One area should be designated just for preparation of the master mix and another area designated just for adding the template. This is crucial to avoid contamination of the RT-PCR reagents with aerosols containing the template. A third area, remote from the other two, should be used for running gels and visualizing RT-PCR products. Pipette sets must be dedicated for use in these areas and must not be moved between areas.
3. A selection of 1,101 full-length segment 6 sequences from all nine subtypes, of a range of host species and from different geographical locations were retrieved and aligned.
4. GAP is the number of viruses for which information is lacking regarding nucleotide composition at a particular position of a nucleic acid alignment.
5. Multiplicity of primers for an RT-PCR reaction was calculated by multiplying the number of different bases possible at each position in a primer sequence.
6. In oligonucleotide duplexes, deoxyinosine forms hydrogen bonds with the complementary base in the following order of stability dC > dA > dG > dT > dU (11).
7. Minimum, maximum, and a mean melting temperature of each chosen semi-conserved sequence region with multiplicity <195 was calculated using the OligoAnalyzer 3.1 software from Integrated DNA technologies (available from http://www.idtdna.com/analyzer/applications/oligoanalyzer/). The *range* of melting temperatures for each semi-conserved region was used to determine the approximate temperatures for the touchdown PCR cycling. The lowest annealing temperature for the RT-PCR cycling was calculated based on

the lowest mean melting temperature of the pair of sequence regions that met specifications (e.g. multiplicity, size, GAP, and Tm) –5°C.

8. Carrying out the alignment of potential primer sequences against the sequences grouped into subtype specific folders facilitates faster alignment and allows assessment of the quality of primer matches with each subtype. A mismatch of 2 or more bases of the chosen oligonucleotide with the aligned sequences in the last five 3′ terminal bases suggests that the RT-PCR using this primer might not amplify the samples that contain sequences with those mismatch bases.
9. Shorter products are more easily amplifiable from samples in which RNA may be partially degraded (eg. from wildlife samples). Hence, in this case, two sets of primers were designed. One set (PB2(1)) was designed to amplify an approximately 589 base pair product from the 5′ end of the PB2 gene segment. The second set (PB2(2)) was designed to amplify an approximately 395 base pair product. The external primers from each set could be used to amplify a single long product of approximately 986 base pairs from RNA from cultured viruses.
10. Researchers are urged to familiarize themselves with the applicable national standards and regulations for containment and handling of influenza A virus and for safety in microbiology laboratories. For example, practice in Australia is guided by ref. 12.
11. This was carried out according to the OIE (World Organisation for Animal Health) Manual of Diagnostic Tests and Vaccines for Terrestrial Animals (8). Detection of HA activity, in bacteria-free amnio–allantoic fluids, indicates a high probability of the presence of an influenza A virus or of an avian paramyxovirus.
12. MagNA Pure LC Total Nucleic Acid islation kit, version July, 2007, cat, no. 03 038 505 001. The external lysis protocol is described in detail at www.roche-applied-science.com.
13. Since the primers have moderate multiplicity, the concentration of primer needs to be greater than in a standard PCR. This is because only a small fraction of the primer may be exactly complementary to the target sequence in the first few rounds of the RT-PCR. This fraction is potentially as small as the reciprocal of the primer multiplicity.
14. Note X: **R** = A,G **Y** = C,T **M** = A,C **K** = G,T **S** = C,G **W** = A,T **H** = A,C,T **B** = C,G,T **V** = A,C,G **D** = A,G,T **I** = Inosine.
15. These primer sequences are claimed in patent number WO 2008/000023.

16. The NA8F-M13 and NA10R-M13 primers are identical to the NA8F and NA10R except they have an M13 sequence appended to the 5′ ends. The 5′ extension is referred to as a "tag".
17. The temperature of 68°C is used for extension. This is the temperature recommended for the Platinum® *Taq* DNA polymerase.
18. For longer PCR products (eg. The 986 base pair PB2 product), 1 min 15 s extension time is used. For shorter products (e.g. those generated using the combination of primers PB2(1) F and PB2(1) R (approx 589 bp) or PB2(2) F and PB2(2) R (approx 395 bp)), a shorter extension time of 40 s can be used.
19. The failure of amplification for H7N4 in the experiment shown could be due to RNA degradation, because shorter PB2 products of 395 and 589 bp were amplifiable from this same sample under similar conditions using PB2(2)F and R or the PB2(1) F and R sets of primers respectively.
20. See detailed protocol in QIAquick® Spin Handbook, July 2002, page 23: QIAquick Gel Extraction Kit Protocol: htttp://www1.qiagen.com/literature/handbooks/PDF/DNACleanupAndConcentration/QQ_Spin/1021422_HBQQSpin_072002WW.pdf
21. CSIRO-AAHL: Commonwealth Scientific Industrial Research Organisation, Australian Animal Health Laboratories; Private Bag 24, Geelong, VIC 3220. AGRF: Australian Genome Research Facility; Level 5 Gehrmann Laboratories University of Queensland, Research Road, St Lucia, QLD, 4072, Australia.
22. Take care with pipetting (i.e. pipette the sample no more than twice to avoid shearing the RNA). It is very important to change filtered tips when making each serial dilution. Carryover of RNA/DNA can occur if tips are not changed between dilutions, and an incorrect result can be obtained for the detection limit. In addition, it is important that when adding the template to the RT-PCR mix, the experimenter should start from the lowest serial dilution to the higher serial dilution (e.g. smallest in concentration to higher concentration) to avoid contaminating the lower concentration with aerosols that can be generated during pipetting.
23. The detection limit as determined by the presence of a clear band in ethidium bromide stained agarose gel varied depending on the N type. For cultured viruses H5N1 and H3N8, a band of the expected size was visible for 10^3 and 10^4 copies respectively. Such variation is not unexpected because the sequence of the regions complementary to the PCR primers will be different for different subtypes.

Acknowledgments

We thank Dr Wen-Bin Chen of Pathology, Queensland for assisting with samples provided for this work. We thank Elena Virtue and Tomoko Nagasaki for technical assistance during the testing of the Neuraminidase RT-PCR and PB2 RT-PCR respectively, and Paul Selleck for culture of the viruses. We thank Dr. Mark Gibbs and Prof. Adrian Gibbs for their advice on bioinformatics. This work was supported by Biochip Innovations Ltd., the Australian Biosecurity CRC for Emerging Infectious Diseases, the University of Queensland and CSIRO. Parts of this work have been published as a research article in the open access *Virology Journal* (10).

References

1. Drake JW, Holland JJ (1999) Mutation rates among RNA viruses. Proc Natl Acad Sci USA 96:13910–13910
2. Don RH, Cox PT, Wainwright BJ, Baker K, Mattick JS (1991) 'Touchdown' PCR to circumvent spurious priming during gene amplification. Nucleic Acids Res 19:4008
3. World Health Organization (2008) Influenza A (H1N1) virus resistance to oseltamivir – last quarter 2007 to 4 April 2008. http://www.who.int/csr/disease/influenza/h1n1_table/en/index.html
4. Yao Y, Mingay LJ, McCauley JW, Barclay WS (2001) Sequences in influenza A virus PB2 protein that determine productive infection for an avian influenza virus in mouse and human cell lines. J Virol 75:5410–5415
5. National Center for Biotechnology Information (2008) Influenza virus resource database (IVRD). http://www.ncbi.nlm.nih.gov/genomes/FLU/
6. Shaner MC, Blair IM, Schneider TD (1993) Sequence logos: a powerful, yet simple tool. In: Mudge TN, Milutinovic V, Hunter L (eds) Proceedings of the twenty-sixth annual hawaii international conference on system sciences. Architecture and biotechnology computing volume 1. IEEE Computer Society Press, Los Alamitos, CA, pp 813–821
7. Hall T (2007) Biological sequence alignment editor for Win95/98/n2/2K/XP (BioEdit). http://www.mbio.ncsu.edu/BioEdit/bioedit.html
8. The World Organisation for Animal Health (2008) OIE terrestrial manual, pp 467–470
9. Altschul SF, Madden TL, Schaffer AA, Zhang J, Zhang Z, Miller W, Lipman DJ (1997) Gapped BLAST and PSI-BLAST: a new generation of protein database search programs. Nucleic Acids Res 25:3389–3402
10. Alvarez AC, Brunck MEG, Boyd V, Lai R, Virtue E, Chen W, Bletchly C, Heine HG, Barnard R (2008) A broad spectrum, one-step reverse-transcription PCR amplification of the neuraminidase gene from multiple subtypes of influenza A virus. Virol J 5:77
11. Kawase Y, Iwai S, Inoue H, Miura K, Ohtsuka E (1986) Studies on nucleic acid interactions I. Stabilities of mini duplexes (dG2A4XA4G2*dC2T4YT4C2) and self-complementary d(GGGAAXYTTCCC) containing deoxyinosine and other mismatched bases. Nucleic Acids Res 14:7727–7736
12. Standards Australia (2002) AS/NZS 2243.3:2002 Safety in laboratories: Part. 3 microbiological aspects and containment facilities

Chapter 6

Detection and Identification of CD46 Splicing Isoforms by Nested RT-PCR

Anita Szalmás, József Kónya, István Sziklai, and Tamás Karosi

Abstract

CD46 (Membrane Cofactor Protein, MCP) is a transmembrane glycoprotein, which is expressed by all nucleated human cells whose purpose is to protect against autologous complement attack. In addition, CD46 can serve as a receptor for several viruses and bacteria and as a potent regulator of the inflammatory response by affecting T cell differentiation. Multiple isoforms of CD46 exist due to alternative splicing and are coexpressed in human cells in various patterns and expression levels. However, specific diseases have not been associated with isoform coexpression. We applied a nested RT-PCR method to investigate the coexpression pattern of CD46 splicing variants in otosclerotic and normal stapes footplate specimens. Using this method, we detected an altered isoform expression pattern and identified four novel CD46 splicing variants overexpressed in otosclerotic bone. This study is the first comprehensive report to provide evidence for disease associated alternative splicing of CD46.

Key words: Alternative splicing, CD46, Measles virus, Nested RT-PCR, Otosclerosis

1. Introduction

Otosclerosis (otospongiosis) is a complex bone remodeling disorder of the human otic capsule, which is associated with persistent measles virus infection (1). Otosclerosis results in bony fixation of the stapes (the third hearing ossicle) into the oval window niche, causing conductive and/or sensorineural hearing loss (1). Until now, the exact molecular pathogenesis of otosclerosis remains unclear. However, we hypothesized that altered paramyxovirus favoring receptor expression pattern of the otic capsule could form the pathologic background of otosclerotic bone remodeling (1).

Nicola King (ed.), *RT-PCR Protocols: Second Edition*, Methods in Molecular Biology, vol. 630,
DOI 10.1007/978-1-60761-629-0_6, © Springer Science+Business Media, LLC 2010

The main cellular receptor for the measles virus is the CD46 molecule, which is also known as human membrane cofactor protein (MCP) and serves as a cofactor for inactivation of complement components C3b and C4b by serum factor I (2, 3). The primary role of CD46 is the protection of host cells from complement activation-mediated damage by inactivation of C3b/C4b complexes deposited on the cellular membrane (3). CD46 has also been referred as a "pathogen magnet" because it serves as a receptor for different viruses (e.g. measles virus, HHV-6) and also for bacteria (*Neisseria* species) (2, 4, 5).

The gene of CD46 has 14 known splicing isoforms, which are expressed in various patterns and expression levels on the surface of all nucleated human cells (6). The mRNA of CD46 is translated from a single gene (1q32) comprising 14 exons (7, 8) (Fig. 1). Exons 1–4 encode the extracellular Complement Control Protein Repeats (CCP) region of the molecule (5, 8) (Fig. 1). CCPs are followed by the Short Consensus Repeats (SCR, exons 5–6) and the Serine-Threonine-Proline (STP) rich coding regions (exons 7–9) (5, 8) (Fig. 1). Exons 11 and 12 encode the transmembrane domain (TM), while exons 13 and 14 encode the cytoplasmic tail (CYT) (5, 8) (Fig. 1). Exons 1–6, 10, 11, and 14 are highly conserved regions (5, 8) (Fig. 1). Exons 7–9, and 13 form the hypervariable region, supporting the molecular heterogeneity (5, 8). Common isoforms (a-l) of CD46 are featured by two types of cytoplasmic domains: a shorter *CYT-1* (exon 13 (stop codon) and 14; 17 amino-acids) and a longer *CYT-2* (exon 14; 24 amino-acids) (9) (Fig. 1). However, isoforms "m" and "n" are featured by two quite rare types of cytoplasmic tail with an unclear signaling process: *CYT3* (exon 14; 25 amino acids) and *CYT4* (exon 14; 25 amino acids) (9, 10) (Fig. 1).

Signals mediated by CD46 result in TCR (T-cell receptor) dependent proliferation and induction of primary human $CD4^+$ T-cells, and also the genesis of regulatory T1-cells (9–11). It should be established, that signals mediated by CD46 have greater effectiveness on T-cell activation than of those mediated by CD3 and CD28 molecules (9, 10, 12).

Despite its essential role in complement regulation and pathogen invasion, specific diseases have not yet been attributed to CD46 isoform coexpression (6). We have previously shown that otosclerotic bone is featured by a unique CD46 isoform coexpression pattern, which could be the basis of the pathogenesis of otosclerosis (1).

This study presents a nonconventional nested RT-PCR technique for the simple detection and identification of CD46 splicing variants in bone specimens. Using this method, four novel and disease associated CD46 variants were detected in otosclerotic bone (1).

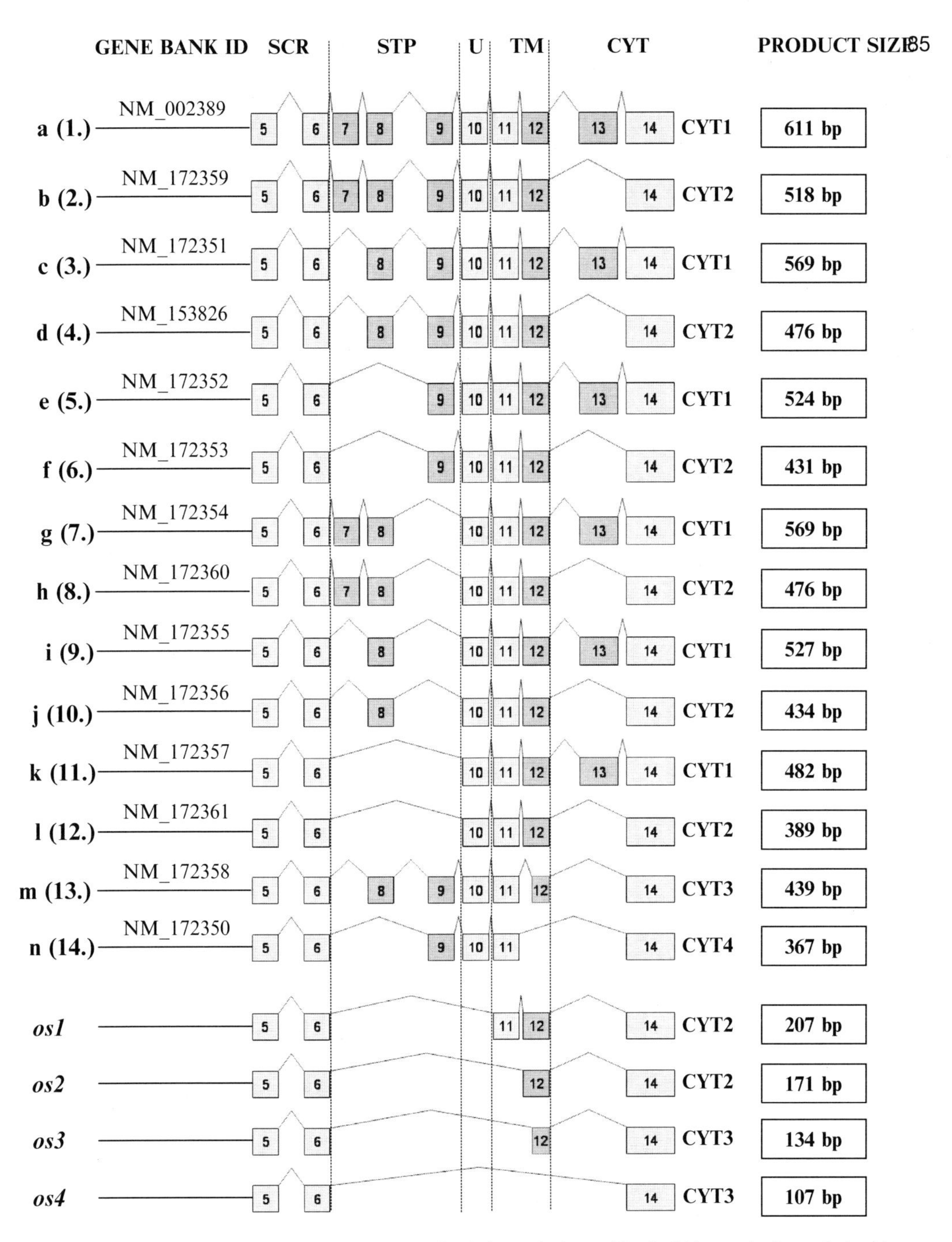

Fig. 1. The mRNA of CD46 has 14 exons (*grey boxes*) spliced alternatively resulting in 14 known isoforms derived from a single gene (1q32). Exons 1–4 encoding the CCPs region (Complement Control Protein repeats) are not shown, because this is highly conserved region. The CCPs are followed by the SCR- (Short Consensus Repeat; exons 5–6) and the heavily O-glycosylated STP-coding regions (Serine-Threonine-Proline rich; exons 7–9). STP is followed by the transmembrane domain (TM; exons 11–12) and cytoplasmic tail (CYT; exons 13–14). Exon 10 encodes a region of unknown (U) significance. There are four known types of cytoplasmic domains: *CYT-1* (exons 13 and 14; 17 amino acids), *CYT-2* (exon 14; 24 amino acids), *CYT3* (frame shift of exon 14; 25 amino acids) and *CYT4* (frame shift of exon 14; 25 amino acids). Four novel CD46 splicing variants were detected in each of the otosclerotic specimens named *os1, os2, os3* and *os4.* The transmembrane domain was partially missed from the newly described CD46 variants, whereas the virus receptor region was conservatively present. Isoform *os4* may be soluble and not expressed on the cell surface. Variants *os1* and *os2* are featured by CYT2 type cytoplasmic tail, while *os3* and *os4* contain the quite rare CYT3 type cytoplasmic end

2. Materials

2.1. Sample Preparation

1. Stapes footplate specimens removed during stapes surgery (stapedectomy) *or*
2. Alternative tissue samples (approximately 50 mg) from different organs (surgical biopsy, surgical resection, fine needle aspiration biopsy (FNAB) etc.).
3. 10% (w/v) buffered formaldehyde solution.
4. 0.5 M sodium-ethylene-diamino-tetraacetate (Na-EDTA, pH 8.0, with sodium-hydroxide).
5. 15 % (w/v) purified gelatine suspension.
6. 4% (w/v) paraformaldehyde solution.
7. 20% (w/v) saccharose solution.
8. Phosphate buffered saline (PBS, pH 7.4) containing sodium-azide 1.5% (w/v).

2.2. RNA Extraction and Nested RT-PCR

1. TRI Reagent™
2. Chloroform prechilled to 4°C.
3. Isopropanol prechilled to 4°C.
4. 75% (v/v) ethanol, prechilled to 4°C.
5. Diethyl-pyrocarbonate (DEPC) treated distilled H_2O: add 0.2 mL DEPC to 100 ml of distilled water (see Note 1).
6. Eppendorf centrifuge (preferably refrigerated).
7. High-Capacity cDNA Reverse Transcription Kit™ (Applied Biosystems).
8. AccuTaq™ LA DNA Polymerase (Sigma-Aldrich) with 10× PCR buffer.
9. dNTPs.
10. External primers (10 μM) (Table 1) (see Note 2):
    (a) CD46Kfor (5′-GAGTGTAAAGTGGTCAAATGTCG-3′)
    (b) CD46Krev(5′-GGAGTGGTTGATTTAGTCTGGTAA-3′).
11. Thermal cycler.

2.3. Detection and Purification of DNA Using Nondenaturing Polyacrylamide Gel Electrophoresis

1. 5× TBE electrophoresis buffer: 54 g of Tris-base, 27.5 g of boric acid, 20 mL of 0.5 M Na-EDTA (pH 8.0, with sodium-hydroxide) dissolved in 800 mL of distilled water and then made up to 1 L with distilled water.
2. 40% (w/v) acryl-amide (38 g of acryl-amide, 2 g of *bis*-acryl-amide) in distilled H_2O.

Table 1
Details of CD46 mRNA-specific oligonucleotide primers for nested RT-PCR

Primer	Sequence (5′–3′)	Tm (°C)[a]	Exons in CD46 mRNA	Annealing site in CD46 gene
46Kfor (+)	GAG TGT AAA GTG GTC AAA TGT CG	62.3	5–6 border	+1,432 to +1,470
46Krev (–)	GGA GTG GTT GAT TTA GTC TGG TAA	62.0	14	+3,394 to +3,417
46Bfor (+)	TCG ATG GCA GCG ACA CAA	68.2	6	+1,567 to +1,584
46Brev (–)	GGT AAG TGG CAT ATT CAG CTC CA	65.9	14	+3,376 to +3,398

[a] *Tm* melting temperature (see Note 4)

3. 10% (w/v) ammonium-persulfate (APS, freshly made).
4. TEMED.
5. Gel loading buffer: 0.25% (w/v) bromophenol blue, 0.25% (w/v) xylene-cyanol, and 30% (v/v) glycerol in distilled water.
6. Molecular weight markers.
7. *Fixation solution*: 10% (v/v) ethanol, and 0.5% (v/v) acetic acid in H_2O; 0.2% (w/v) silver-nitrate in H_2O.
8. *Developing solution*: 1.5% (w/v) sodium-hydroxide, and 0.05% (w/v) formaldehyde in H_2O.
9. 0.5 μg/mL of ethidium-bromide.
10. Elution buffer for DNA recovery from poly-acryl-amide gels: 0.5 M ammonium-acetate, 1 mM EDTA (pH 8.0, with sodium-hydroxide) (see Note 3)
11. 100% and 70% (v/v) ethanol.
12. 2.5 M sodium-acetate.
13. TE buffer: 10 mL 1 M Tris-Cl stock solution (60.57 g of Tris (hydroxyl-methyl amino-methane) dissolved in 0.5 L distilled water, pH adjusted to 7.5 using HCl) and 2 mL 500 mM Na-EDTA stock solution (18.6 g EDTA dissolved in 100 mL distilled water, pH adjusted to 8.0 using sodium-hydroxide) made up to 1 L with distilled water.
14. Sterile scalpel blades.
15. UV transluminator.

16. Vertical gel electrophoresis system.
17. Heating block.
18. Eppendorf centrifuge.

2.4. Nested PCR: Secondary Amplification and DNA Purification before Sequencing

1. AccuTaq™ LA DNA Polymerase (Sigma-Aldrich) with 10× PCR buffer.
2. dNTPs.
3. Internal primers (10 μM) (Table 1) (see Note 2):
   (a) CD46Bfor (5′-TCGATGGCAGCGACACAA-3′)
   (b) CD46Brev (5′-GGTAAGTGGCATATTCAGCTCCA-3′).
4. Sodium-acetate (2.5 M).
5. 96% (v/v) ethanol, prechilled to 4°C.
6. 70% (v/v) ethanol, prechilled to 4°C.

2.5. Sequencing

1. Big Dye terminator 3.1 sequencing kit™ (Applied Biosystems).
2. ABI PRISM 3100-Avant Genetic Analyzer automated DNA sequencing system™ (Applied Biosystems).

3. Methods

3.1. Sample Preparation for RNA Extraction and Histological Analysis

1. Fix the (stapes) specimens in 5 ml of 10% buffered formaldehyde solution (see Note 5).
2. If necessary, osseous tissues should be decalcified in 10 mL of 0.5 M Na-EDTA (72 h, 4°C).
3. Embed the specimens in 15% purified gelatine (24 h, 56°C).
4. Refrigerate the gelatine-embedded specimens (4 h, 4°C) and form a cubical shaped gelatine block using a scalpel.
5. Refix the gelatine-embedded specimens in 5 mL of 4% paraformaldehyde (24 h, 20°C).
6. Cryoprotect the blocks in 10 mL of 20% saccharose solution (2 h, 4°C).
7. Cut the blocks by a cryomicrotome into 10 μm slides (–25°C).
8. Store slides in 5 mL of 0.1 M PBS (phosphate-buffered saline) containing 1.5% (w/v) sodium-azide at 4°C.
9. Consecutive 10 μm frozen cut sections could be examined as follows: (1) Nucleic acid extraction and nested RT-PCR amplification; (2) Staining with hematoxylin and eosin; (3) CD46 specific immunohistochemistry.

3.2. RNA Extraction

1. Homogenize the prepared sample in a 1.5 mL Eppendorf tube with 1 mL TRI Reagent® (see Note 6). Lyse cells by incubation for 10 min at room temperature.
2. Add 200 μL of chloroform to the lysate, and shake the tightly covered sample for 10–15 s, leave at room temperature for 5–10 min, and centrifuge at 12,500 × *g* for 15 min at 4°C. Following centrifugation, the mixture separates into a lower red phase, interphase, and an upper colorless phase. The upper aqueous phase contains the RNA.
3. For RNA precipitation, transfer the upper aqueous phase to a fresh 1.5 mL sterile tube and add 500 μL of isopropanol. Leave the sample at room temperature for 5–10 min and centrifuge at 12,500 × *g* for 15 min at 4°C.
4. Remove the supernatant carefully, wash the RNA pellet with 1 mL of 75% ethanol, and centrifuge at 12,500 × *g* for 5 min at 4°C.
5. Remove the ethanol very carefully as much as possible. Dry RNA pellet for 5 min at 37°C and resuspend in 20 μL of DEPC-treated distilled water. Incubate the RNA solution at 55°C for 15 min and then store at –70°C.

3.3. RT-PCR

1. Perform reverse transcription (RT step) using the High-Capacity cDNA Reverse Transcription Kit in a final volume of 20 μL containing: 100 ng of total RNA, 50 U of MultiScribe RT enzyme, 2 μL of 10× RT buffer, 10 U of RNase inhibitor, CD46 sequence-specific antisense primer (CD46Krev) at a final concentration of 1 μM, and dNTP mix at a final concentration of 100 μM for each deoxyribonucleotide, make up volume to 20 μL using RNase free water (see Note 7–9).
2. Incubate the reaction for 10 min at 25°C, for 120 min at 37°C, and finally for 5 s at 85°C. Store the cDNA product at -20°C.
3. Prepare a PCR reaction mix (25 μL) containing: 4 μL of the RT reaction mixture, 2.5 IU of AccuTaq™ LA DNA Polymerase, 2.5 μL of 10× PCR buffer, 0.2 μM of each primer, and 200 μM dNTP (50 μM each), make up volume to 25 μL using RNase free water.
4. Perform thermal cycling as follows: 2 min of initial denaturation at 94°C, followed by 35 cycles of denaturation for 15 s at 94°C, annealing for 30 s at 60°C, and extension for 1 min at 68°C followed by a final extension step at 68°C for 5 min.

3.4. Detection and Purification of DNA Using Nondenaturing Polyacrylamide Gel Electrophoresis (PAGE): Isolation of CD46 Splicing Variants

1. Prepare a nondenaturing poly-acryl-amide gel (10%) by mixing 1.25 mL of 40% acryl-amide, 1 mL of 5× TBE, and 2.75 mL of distilled water (5 mL total, sufficient for a 10×8 cm gel). Add 2.5 μL of TEMED and 25 μL of 10% APS to the gel solution immediately before pouring. Insert combs and leave the gel to polymerize for 20 min (see Note 10).
2. Load samples (approx. 5 μL) into the wells and perform electrophoresis at 110 Volts (constant, room temperature) for approx. 1.5 h along with a molecular weight marker. Run each sample in two gels concurrently.
3. After electrophoresis, stain the first gel with silver staining: 2× for 5 min with Fixation solution, 1× for 10 min with 0.2% silver-nitrate solution, then 1× for 5 min with Developing solution without formaldehyde, and finally 1× with Developing solution containing 0.05% formaldehyde until DNA bands become visible. Wash the gel with distilled water and analyze immediately (see Note 11) (see Fig. 2).
4. Stain the second gel with ethidium-bromide for 20 min, visualize the DNA bands of interest under ultraviolet light, and excise using a scalpel (see Note 12).
5. Transfer gel fragments into 1.5 mL Eppendorf tubes and treat with 3 volumes of Elution buffer overnight at 37°C.

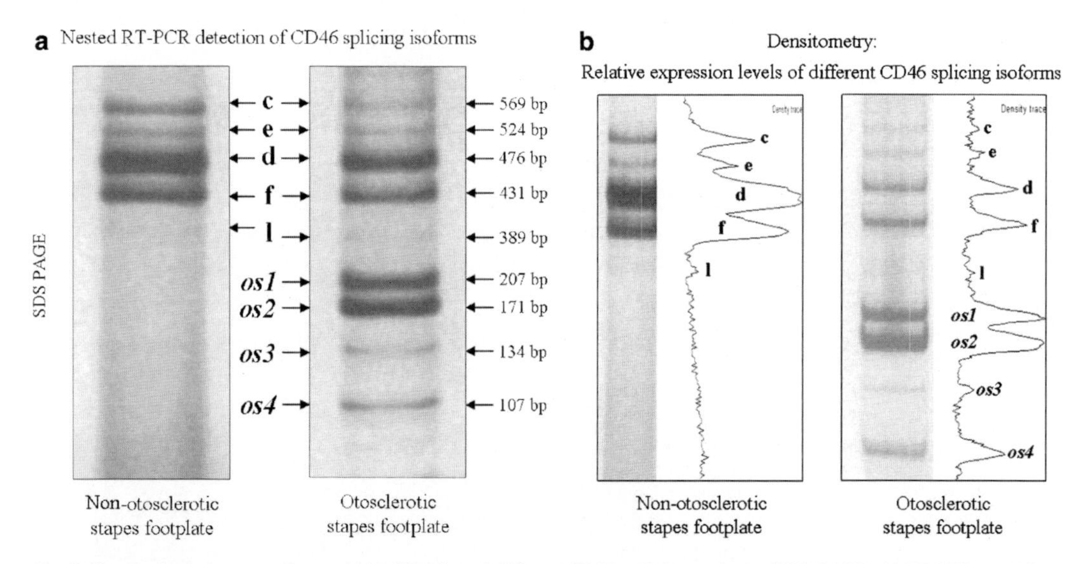

Fig. 2. Electrophoresis separation and identification of different CD46 splicing variants (SDS-PAGE). (**a**) RT-PCR amplification of CD46 mRNA in nonotosclerotic and in otosclerotic stapes footplates. Coexpression of "c", "d", "e", "f", and "l" isoforms was proved in all specimens, however, in otosclerosis, four short, unknown splicing variants were detected. (**b**) Densitometric representation of relative expression levels of different splicing variants of CD46 mRNA in non-otosclerotic- and in otosclerotic stapes footplates. In case of nonotosclerotic, ankylotic stapes footplate, variant "d" was expressed in the largest level which was followed by "c", "f", "e", and "l" isoforms. In otosclerosis, variants *os1* and *os2* show the highest expression level

6. After incubation, spin down the gel fragment by centrifugation at room temperature for 30 s at 5,000 × *g*, and then transfer the supernatant into a sterile Eppendorf tube (see Note 13).
7. Extract DNA from the supernatant with 2 volumes of ice-cold 100% ethanol and 1/10 volume of 2.5 M sodium-acetate, incubate in a –70°C freezer for 15 min or longer, or in a –20°C freezer for at least 30 min, and then centrifuge at 12,500 × *g* for 15 min at 4°C (see Note 14).
8. Remove the supernatant carefully, wash the DNA pellet with 1 mL of 70% ethanol, and centrifuge at 12,500 × *g* for 5 min at 4°C.
9. Decant supernatant, dry pellet, and resuspend in 20 μL of TE buffer or distilled water.

3.5. Nested PCR: Secondary Amplification before Sequencing

1. Prepare a PCR reaction mix (50 μL) containing: 5 μL of the purified primary PCR product, 2.5 IU of AccuTaq™ LA DNA Polymerase, 5 μL of 10× PCR buffer, 0.2 μM of both primer, and 200 μM dNTP (50 μM each), make up volume to 50 μL with RNase free distilled water (see Note 15).
2. Perform thermal cycling as follows: 2 min of initial denaturation at 94°C, followed by 40 cycles of denaturation for 15 s at 94°C, annealing for 30 s at 64°C, and extension for 1 min at 68°C followed by a final extension step at 68°C for 5 min.
3. Analyze the PCR products by running on a 0.5× TBE, 1.5% agarose gel.

3.6. Sequencing: Confirmation of PCR Product Specificity and Identification of Splicing Variants

1. Extract PCR products by adding 1/10 volume of 2.5 M sodium-acetate and 2.5 volumes of ice-cold 96% ethanol to the PCR mixtures, incubate in a –70°C freezer for 15 min or longer, or in a –20°C freezer for at least 30 min, and centrifuge at 12,500 × *g* for 15 min at 4°C (see Note 16).
2. Remove the supernatant carefully, wash the pellet with 100 μL of 70% ethanol, and centrifuge at 12,500 × *g* for 5 min at 4°C.
3. Remove the ethanol very carefully as much as possible. Dry pellet for 5 min at 37°C and resuspend in 20 μL of TE buffer or distilled water, and then store at –20°C.
4. Perform sequencing by using the Big Dye terminator 3.1 sequencing kit according to the manufacturer's instructions (see Note 17).
5. Analyze the sequencing products (see Note 18) (see Fig. 3).

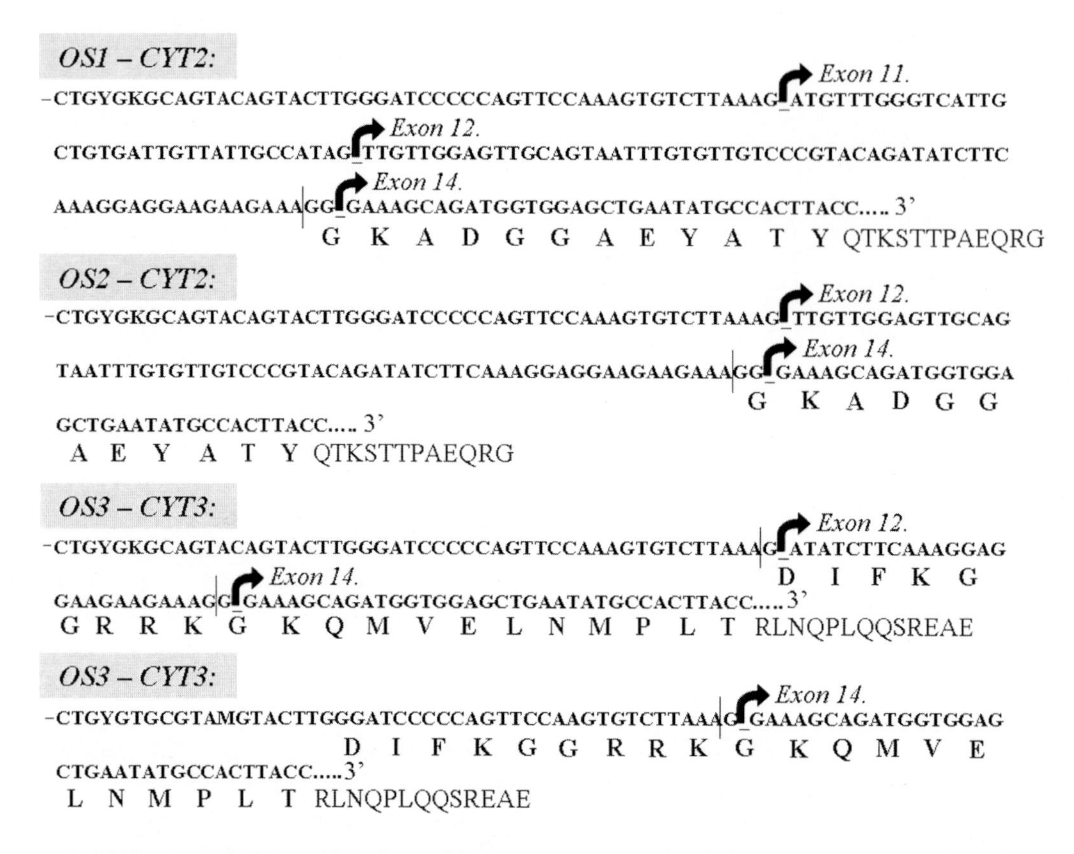

Fig. 3. Sequence analysis of the cDNA of novel CD46 splicing variants associated with otosclerosis. Isoforms *os1* and *os2* encode the CYT2 type cytoplasmic tail, whereas *os3* and *os4* encode the CYT3 type cytoplasmic end due to a frame shift at the beginning of exon 14 caused by alternative splicing. Starting nucleotides of the numbered exons are indicated by *curved arrows*

3.7. Conclusion

Although specific diseases have not yet been attributed to CD46 isoform co-expression, present results supply essential information about novel splicing variants of CD46 emphasizing specific association between isoform coexpression and pathogenesis of otosclerosis (1).

4. Notes

1. When preparing DEPC distilled water, shake vigorously to get the DEPC into solution. Autoclave the DEPC treated distilled water to inactivate remaining DEPC. Wear gloves and use a fume hood when using DEPC, as it is suspected to be carcinogenic.
2. Although RT-PCR is a sensitive and robust method for examination of the presence of CD46 variants within the transcribed cDNA, several disadvantages may occur. In our study,

the various kinds of similar-sized amplicons due to alternative splicing were relatively difficult to identify. Therefore, we invented a nested RT-PCR method with a selective purification step between the first and the second round PCR amplifications.

3. The elution buffer should be protected from light, and it is stable for several months at room temperature.
4. The high and uniform melting temperature (Tm) of the primer pairs is important to avoid unspecific annealing of one primer at a lower temperature demanded by the other primer.
5. If you do not intend to perform histological analysis, the following steps are not necessary. In this case, homogenize the tissue in a sterile dry rubber cup and make a suspension by adding DEPC distilled water and just follow the RNA extraction using the methods described under Subheading 3.2.
6. Several kits are also commercially available for quick RNA isolation.
7. The cDNA sequences for known human CD46 isovariants can be retrieved from the EMBL GenBank database (http://www.ncbi.nlm.nih.gov/sites/entrez?db=nucleotide). Using these sequences, external primers were designed to annelate conservative sequences of CD46 cDNA, and the forward external primer was designed to span an intron in its genomic sequence, thus ensuring mRNA specific amplification. Primers span the entire variable molecule region affected by alternative slicing, thus generating different sized PCR products for the different splice variants.
8. When dealing with a small amount or partially degraded RNA, it is essential to choose an appropriate RT-PCR kit.
9. Either random hexamers or gene sequence-specific primers can be used in cDNA synthesis. The use of a specific primer for RT step leads to higher sensitivity and specificity via generating cDNA that is enriched for the specific target.
10. PAGE was used for visualization and preparative separation of cDNA PCR products because polyacrylamide gels provide better resolution than agarose gels. Agarose gel electrophoresis is inappropriate for detection of under-represented splice variants, and for differentiation between similar-sized PCR products.
11. Silver staining method is widely used to visualize DNA fragments and other organic molecules with unsurpassed detail following traditional PAGE. To achieve the best sensitivity and visual clarity, clean handling is important. In our study, densitometry software was used for determining the band

intensity of PCR products on silver-stained poly-acryl-amide gels. In view of the concurrent amplification of competing PCR products, the evaluation of signals by densitometry is semi-quantitative.

12. Remove the band from the gel as cleanly as possible. Repeat purification of amplimers until obtaining single bands. Ultraviolet light produces strand breaks in DNA molecules. Ensure quick recovery of the bands. When we recover multiple bands, we block the UV light on other parts of the gel using cardboard under the gel support. Addition of 1 mM cytidine or guanosine to the gel and electrophoresis buffers could protect DNA during translumination.
13. An additional washing step can be included at this point in the protocol. Wash the gel with 100 μL of Elution solution, spin down by centrifugation, and remove supernatant to a separate 1.5 mL tube.
14. For efficient DNA extraction, it is important to use ice-cold 100% or 96% ethanol and subsequent incubation in a freezer (at –70°C for at least 15 min, or at –20°C for at least 30 min).
15. Using a nested primer pair for a second amplification further increases sensitivity and results in a high amount of DNA eligible for sequencing. Thus, oligonucleotides that also annelate the conservative regions of the CD46 gene and amplify a sequence internal to the primary PCR products were selected. This nested primer pair (CD46B for and CD46 Brev) also span the variable region of CD46 mRNA and is therefore able to amplify all purified first round PCR products. These nested primers amplify a 140-bp shorter sequence compared to the first round primers.
16. The PCR products need to be purified before the sequencing reaction in order to remove the excess primer and dNTPs left from the PCR, which can interfere with the activity of Taq DNA Polymerase in the sequencing reaction. This purification will also remove the excess salts and proteins, which can interfere with the capillary electrophoresis. Before sequencing, always check the purified PCR fragments on an agarose gel.
17. Specificity of PCR products can be confirmed by direct nucleic acid sequencing using the PCR oligonucleotides as primers. Restriction mapping can also be used to identify alternative splicing variants. In our case, detection of the presence of given exons is possible using specific restriction endonucleases. The amplified CD46 sequence contains unique restriction sites for *Hinf*I in exon 7, *Acc*I in exon 8, *Stu*I in

exon 9, *Hpy*CH4V in exon 12, and *Bsa*I in exon 13. However, novel splice variants cannot be identified by this method.

18. Known and novel CD46 splicing variants could be successfully identified by comparing the sequencing results to CD46 cDNA reference sequences in the European Molecular Biology Laboratories (EMBL) GenBank nucleotide sequence database.

Acknowledgments

This work was supported by grants from Hungarian Scientific Research Fund (OTKA PD75371) and Mecenatúra Fund of Debrecen University (DE OEC Mec 17/2008).

References

1. Karosi T, Szalmás A, Csomor P, Kónya J, Petkó M, Sziklai I (2008) Disease-associated novel CD46 splicing variants and pathologic bone remodeling in otosclerosis. Laryngoscope 118(9):1669–1676
2. Liszewski MK, Kemper C, Price JD, Atkinson JP (2005) Emerging roles and new functions of CD46. Springer Semin Immunopathol 27(3):345–358
3. Riley-Vargas RC, Gill DB, Kemper C, Liszewski MK, Atkinson JP (2004) CD46: expanding beyond complement regulation. Trends in Immunol 25(9):496–503
4. Erlenhofer C, Duprex WP, Rima BK, ter Meulen V, Schneider-Schaulies J (2002) Analysis of receptor (CD46, CD150) usage by measles virus. J Gen Virol 83(6): 1431–1436
5. Dhiman N, Jacobson RM, Poland GA (2004) Measles virus receptors: SLAM and CD46. Rev Med Virol 14(4):217–229
6. Riley RC, Tannenbaum PL, Abbott DH, Atkinson JP (2002) Cutting edge: inhibiting measles virus infection but promoting reproduction: an explanation for splicing and tissue-specific expression of CD46. J Immunol 169(10):5405–5409
7. Schneider-Schaulies J, Dunster LM, Kobune F, Rima B, ter Meulen V (1995) Differential downregulation of CD46 by measles virus strains. J Virol 69(11):7257–7259
8. Manchester M, Naniche D, Stehle T (2000) CD46 as a measles receptor: form follows function. Virology 274(1):5–10
9. Zaffran Y, Destaing O, Roux A, Ory S, Nheu T, Jurdic P, Rabourdin-Combe C, Astier AL (2001) CD46/CD3 costimulation induces morphological changes of human T cells and activation of Vav, Rac and extracellular signal-regulated kinase mitogen-activated protein kinase. J Immunol 167(12): 6780–6785
10. Kemper C, Chan AC, Green JM, Brett KA, Murphy KM, Atkinson JP (2003) Activation of human CD4+ cells with CD3 and CD46 induces a T-regulatory cell 1 phenotype. Nature 421(6921):388–392
11. Astier A, Trescol-Biemont MC, Azocar O, Lamouille B, Rabourdin-Combe C (2000) Cutting edge: CD46, a new costimulatory molecular way for T cells that induces p120CBL and LAT phosphorylation. J Immunol 164(12):6091–6095
12. Kurita-Taniguchi M, Fukui A, Hazeki K, Hirano A, Tsuji S, Matsumoto M, Watanabe M, Ueda S, Seya T (2000) Functional modulation of human macrophages through CD46 (measles virus receptor): production of IL-12 p40 and nitric oxide in association with recruitment of protein-tyrosine phosphatase SHP-1 to CD46. J Immunol 165(9): 5143–5152

Chapter 7

Simultaneous Detection of Bluetongue Virus RNA, Internal Control GAPDH mRNA, and External Control Synthetic RNA by Multiplex Real-Time PCR

Frank Vandenbussche, Elise Vandemeulebroucke, and Kris De Clercq

Abstract

Bluetongue is an insect-borne disease of domestic and wild ruminants that requires strict monitoring by sensitive, reproducible and robust methods. Real-time reverse transcription polymerase chain reaction (RT-qPCR) analysis has become the method of choice for routine viral diagnosis. As false-negative test results can have serious implications; an internal/external control system should be incorporated in each analysis to detect RT-qPCR failure due to poor sample quality, improper nucleic acid extraction and/or PCR inhibition. To increase the diagnostic capacity and reduce costs, it is recommended to use a multiplex strategy which enables the amplification of multiple targets in a single reaction. This chapter describes the application of a triplex RT-qPCR for the simultaneous detection of bluetongue viral RNA, an internal control and an external control. The primer and probe sequences of the BTV RT-qPCR were taken from Toussaint et al. (J Virol Methods 140:115–123, 2007), whereas the internal and external RT-qPCRs were specifically designed to detect endogenous glyceraldehyde-3-phosphate dehydrogenase mRNA and a synthetic RNA, respectively. To maximize the sensitivity of the assay, the primer concentrations of the internal/external control reactions were limited and the amount of *Taq* DNA polymerase was increased. A comparison of the singleplex versus triplex RT-qPCR indicated that the triplex RT-qPCR exhibits a higher analytical sensitivity. Due to the incorporation of the internal/external control system, the triplex RT-qPCR allows an even more reliable and rapid diagnosis of bluetongue than the previously described singleplex RT-qPCR (J Virol Methods 140:115–123, 2007).

Key words: Multiplex, RT-qPCR, Bluetongue virus, Glyceraldehyde-3-phosphate dehydrogenase mRNA, Synthetic RNA, Internal control, External control

1. Introduction

Bluetongue (BT) is an insect-borne disease of domestic and wild ruminants that induces variable clinical signs depending on the species and the breed (1, 2). The disease is caused by the bluetongue

Nicola King (ed.), *RT-PCR Protocols: Second Edition*, Methods in Molecular Biology, vol. 630,
DOI 10.1007/978-1-60761-629-0_7, © Springer Science+Business Media, LLC 2010

virus (BTV), which is a member of the genus *Orbivirus* within the family *Reoviridae* (3).

Since an introduction of BTV is associated with substantial economic losses, the disease has to be strictly monitored by sensitive, reproducible and robust methods (4). Although the World Organization for Animal Health (Office international des epizooties, OIE) acknowledges various tests, real-time reverse transcription polymerase chain reaction (RT-qPCR) analysis has become the method of choice to demonstrate the absence or presence of viral nucleic acids. Different sample types, such as EDTA blood samples, spleen, brain or lungs, can be used for the diagnosis of BTV, with the first being the preferred sample type.

This chapter describes the application of a triplex RT-qPCR for the simultaneous detection of viral RNA (i.e., bluetongue virus), an internal control (IC, i.e., glyceraldehyde-3-phosphate dehydrogenase mRNA (IC-GAPDH)) and an external control (EC, i.e., synthetic RNA (EC-EXTR)). Since the IC-GAPDH consists of endogenous mRNA, a negative IC-GAPDH test result is indicative of poor sample quality. Absence of IC-GAPDH amplification can, however, also be due to an improper nucleic acid extraction and/or PCR inhibition. To exclude this possibility, a fixed amount of EC-EXTR is added to each sample prior to extraction, which allows monitoring of the extraction and RT-qPCR efficiency in more detail (5–7). To increase the diagnostic capacity, a multiplex strategy, which enables the amplification of multiple targets in a single reaction, is used (8). Although the number of fluorophores that can be simultaneously detected is limited due to spectral overlap, the multiplex strategy has been successfully transferred to real-time PCR and is widely used in clinical diagnostics (7, 9). The results presented here, clearly demonstrate that the use of a multiplex strategy is not necessarily accompanied by a reduction in sensitivity and can greatly improve routine diagnosis (10).

2. Materials

2.1. Equipments and Disposables

1. Pipettes capable of dispensing volumes <1 μL
2. Multichannel pipette capable of dispensing volumes of 4 μL
3. Sterile filter pipette tips
4. Sterile 1.5 mL low-retention Eppendorf tubes
5. 96-well PCR plates
6. Sealing foil
7. LightCycler® 480 Multiwell Plate 96
8. LightCycler® 480 Sealing Foil

9. Centrifuge for 96-well PCR plates
10. Spectrophotometer
11. Thermocycler
12. Real-time PCR cycler
13. PCR laminar flow cabinets
14. –20°C freezer
15. –80°C freezer

2.2. Solutions, Kits and Reagents

1. AmpliTaq Gold® PCR Master Mix (Applied Biosystems)
2. Tris–EDTA buffer (10 mM Tris–HCl, 1 mM EDTA; pH 8.0)
3. TOPO® TA Cloning® Kit (Invitrogen)
4. NucleoBond PC 20 (Macherey–Nagel)
5. RNase-free water
6. Riboprobe® System-T7 (Promega)
7. Recombinant, RNase-free DNase I
8. Sodium acetate solution (3M; pH 5.2)
9. Acid-phenol:chloroform:isoamyl alcohol solution (125:24:1; pH 4.5)
10. Chloroform:isoamyl alcohol solution (24:1)
11. RNA Storage Solution
12. NucleoSpin RNA Virus kit (Macherey–Nagel)
13. Proteinase K
14. Dimethyl sulfoxide (DMSO)
15. RNA UltraSense™ One-step Quantitative RT-PCR System (Invitrogen)
16. Platinum® *Taq* DNA polymerase (Invitrogen)
17. Platinum qPCR Supermix UDG

3. Methods

3.1. Preparation of EC-EXTR Synthetic RNA by Primer Overlap Extension

1. Prepare a 50 μL reaction containing 25 μL 2× AmpliTaq Gold® PCR Master Mix, 5 μL of each overlapping primer and 15 μL RNase-free water (see Notes 1–3).
2. Run the following program on a thermocycler: denaturation at 95°C for 5 min, 10 cycles of denaturation at 95°C for 15 s, annealing at 50°C for 15 s and elongation at 72°C for 15 s. Verify the size and yield of the obtained PCR product by gel electrophoresis.

3. Clone the PCR product into a pCR®2.1-TOPO® vector and purify the resulting plasmid using the NucleoBond PC 20 kit (see Note 4).
4. Reamplify the amplicon in a 50 μL reaction containing 25 μL 2× AmpliTaq Gold® Master Mix, 5 μL of each insert-overspanning M13 primer, 12.5 μL RNase-free water and 2.5 μL purified plasmid and run the following program on a thermocycler: 95°C during 5 min, 35 cycles of 95°C for 15 s, 50°C for 15 s and 72°C for 30 s and final extension at 72°C for 7 min (see Note 5).
5. Verify the size and yield of the obtained PCR product by gel electrophoresis.
6. Use 1 μL of the resulting M13 amplicon to synthesize the EC-EXTR synthetic RNA using the Riboprobe® System-T7 (see Note 4).
7. After transcription, digest the template DNA from the transcription reaction by adding 40 μL of recombinant, RNase-free DNase I to the transcription reaction mixture (see Note 4).
8. Remove proteins and free nucleotides by an acid-phenol:chloroform:isoamyl alcohol (125:24:1; pH 4.5) extraction and ethanol precipitation. Add 75 μL RNase-free water and 15 μL of sodium acetate solution (3M; pH 5.2) to a 60 μL reaction mixture and mix thoroughly. Extract once with an equal volume of acid-phenol:chloroform:isoamyl alcohol (125:24:1; pH 4.5) and then twice with an equal volume of chloroform/isoamylalcohol (24:1). Collect the aqueous phase and transfer to a new tube. Precipitate the RNA by adding two volumes of 100% ethanol and incubate for at least 30 min at –80°C. Collect the RNA by centrifugation at 18,000×*g* for 10 min. Remove the supernatant and rinse the pellet with 500 μL of cold 70% ethanol.
9. Dissolve the resulting EC-EXTR synthetic RNA in RNA Storage Solution and quantify by spectrophotometry. Convert the obtained concentration from g/L to copies/L according to the following formula:

$$\frac{\text{copies}}{\text{L}} = \frac{\text{concentration}\,[\text{g/L}]}{\text{molecular weight}\,[\text{g/mol}]} \times \frac{6.023 \times 10^{23}\,[\text{copies}]}{1\,[\text{mol}]}$$

$$\text{MW} = (A_n \times 329.21) + (U_n \times 306.17) + (C_n \times 305.18) + (G_n \times 345.21) + 159.0$$

where A_n, U_n, C_n and G_n is the number of each respective nucleotide.

10. Prepare a tenfold dilution series of the EC-EXTR synthetic RNA in a constant background of RNA from BTV-negative blood samples extracted as described under subheading 3.3. Analyse the dilution series by triplex RT-qPCR as described under subheading 3.4.
11. Select the dilution which results in a Cp-value of ± 30 and prepare 100× concentrated aliquots (see Note 6).
12. Verify absence of DNA by a singleplex EC-EXTR qPCR using the Platinum qPCR Supermix UDG. The PCR mix consists of 10 μL Platinum qPCR Supermix UDG, 2 μL of both forward and reverse primer EC-EXTR and 1 μL of EC-EXTR probe in a total volume of 15 μL (see Notes 7 and 8). Add 5 μL of synthetic RNA and run the following program on a LightCycler® 480 Instrument:
    (a) Activation: 95°C for 2 min
    (b) Amplification: 45 cycles of 95°C for 15 s, 60°C for 30 s with a single fluorescence acquisition at 60°C in each cycle.
    (c) Cooling: 40°C for 10 s

3.2. Preparation of BTV-S5 Synthetic RNA Standard Curve

1. Prepare a BTV singleplex RT-qPCR in a 20 μL reaction containing 1 μL RNA UltraSense™ Enzyme Mix, 4 μL RNA UltraSense™ 5× Reaction Mix, 0.75 μL forward primer BTV-S5, 1 μL reverse primer BTV-S5, 1.4 μL BTV-S5 probe, 7 μL BTV RNA and 4.85 μL RNase-free water (see Notes 3, 7, and 8).
2. Run the following program on a LightCycler® 480 Instrument:
    (a) Reverse transcription: 55°C for 15 min
    (b) (in)activation: 95°C for 2 min
    (c) Amplification: 45 cycles of 95°C for 10 s, 60°C for 30 s with a single fluorescence acquisition at 60°C in each cycle.
    (d) Cooling: 40°C for 10 s
3. Clone the BTV-S5 amplicon into a pCR2.1®-TOPO® vector as described under subheading 3.1, step 3.
4. Synthesize BTV-S5 synthetic RNA and quantify the concentration by spectrophotometry as described in Subheading 3.1, steps 4–8. Convert the obtained concentration from g/L to copies/L according to the formula described under subheading 3.1, step 9.
5. Prepare a tenfold dilution series, based on these calculations ranging from 10^7 to 10^1 copies/μL in a constant background of RNA from BTV-negative blood samples extracted as described under subheading 3.3. (see Note 9).

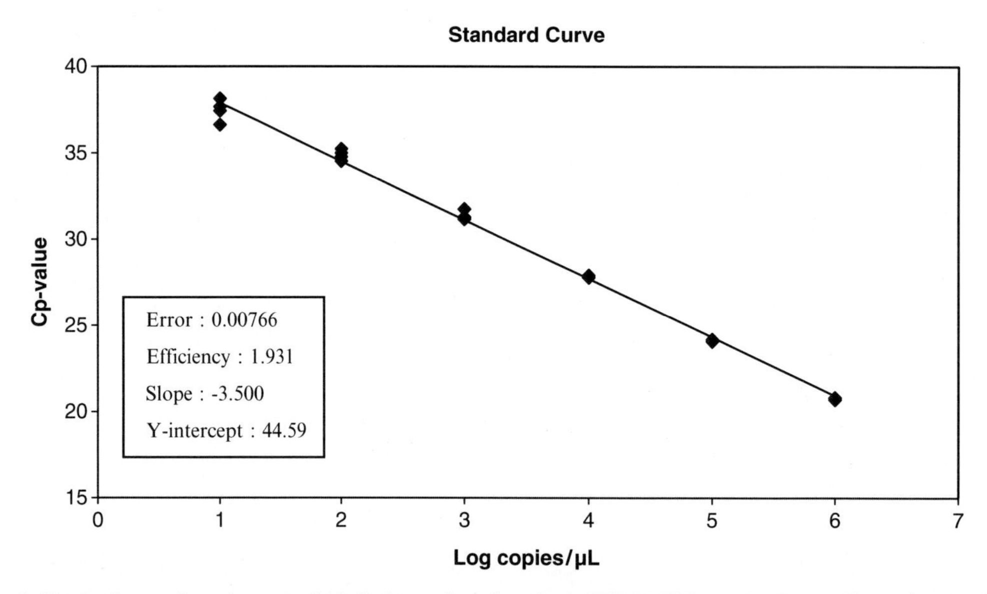

Fig. 1. Standard curve based on a tenfold dilution series of synthetic BTV-S5 RNA ranging from 10^7 to 10^1 copies/µL. Cp-values from five independent experiments are plotted against the log-transformed copy number. The box summarizes the correlation between Cp-value and copy number

Analyse the dilution series by triplex RT-qPCR as described under subheading 3.4 (see Note 10) (see Fig. 1).

3.3. Nucleic Acid Extraction

1. Extract nucleic acids using the NucleoSpin RNA Virus kit according to the support protocol for the simultaneous extraction of viral RNA and DNA with slight modifications (see Notes 3 and 11). Prior to extraction, add one aliquot EC-EXTR RNA to one bottle of RAV1 lysis buffer. Use 150 µL of EDTA-treated whole blood as starting material. Elute nucleic acids into 100 µL preheated RNase-free water containing 10% (v/v) DMSO and store at –80°C until use (see Notes 6 and 12).

3.4. Detection of Bluetongue RNA, IC-GAPDH and EC-EXTR by Multiplex RT-qPCR

1. Prepare a master mix which contains 1 µL RNA UltraSense™ Enzyme Mix, 4 µL RNA UltraSense™ 5× Reaction Mix, 1.4 µL BTV-S5 probe, 0.75 µL IC-GAPDH probe, 0.4 µL EC-EXTR probe, 0.3 µL RNase-free water and 0.8 µL Platinum® *Taq* DNA polymerase per sample (see Notes 3, 4, 7, 13–16).
2. Pipette 8.65 µL of the master mix into each well of a LightCycler® 480 Multiwell Plate 96.
3. Prepare a primer mix which contains, for each sample, 0.75 µL forward primer BTV-S5, 1 µL reverse primer BTV-S5, 1 µL

forward primer IC-GAPDH, 1 μL reverse primer IC-GAPDH, 0.3 μL forward primer EC-EXTR and 0.3 μL reverse primer EC-EXTR (see Note 8, 14, 17 and 18).

4. Pipette 4.35 μL of the primer mix and 7 μL of the RNA sample into each well of a 96-well PCR plate (see Note 14).
5. Seal the 96-well plate with a Sealing Foil and centrifuge at 1,800×*g* for 1 min.
6. Place the 96-well plate in a thermocycler and run the following program: 95°C for 3 min and hold at 4°C until use.
7. Finally, add 11.35 μL of the sample/primer mix to the master mix in the LightCycler® 480 Multiwell Plate 96 (see Note 19).
8. Seal the LightCycler® 480 Multiwell Plate 96 with a LightCycler® 480 Sealing Foil and centrifuge at 1,800×*g* during 2 min.
9. Place the LightCycler® 480 Multiwell Plate 96 in the LightCycler® 480 Instrument and run the following program (see Note 3):
   (a) Reverse transcription: 55°C for 15 min.
   (b) (in)activation: 95°C for 2 min.
   (c) Amplification: 45 cycles of 95°C for 10 s, 60°C for 30 s with a single fluorescence acquisition at 60°C in each cycle.
   (d) Cooling: 40°C for 10 s

3.5. Analysis

1. Prior to analysis, create a Color Compensation (CC) object as described in the LightCycler® 480 Operator's Manual to eliminate fluorescence cross-talk between the different channels (see Note 20).
2. Perform an Absolute Quantification Analysis using the BTV-S5 synthetic RNA standard curve as external standard. Analyse the data using the Second Derivative Maximum Method of the LightCycler® 480 Software after applying the above-described CC object.
3. For each sample, compare the obtained Cp-values for IC-GAPDH and EC-EXTR with the acceptance criteria (see Note 21).
4. Determine the number of BTV-S5 copies present in each sample by use of the following formula:

   viral load [copies/mL blood] = conv × viral load [copies/PCR reaction] where conv is the conversion factor relating BTV-S5 copies/PCR reaction to BTV-S5 copies/mL of EDTA-treated whole blood.

4. Notes

1. The primers were received lyophilized and dissolved in Tris–EDTA buffer to obtain a concentrated stock of 100 μM. This stock was further diluted in Tris–EDTA buffer to a concentration of 5 μM.
2. The primers used in the overlap extension PCR had the following sequences: forward primer EC-EXTR (5′–3′) tttccagttgctaccgattttacatagagagaaacaccctctcctgcccatcatg and reverse primer EC-EXTR (5′–3′) aaaggacgtttgtaatgtccgctcagtcatagatcatgatgggcaggagagggtg.
3. In order to avoid cross-contamination, the preparation of synthetic RNA, nucleic acid extraction, RT-qPCR setup and RT-qPCR amplification were performed in four separate rooms. Tubes were kept closed whenever possible and always centrifuged briefly before being opened. Latex or vinyl gloves were worn throughout the complete procedure to prevent contamination with RNases from the skin. Only sterile, RNase- and DNase-free plastic disposables were used.
4. The protocol was executed according to the kits' instructions.
5. The insert-overspanning M13 primers had the following sequences: forward M13 primer (5′–3′) gtaaaacgacggccag, reverse M13 primer (5′–3′) caggaaacagctatgac.
6. RNA was stored at −80°C and repeated freezing/thawing cycles were avoided as much as possible to minimize degradation.
7. The probes were received lyophilized, from which a concentrated stock of 100 μM was prepared by dissolving them in Tris–EDTA buffer. The use of RNase-free water is not recommended as some fluorophores are unstable under acidic conditions. The concentrated stock was further diluted in Tris–EDTA buffer before use. The BTV-S5, IC-GAPDH and EC-EXTR probes were diluted to a concentration of 5 μM, 2 μM and 5 μM, respectively. All stocks were stored at −20°C.
8. The primers were received lyophilized, from which a concentrated stock of 100 μM was prepared by dissolving them in Tris–EDTA buffer. This concentrated stock was further

diluted in Tris–EDTA buffer before use. The BTV-S5, IC-GAPDH and EC-EXTR primers were diluted to a concentration of 20 μM, 1 μM and 10 μM, respectively. All stocks were stored at −20°C.

9. Low-retention Eppendorf tubes were used to improve the accuracy of the highest dilutions. RNA from BTV-negative blood samples was used as diluent to avoid incorrect quantification due to differences in the BTV-S5 RT-qPCR efficiency in the absence or presence of IC-GAPDH and EC-EXTR amplification.

10. A standard curve should be included in each experiment in order to quantify the BTV RNA in each sample.

11. Although the NucleoSpin RNA Virus kit is originally designed for the extraction of nucleic acids from cell-free biological fluids, it can also be used for the isolation of BTV RNA from EDTA-treated whole blood.

12. In order to increase the sensitivity of the RT-qPCR, the RNA was denatured prior to analysis by heating in the presence of DMSO. To minimize the number of manipulations, DMSO was added directly to the elution buffer.

13. The BTV-S5 primers/probe sequences are identical to the ones described by Toussaint et al. (11). The IC-GAPDH and EC-EXTR primers/probe sequences are as follows: forward primer_IC-GAPDH (5′–3′) tcaccatcttccaggagcgag, reverse primer_IC-GAPDH (5′–3′) aaggtgcagagatgatgaccctc, probe_IC-GAPDH (5′–3′) YakimaYellow-caagtggggtgatgctggtgct gagta-BHQ1, forward primer_EC-EXTR (5′–3′) gacgtttg-taatgtccgctc, reverse primer_EC-EXTR (5′–3′) ccagttgctac-cgattttacata, probe_EC-EXTR (5′–3′) Texas Red-acaccct CTCCTGCCCATCATGATC-BHQ2.

14. To prepare a master or primer mix for more than one sample, the indicated volumes were multiplied by the number of samples to be analysed and an additional 10% (v/v) was added to facilitate pipetting small volumes.

15. The detection of several probes in a single PCR reaction is possible by labelling each probe with a different fluorophore. To avoid overlap of the signal, fluorophores with a good spectral separation were chosen. In this assay the BTV-S5, IC-GAPDH and EC-EXTR probes were labelled with FAM, Yakima Yellow and Texas Red at their 5′-end, respectively. To reduce background signals, dark quencher molecules were used (i.e., molecules which quench the fluorescence of a fluorophore without emitting fluorescence themselves). The BTV-S5 and IC-GAPDH probes contained

a BHQ1 molecule, whereas the EC-EXTR probe contained a BHQ2 molecule at their 3′-end (12).

16. The additional Platinum® *Taq* DNA polymerase is the same enzyme as the one contained in the RNA Ultrasense™ Enzyme Mix. The addition of supplemental *Taq* DNA polymerase is needed to overcome the competition of the various PCRs and to increase the assay sensitivity.
17. To avoid contamination, the master mix and sample/primer mix were prepared in separate laminar flow cabinets using dedicated sets of pipettes. Laminar flow cabinets were cleaned with DNA Remover™ after each reaction setup and daily UV-irradiated for 30 min.
18. The primer and probe concentrations of the IC-GAPDH and EC-EXTR reactions were severely limited to avoid competition with the BTV-S5 reaction for common reagents. The effect of multiplexing on the sensitivity of the BTV-S5 reaction was assessed by preparing a tenfold dilution series of a strongly BTV-positive field sample using RNA from BTV-negative blood samples as diluent. The dilution series was tested in parallel with the BTV singleplex and the BTV/IC/EC triplex RT-qPCRs. The BTV singleplex RT-qPCR was performed as described above but without the IC-GAPDH or EC-EXTR primers/probes (see Fig. 2).
19. To avoid renaturation of the double-stranded BTV RNA, the sample/primer mix was added to the master mix immediately after denaturation.
20. The use of a CC object is a prerequisite for the accurate detection/quantification of the number of BTV-S5 copies in the sample. Although the emission filters of the LightCycler® 480 Instrument are optimized for a maximum specific emission, all currently available fluorescent dyes exhibit emission spectra with long "tails", which leads to spectral overlap. To eliminate the cross-talk between the different fluorophores, the instrument has to be calibrated by running a CC experiment. During such an experiment, the fluorescence of each dye is measured in all channels and a CC object is generated. In subsequent experiments, this object is used to reassign the fluorescence in each channel to the appropriate dye.
21. During the validation process, the triplex RT-qPCR was performed on a large number of samples of good quality. Based on these results, acceptance criteria for the IC-GAPDH and EC-EXTR reactions were defined as the mean Cp-value ± 2× the standard deviation. BTV-negative test results are considered not valid whenever the IC/EC reactions exhibit a Cp-value higher than the acceptance criteria. Since the primer/probe concentrations of the IC/EC reactions are

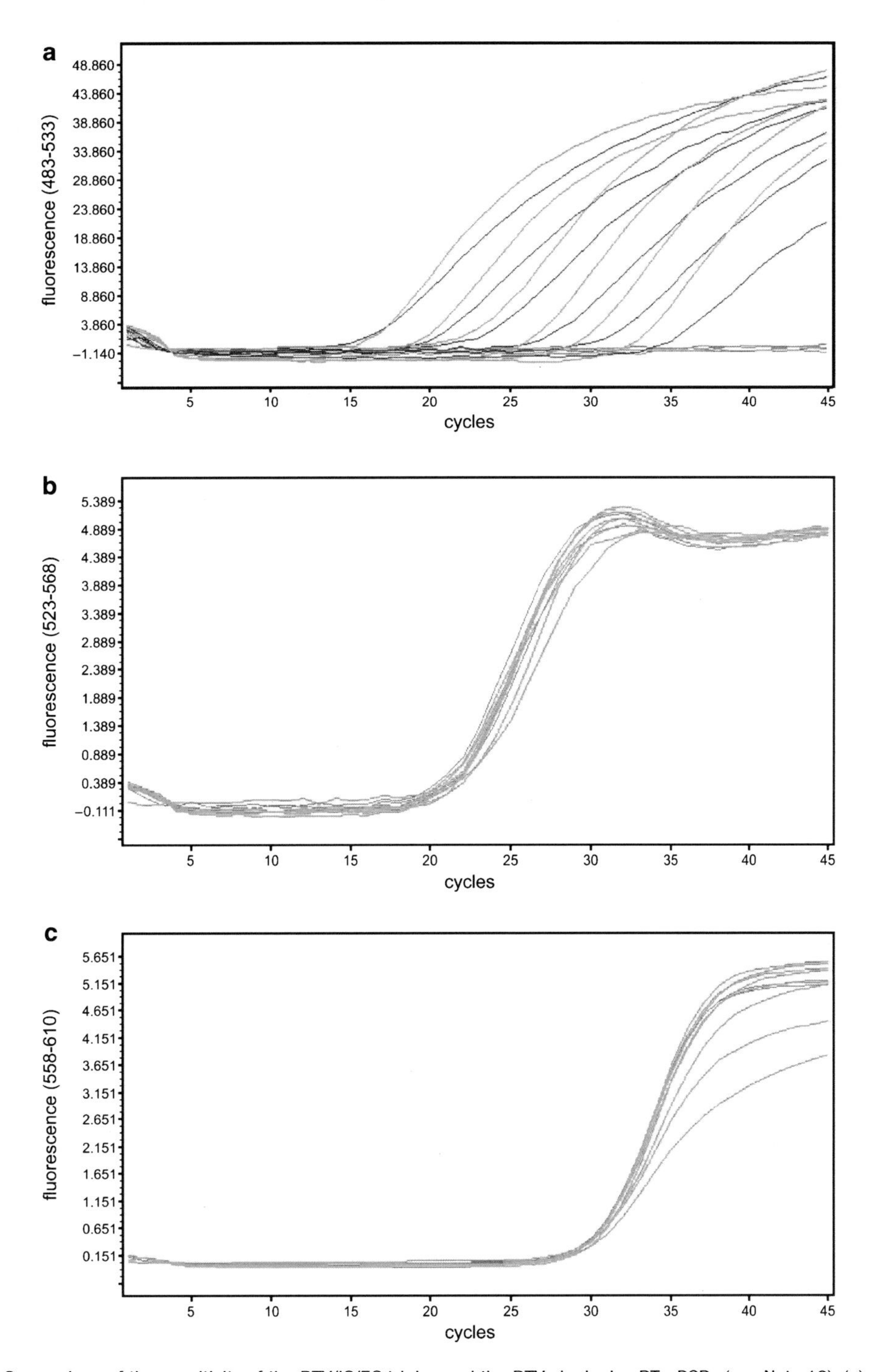

Fig. 2. Comparison of the sensitivity of the BTV/IC/EC triplex and the BTV singleplex RT-qPCRs (see Note 18). (**a**) BTV-S5 amplification plots for both RT-qPCRs from a tenfold dilution series of a strong positive field sample in the FAM-channel (*red lines* = singleplex RT-qPCR; *green lines* = triplex RT-qPCR). (**b**) IC-GAPDH and (**c**) EC-EXTR amplification plots for the triplex RT-qPCR in the HEX- and *Red* 610-channel

severely limited, the IC/EC controls may not be amplified in BTV-positive samples with a higher viral load. In these cases the BTV-positive test results are considered valid because the IC/EC results are irrelevant.

Acknowledgments

The study was supported by grants from DG4-6406 Bluetongue-RF 6187; the Belgian Federal Agency for the Safety of the Food Chain; EC-BTVAC SSPE-CT-2006-044211; EC-MedReoNet-SSPE-CT-2006-04428; EC-EPIZONE FOOD-CT-2006-016236 and the Veterinary and Agrochemical Research Centre. We wish to thank all technicians of the VAR who contributed to this study.

References

1. MacLachlan NJ (1994) The pathogenesis and immunology of bluetongue virus infection of ruminants. Comp Immunol Microbiol Infect Dis 17:197–206
2. Verwoerd DE (2004) Bluetongue. In: Tustin RC, Coetzer JA (eds) Infectious diseases of livestock. Oxford University Press, Cape Town, pp 1201–1220
3. Mertens PP, Diprose J, Maan S, Singh KP, Attoui H, Samuel AR (2004) Bluetongue virus replication, molecular and structural biology. Vet Ital 40:426–437
4. MacLachlan NJ, Osburn BI (2006) Impact of bluetongue virus infection on the international movement and trade of ruminants. J Am Vet Med Assoc 228:1346–1349
5. Rosenstraus M, Wang Z, Chang S-Y, DeBonville D, Spadoro JP (1998) An internal control for routine diagnostic PCR: design, properties, and effect on clinical performance. J Clin Microbiol 36:191–197
6. Hoorfar J, Malorny B, Abdulmawjood A, Cook N, Wagner M, Fach P (2004) Practical considerations in design of internal amplification controls for diagnostic PCR assays. J Clin Microbiol 42:1863–1868
7. Espy MJ, Uhl JR, Sloan LM, Buckwalter SP, Jones MF, Vetter EA, Yao JDC, Wengenack NL, Rosenblatt JE, Cockerill FR III, Smith TF (2006) Real-time PCR in clinical microbiology: applications for routine laboratory testing. Clin Microbiol Rev 19:165–256
8. Chamberlain JS, Gibbs RA, Ranier JN, Nguyen PN, Caskey CT (1988) Deletion screening of the Duchenne muscular dystrophy locus via multiplex DNA amplification. Nucl Acids Res 16:11141–11156
9. Elnifro EM, Ashshi AM, Cooper RJ, Klapper PE (2000) Multiplex PCR: optimization and application in diagnostic virology. Clin Microbiol Rev 13:559–570
10. Vandenbussche F, Vanbinst T, Vandemeulebroucke E, Goris N, Sailleau C, Zientara S, De Clercq K (2008) Effect of pooling and multiplexing on the detection of bluetongue virus RNA by real-time RT-PCR. J Virol Methods 152:13–27
11. Toussaint JF, Sailleau C, Breard E, Zientara S, De Clercq K (2007) Bluetongue virus detection by two real-time RT-qPCRs targeting two different genomic segments. J Virol Methods 140:115–123
12. Holland PM, Abramson RD, Watson R, Gelfand DH (1991) Detection of specific polymerase chain reaction product by utilizing the 5′→3′ exonuclease activity of thermus aquaticus DNA polymerase. Proc Natl Acad Sci 88:7276–7280

Chapter 8

Detection of West Nile Viral RNA from Field-Collected Mosquitoes in Tropical Regions by Conventional and Real-Time RT-PCR

Ana Silvia González-Reiche, María de Lourdes Monzón-Pineda, Barbara W. Johnson, and María Eugenia Morales-Betoulle

Abstract

West Nile virus (WNV) is an emerging mosquito-borne flavivirus, which has rapidly spread and is currently widely distributed. Therefore, efforts for WNV early detection and ecological surveillance of this disease agent have been increased around the world. Although virus isolation is known to be the standard method for detection and identification of viruses, the use of RT-PCR assays as routine laboratory tests provides a rapid alterative suitable for the detection of viral RNA on field-collected samples. A method for WNV RNA genome detection in field-collected mosquitoes is presented in this chapter. This method has been designed for virus surveillance in tropical regions endemic for other flaviviruses. Reverse Transcriptase-PCR (RT-PCR) assays, both standard and real time, to detect WNV and other flaviviruses are described. A first screening for flavivirus RNA detection is performed using a conventional RT-PCR with two different sets of flavivirus consensus primers. Mosquito samples are then tested for WNV RNA by a real-time (TaqMan) RT-PCR assay. Sample preparation and RNA extraction procedures are also described.

Key words: RT-PCR, Real-time RT-PCR, Viral RNA, Flavivirus, West Nile virus, Field-collected mosquitoes

1. Introduction

West Nile virus (WNV; *Flavivirus*, *Flaviviridae*) is considered an emerging mosquito-borne infectious agent in the New World. It was described for the first time in New York, United States in 1999 (1–3). Since then, the geographical range of WNV has increased throughout the United States, Canada, and Mexico. In Latin America, circulation of WNV in tropical and temperate regions was described by serological evidence after 2001 in several countries (4–10). Countries that reported viral isolates

Nicola King (ed.), *RT-PCR Protocols: Second Edition*, Methods in Molecular Biology, vol. 630,
DOI 10.1007/978-1-60761-629-0_8, © Springer Science+Business Media, LLC 2010

include Mexico, Argentina, Puerto Rico, and Guatemala (11–13) (Morales–Betoulle et al. unpublished data). Most of these countries, particularly those with tropical ecosystems, are endemic for several flaviviruses. From the approximately 70 known taxa of flaviviruses, at least 16 have been reported in Latin America, including human pathogens such as Dengue and recently described insect flaviviruses (14–17). Under these conditions, the significance of emerging flaviviruses such as WNV on public health remains unclear. It is possible that transmission of other flaviviruses, especially dengue virus, may provide some degree of cross-protection against WNV. Characterization of viral strains could allow better understanding of the evolution, behavior, and possible adaptation to humans of WNV in tropical countries.

West Nile Virus is a positive-sense RNA virus, which is transmitted principally by mosquitoes of the genus *Culex* (15). Thus, to actively monitor WNV, field-collected mosquitoes can be tested for viral RNA. Several techniques can be applied for this assessment. Although virus isolation is the "gold standard" technique used to characterize viruses, it is time consuming, relatively expensive, and requires specialized laboratories with tissue culture and biosafety level 3 conditions. An alternative to virus isolation is to use molecular techniques to detect viral genomic RNA. Molecular-based assays are highly sensitive and specific and are readily standardized with good reproducibility within and between laboratories. As a result, nucleic acid detection assays have become acceptable alternatives to virus isolation and important tools for diagnosis and research in virology (18). Several RT-PCR assays for WNV detection have been reported (19–22).

This chapter focuses on the use of Reverse Transcriptase–Polymerase Chain Reaction (RT-PCR) assays, both standard and real-time, to detect WNV and other flaviviruses in field-collected mosquitoes. The strategy described here was developed to be used in regions with possible co-circulation of several flaviviruses.

After collection, mosquitoes are frozen and transported to the laboratory, where they are identified and sorted by genus and species (when possible). Mosquitoes of the same type are pooled and then ground in a diluent with antibiotics to obtain a tissue homogenate. Because field-collected samples are mixtures of several types of cells and other material (host cells, bacteria, virus, etc.), RNA is extracted using a silica-gel-based membrane. This membrane binds nucleic acid, so it is possible to remove other material that can interfere with molecular detection. After RNA extraction the samples are screened in a one-step RT-PCR reaction, using primers designed to amplify a consensus region of the flavivirus non-structural 5 gene (14).The FU1/cFD2 primer pairs are generally more sensitive than the FU2/cFD3 primers. However, they lack specificity and can produce non-specific bands.

The FU2/cFD3 primers are more specific, and because of this, the flavivirus RT-PCR screening assay is performed with two different sets of primers. In order to confirm the resultvand identify the flavivirus, the RNA should be tested with a specific flavivirus real-time RT-PCR assay as described here for WNV. Alternatively, the correct-sized amplicon resulting from the FU1/cFD2 or FU2/cFD3 RT-PCR reaction should be extracted from the gel and purified, followed by nucleotide sequencing of the DNA product using the same primers as those in the initial RT-PCR procedure (Fig. 1). Virus identification or association can be made by comparing the nucleotide sequence with the Genbank database using the BLAST program.

In addition to this initial screening, the samples are tested by a WNV-specific one-step real-time RT-PCR assay, designed to detect the 3′ non-coding region of the WNV NY99 genome following the technique described by Lanciotti et al. (19). Specific real-time RT-PCR assays to detect SLEV RNA in *Culex* mosquitoes can also be used (19).

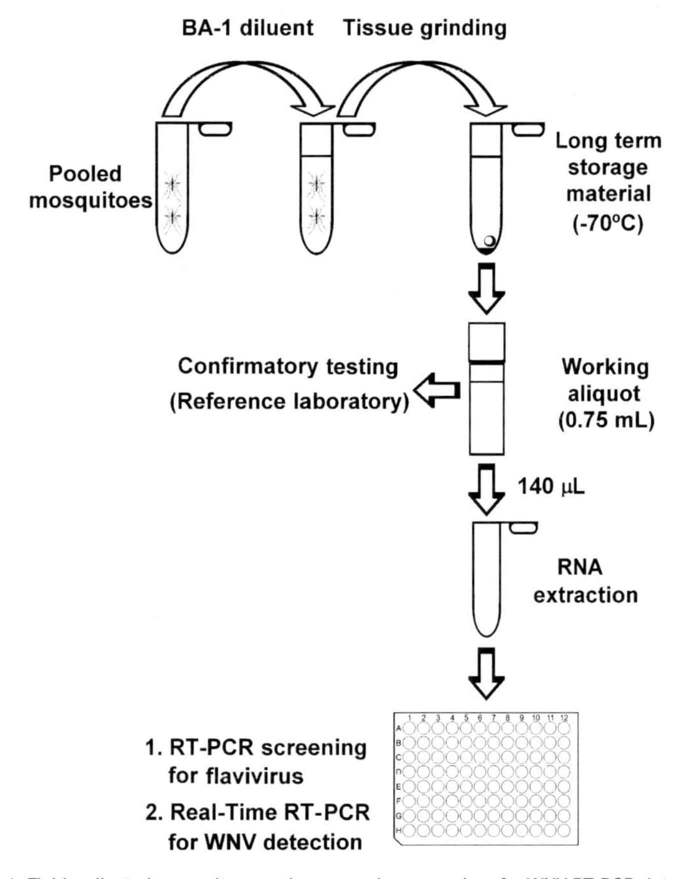

Fig. 1. Field-collected mosquito sample processing procedure for WNV RT-PCR detection

Because several sets of primers and RT-PCR protocols are recommended here, this procedure can be applied to detect not only WNV but also other flaviviruses. Therefore, in addition to human pathogens, novel flaviviruses (e.g. insect flaviviruses) can also be identified. Despite specific and sensitive molecular techniques, virus isolation by cell culture isolation is always recommended.

2. Materials

Since all procedures described here involve working with RNA, some general precautions are required to minimize contamination and exposure to RNases: use of nuclease-free reagents, supplies, and aerosol block pipette tips; a laboratory area exclusively used for RNA work, including dedicated equipment, instruments, and pipettes; and use of gloves at all times, with glove changes between steps to avoid RNA carry over.

2.1. Sample Preparation

2.1.1. Mosquito Grinding

1. BA-1 diluent: 1× medium 199 with Hanks' balanced salt solution, 2 mM L-Glutamine, 0.05 M Tris–HCl buffer, pH 7.6, 1% bovine serum albumin, 0.35 g/L sodium bicarbonate, 100 units/mL penicillin, 100 μg/mL streptomycin, 1 mg/mL amphotericin B.
2. Copper-coated steel Ball Baring (BBs) air gun pellets (Crosman Corporation, East Bloomfield, NY), 1 per sample.
3. Nuclease free safe-lock microcentrifuge tubes (2.0 mL).
4. Sterile and nuclease-free aerosol-resistant pipette tips.
5. Mixer mill (see Note 1).
6. Sample material (see Note 2): field-collected mosquitoes.

2.1.2. RNA Extraction

1. Qiagen QIAamp viral RNA kit or equivalent (see Note 3).
2. 100% Ethanol (absolute GR for analysis).
3. Nuclease-free microcentrifuge tubes (1.5 mL).

2.2. RT-PCR for Flavivirus Detection

2.2.1. Primer Reconstitution

1. Lyophilized primers (Table 1).
2. DEPC treated H_2O (for primer reconstitution) (see Note 4).

2.2.2. Additional RT-PCR Materials

1. Thermal cycler (see Note 5).
2. Qiagen™ OneStep RT-PCR kit or equivalent reagents (see Note 6).
3. Positive control RNA.

Table 1
Flavivirus and West Nile virus specific primers and probe sequences and expected product sizes

Specificity	Name	Sequence (5′–3′)	Genomic position	Expected product size
Flavivirus (14)	FU1 (forward)	TACAACATGATGGGAAA-GAGAGAGAA	8993[a]	250 bp with cFD2
	cFD2 (reverse)	GTGTCCCAGCCGGCG-GTGTCATCAGC	9258[a]	–
	FU2 (forward)	GCTGATGACACCGCC-GGCTGGGACAC	9233[a]	850 bp with cFD3
	cFD3 (reverse)	AGCATGTCTTCCGTG-GTCATCCA	10077[a]	–
West Nile Virus (19)	WN3′NC-forward	CAGACCACGCTACGGCG	10668	103 bp with 10770c
	WN3′NC-reverse	CTAGGGCCGCGTGGG	10770	–
	WN3′NC-probe	[FAM]-TCTGCGGAGTGCAGT-TCTGCGAT-[BHQ-1]	10691	–

[a]Position can vary between different flavivirus genomes (see Note 25)

2.2.3. RT-PCR Product Visualization with Agarose Gel Electrophoresis

1. Horizontal agarose gel electrophoresis system.
2. Agarose (molecular grade).
3. TBE (Tris/borate/EDTA) 1× electrophoresis buffer: 89.0 mM Tris base/89.0 mM borate/2 mM EDTA, pH 8.0.
4. 6× loading dye.
5. 100 bp DNA ladder.
6. Ethidium bromide solution (10 mg/mL) (see Note 7).

2.3. Real-Time RT-PCR for West Nile Virus Detection

2.3.1. Primers and Probe Dilution

1. Primers (see Note 8) (Table 1).
2. Dual-labeled probe (see Note 9) (Table 1).
3. DEPC treated H_2O (for primer and probe reconstitution) (see Note 4).

2.3.2. Additional Real-Time RT-PCR Material

1. Real-time PCR instrument (see Note 10).
2. Qiagen™ Quantitect Probe RT-PCR kit.
3. Positive control RNA.
4. 96-well PCR plates with optical seal plates, 0.2 mL low profile PCR tubes or appropriate reaction vessel for the real-time PCR instrument being used.

3. Methods

3.1. Sample Preparation

3.1.1. Mosquito Grinding

1. Perform all steps of this procedure in a Class II Biological safety Cabinet (see Note 11).
2. Place fresh or frozen samples (field collected mosquitoes) (see Note 12) in a 2-mL tube containing one copper-coated steel BB.
3. Add 1.75 mL of BA-1 diluent to each tube using a micropipette with a barrier tip. Change tips between each sample to avoid cross contamination.
4. Grind samples using the mixer mill for 4 min at maximum vibration frequency (25 Hz) (see Note 1).
5. Centrifuge at 13,400 × *g* for 3 min at 4°C to clarify homogenates (see Note 13).

3.1.2. RNA Extraction

1. Prepare all buffers and reagents according to RNA extraction kit protocol.
2. Work in a Class II Biological safety Cabinet (BSC) during all steps prior to viral lysis (steps 3–5).
3. Pipette 140 μL of the supernatant from the mosquito homogenate (see Subheading 3.1.1, step 5 above) into a 1.5 mL sterile, PCR clean (nuclease-free), microcentrifuge tube, being careful to avoid the mosquito pellet.
4. Include a negative extraction control with water nuclease-free, known uninfected mosquito homogenate, or BA-1 diluent and process with the samples.
5. Add 560 μL of kit supplied AVL buffer (lysis buffer) to the supernatant from step 3. Incubate at room temperature for 10 min. Centrifuge briefly (short spin) to remove drops from lid and minimize aerosol generation.
6. Add 560 μL of 100% ethanol and mix by vortexing for 15 s. Centrifuge briefly.
7. Pipette 630 μL of this mixture into a QIAamp mini spin column. Centrifuge at 6,000 × *g* for 1 min at room temperature. Discard the tube containing the filtrate, place the column in a new collection tube (kit supplied) and repeat this step until the entire sample has been applied to the column.
8. Add 500 μL of kit supplied AW1 buffer to the column in a clean collection tube, and centrifuge at 6,000 × *g* for 1 min at room temperature. Discard the tube containing the filtrate and place the column in a new collection tube.
9. Add 500 μL of kit supplied AW2 buffer to the column in the clean collection tube, and centrifuge at 20,000 × *g* for 4 min at room temperature and discard the tube containing the filtrate.

10. Place the column in a 1.5 mL sterile, PCR clean (nuclease-free), microcentrifuge tube and carefully add 60 μL of kit supplied AVE buffer. *Important*: Avoid contact with the column membrane during this final step. Incubate for 1 min and centrifuge at 6,000 × *g* for 1 min at room temperature to elute RNA (see Note 14).

3.2. RT-PCR for Flavivirus Detection

3.2.1. Primer Handling and Dilution

1. Briefly centrifuge (short spin, usually 30 s at 12,000 × *g*) lyophilized primers at room temperature prior to opening the primer tubes to ensure that all DNA is at the bottom of the tube.
2. For both FU1/cFD2 and FU2/cFD3 primer sets, the stock solution should be prepared by adding the required amount of water (nuclease-free, DEPC treated) to obtain a final concentration of 100 μM (see Note 15).
3. Mix gently by inverting the tube several times to allow DNA reconstitution (at least 10 min) before the next step.
4. Pulse centrifuge before opening.
5. Prepare small working aliquots in order to avoid several freeze thaw cycles (see Note 16). It is recommended to prepare 50 μL aliquots, but depends on the number of samples routinely tested at a time.
6. Store working dilution primer aliquots at –20°C (see Note 17).

3.2.2. RT-PCR Reaction

1. Prepare the volume of RT-PCR master mix on ice according to the number of reactions desired (Table 2) (see Note 18).

Table 2
Reaction master mix for Qiagen™ OneStep RT-PCR kit

	Volume (μL) per reaction	Final concentration
OneStep RT-PCR buffer (5×)	5 μl	1×
dNTP mix (containing 10 mM of each dNTP)	1 μl	400 μM of each dNTP
Forward primer (100 μM)	0.25 μl	25 pmol (1 μM)
Reverse primer (100 μM)	0.25 μl	25 pmol (1 μM)
OneStep RT-PCR Enzyme mix	1 μl	40 U
DEPC treated H_2O	7.5 μl	–
Master mix per reaction	15 μl	
RNA template	10 μl	
Total	25 μl	

Pulse centrifuge PCR reagents prior to opening to minimize aerosol generation. It is recommended to add PCR reagents to a clean PCR tube (nuclease-free) in the following order: DEPC treated H_2O, 5× One-Step RT-PCR buffer, dNTP mix, forward primer, reverse primer, and OneStep RT-PCR enzyme mix. Mix by inverting the tube several times and pulse centrifuge before next step.

2. Add 15.0 μL of RT-PCR master mix into each well of the 96-well PCR plate (or PCR tubes). Avoid generating bubbles in this step (see Note 19). Chill PCR plate or tubes on ice until adding to the thermal cycler (see Note 20).
3. Add RNA sample and controls to the corresponding wells of the 96-well PCR plate (or PCR tubes) chilled on ice. Add the negative controls first. Negative controls should include non-template control (nuclease-free H_2O instead of RNA) and negative extraction control. Then, add each RNA sample, and finally add the positive control RNA. Viral RNA extracted from flavivirus (e.g. West Nile virus) infected cell culture supernatant may be used as positive control. Seal the plate or tubes (see Note 21).
4. Run the RT-PCR reaction in the thermal cycler with the conditions described in Table 3.

3.2.3. RT-PCR Product Visualization with Agarose Gel Electrophoresis

1. Prepare a 2% (w/v) agarose gel. Weigh the desired amount of agarose powder and add the needed volume of TBE buffer, mix well and microwave solution at medium power until agarose dissolves completely (approximately 2–3 min). Cool the agarose solution to ~50°C; add the appropriate volume of ethidium bromide solution (1 μL of 10 mg/mL solution per 50 mL of 2% agarose) (see Note 7). Mix well.

Table 3
RT-PCR conditions for flavivirus detection with flavivirus primers

Phase	Number of cycles	Steps	Time	Temperature (°C)
Reverse transcription	1	Reverse transcription	30 min	50
		RT inactivation	15 min	95
PCR	45	Denaturation	30 s	94
		Annealing[a]	1 min	48–50
		Extension	1 min	72
Final extension	1		10 min	72
End	1	Hold	□	4

[a] Annealing temperature for the FU1/cFD2 primer set is 48–50°C, and for the FU2/cFD3 primer set is 50°C

2. Pour the agarose solution into the gel casting platform with desired well combs and allow cooling until the gel is completely polymerized (~30 min depending on gel size).
3. After RT-PCR, a 5 μL portion of each product (samples and controls) is analyzed. The 5 μL portion is mixed with 1 μL of 6× loading dye and loaded into the agarose gel well.
4. Load 5 μL of DNA ladder mixed with 1 μL of 6× loading dye in first and last well beside the first and last sample, respectively.
5. Run electrophoresis at 100 V for 20–30 min until desired resolution is achieved. Visualize PCR products under a UV transilluminator (see Note 22).
6. Check for amplification in the positive control, and lack of contamination in the negative controls (i.e. expected or unspecific products) (Fig. 2). For test samples, the DNA products with same size as positive controls should be considered as a presumptive positive result and retested for further analysis with WNV TaqMan assay described in the following section (see Note 23). For troubleshooting see Note 24.

3.3. Real-Time RT-PCR

3.3.1. Primers and Probe Handling and Dilution

1. West Nile virus specific lyophilized primers should be prepared as described for flavivirus primers in Subheading 3.2.1, to a final concentration of 100 μM.
2. Lyophilized probe should be reconstituted in nuclease-free water to a final concentration of 25 μM. To determine the amount of water needed, divide the amount in pmoles of lyophilized probe (supplied for each oligo by manufacturer)

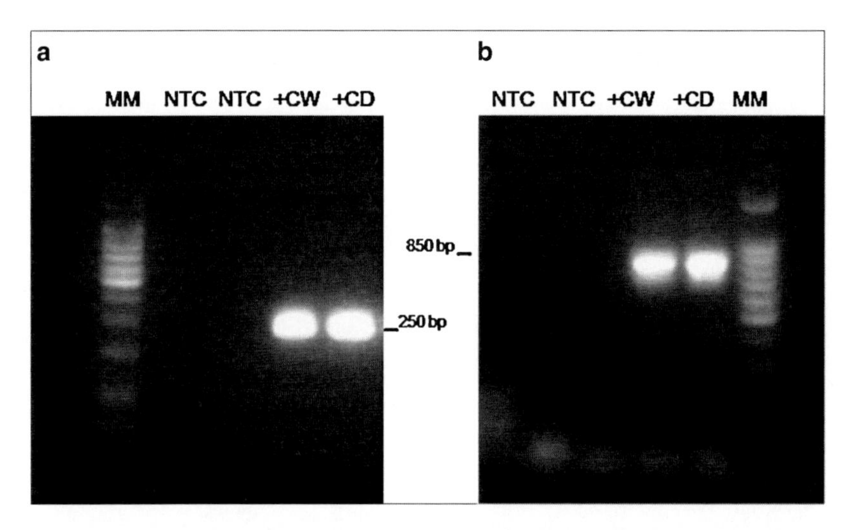

Fig. 2. RT-PCR product using primers (**a**) FU1/cFD2 (250 bp) and (**b**) FU2/cFD3 (850 bp), visualized on a 2% agarose gel. NTC, Non-template control (nuclease-free H_2O); +CW, WNV Positive control; +CD, DENV Positive control; MM, Molecular marker 100 bp.

by the desired concentration (25 pmol/µL) (see example in Note 13). Mix gently by inverting the tube several times and allow probe to reconstitute (10 min) before preparing working aliquots. Store working solution at –20°C and stock solutions at –70°C (see Note 14).

3.3.2. Real-Time RT-PCR Reaction

1. Prepare reaction master mix on ice according to the number of reactions desired (Table 4).
2. Pulse centrifuge PCR reagents prior to opening to minimize aerosol generation. It is recommended to add PCR reagents to a clean PCR tube (nuclease-free) in the following order: DEPC treated H_2O, 2× RT Master Mix, forward primer and reverse primer, probe, then RT enzyme mix.
3. Mix gently by inverting or tapping the tube several times and pulse centrifuge before the next step.
4. Add 10 µL of reaction master mix into each well of a white hard-shell 96-well PCR plate or into individual PCR tubes without creating bubbles. The PCR plate or tubes should be chilled on ice until they are put in the thermal cycler (see Note 20).
5. Add the control and sample RNA to the reaction wells of the plate or tubes. Add the negative control template (nuclease free H_2O) and negative extraction controls first, then add each sample, and without sealing or capping the tubes add the positive control RNA.

Table 4
Reaction master mix for Real-Time RT-PCR with Qiagen™ Quantitect Probe RT-PCR kit

	Volume (µL) per reaction	Final concentration
RT Master Mix (2×)	7.50	1×
Forward primer (100 µM)	0.15	15 pmol (1 µM)
Reverse primer (100 µM)	0.15	15 pmol (1 µM)
RT Mix (enzyme)	0.15	40 U
Probe (25 µM)	0.09	0.15 µM
DEPC treated H_2O	1.96	–
Master mix per reaction	10.0	
RNA template	5.0	
Total	15.0	

6. Viral RNA extracted from WNV-infected cell culture supernatant may be used as positive control.
7. Run the real-time RT-PCR reaction with the conditions described in Table 5.

3.3.3. Analysis of Results

1. Results are obtained as cycle threshold (Ct) value (Fig. 3). This value represents the precise number of PCR cycles at which the fluorescence signal begins to increase exponentially (and therefore the cDNA amount) and crosses the detection threshold (i.e. the background level).

Table 5
Real-time RT-PCR conditions for West Nile virus detection with specific primers

Phase	Number of cycles	Step	Time	Temperature (°C)
Reverse transcription	1	Reverse transcription	30 min	50
		Polymerase heat activation denature RNA:DNA	15 min	95
PCR	45	Denaturation	15 s	95
		Annealing/extension[a]	1 min	60

[a] Perform fluorescence data collection during annealing/extension step

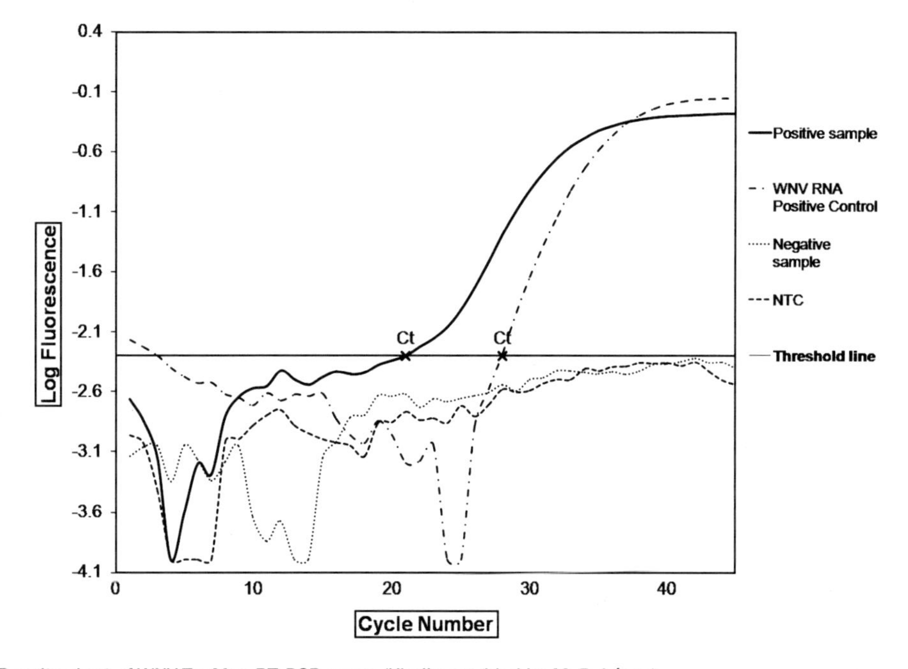

Fig. 3. Results chart of WNV TaqMan RT-PCR assay (Kindly provided by M. R. López)

2. Check for amplification in the positive control, and lack of contamination in the negative controls (i.e. unexpected or non-specific fluorescence) to ensure that reaction conditions were correct (see Note 24).
3. Positive Ct value is ≤38. Maximum positive value is Ct = 39. Samples with Ct value >38 but <39 are considered equivocal or suspect and should be retested for confirmation. Samples with Ct > 39 are considered negative. The cutoff threshold was calculated as the highest Ct (lowest concentration of RNA) at which 100% of the serially diluted reference RNA was detected in 8–10 replicate tests (23, 24).

4. Notes

1. A mixer mill such as the Retsch bench-top ball mill is suitable for disruption of tissues and sample material. If a mixer mill is not available, mosquito samples can also be ground by vortexing for 5 min at maximum speed.
2. *Culex* mosquitoes can be collected using two different traps: a CO_2-baited CDC light trap (John W.Hock Co., Gainesville, FL) and a gravid trap (25). Mosquitoes are killed by CO_2 freezing within their collection bags. They are transferred to collection tubes in the field and transported to the laboratory frozen on dry ice. They should be stored at −70°C until thawed for sorting. Repeated freeze-thaw cycles should be avoided. After identification of adult female mosquitoes, mosquitoes are sorted according to species, location, date and trap type, and pooled into groups of no more than 50 mosquitoes.
3. Alternative methods can be used for RNA extraction. Organic extraction with Trizol LS reagent has been shown to be efficient and provides good quality RNA suitable for this analysis.
4. Prepare DEPC treated water by adding 0.1–0.2 mL diethylpyrocarbonate (DEPC) to each 100 mL of distilled H_2O. Mix well and incubate overnight in a fume hood. Autoclave the solution for 15 min at 121°C at 1 Bar and store at room temperature. Work in a fume hood and wear gloves when using DEPC.
5. Thermocycling conditions for the MJ Research PTC-200 are described in this chapter. If using either this or other thermalcyclers, optimization of RT-PCR conditions in the laboratory must be accomplished prior to the use of this method for diagnostic purposes.

6. Because end point RT-PCR amplification is used for flavivirus detection, products are visualized in agarose gel electrophoresis. Any standard RT-PCR kit can be used in this assay, but reaction conditions such as reagents concentration and amplification parameters must be optimized.
7. Ethidium bromide is toxic and carcinogenic and requires special handling practices. Use an exclusive area and material for ethidium bromide manipulation. Additional protective lab coat and gloves should be used only for this purpose. For safe disposal practices, ethidium bromide stained gels should be destained by placing used gels and buffer with activated charcoal destaining bags into a container exclusive for this purpose. Stir the preparation overnight. Drain the buffer with abundant water and discard gels and bags in the biohazardous waste box.
8. Primers for real-time RT-PCR need to be highly purified. HPLC purification is recommended.
9. Probes are light sensitive and should be stored in amber tubes if available or wrapped in aluminum foil to minimize light exposure.
10. Reaction conditions for the PTC-200 PCR cycler with an optical unit Chromo 4 MJ Research, are described in this chapter; the conditions for the ABI Prism 7700 Sequence Detection System instrument have been reported elsewhere (19).
11. West Nile virus is a Biological safety Level (BSL) 3 agent. Sample material suspected to be infected with WNV should be processed in a Class II Biological safety Cabinet (BSC) until the viral lysis buffer has been added; downstream processing can then be done in BSL-2 conditions.
12. This procedure can also be used for detection of viral RNA in other potentially infected material such as animal tissue or fluid (human or animal serum, CSF).
13. For long-term storage, place a 750-μL aliquot of the supernatant into a 1.5 mL cryogenic vial.
14. Extracted RNA can be stored at 4°C <2 days. If testing cannot be initiated within 48 h, samples should be stored at −70°C. Prepare several aliquots of the sample RNA to avoid freeze and thaw cycles.
15. To calculate the amount of water (in μL), divide the pmoles of lyophilized primers (supplied for each oligo by the manufacturer) into the desired concentration (100 pmol/μL). For example: to reconstitute 31,100 pmol of lyophilized primer to a final concentration of 100 pmol/μL (i.e. 31,100 pmol/100 pmol per μL = 311 μL) add 311 μL of water.
16. Lyophilized oligonucleotides can be handled and reconstituted at room temperature. When reconstituted, several cycles of

freezing and thawing should be avoided. Oligonucleotide primers are more likely to form primer dimers or intrastrand secondary structures after undergoing multiple cycles of warming and slow freezing, which can result in the loss of the ability to amplify DNA (26).

17. To protect primer stocks from untimely demise caused by temperature fluctuations, thaw those aliquots to be used during the course of one or two experiments. For long-term conservation store the primer stocks at –70°C.
18. Maintain separate areas for RNA extraction, RT-PCR master mix preparation, template addition and amplification. Clean all areas, working material, and equipment (pipettes, centrifuges, microtube racks, etc.) with RNase inactivating agents (e.g. 0.1% SDS or RNase inactivating commercial products). Clean benches with 70% ethanol.
19. Air bubble presence and formation during some steps of the PCR thermocycling has been reported as a cause for PCR failure. The bubbles may expel the reagents from the PCR chamber, affecting temperature uniformity and reducing amplification efficiency. In addition, in a real-time PCR reaction, bubble formation may cause increased light scattering that degrades the fluorescent signal reducing the optical detection sensitivity of the instrument (27–29).
20. PCR plate or tubes should be placed on ice and keep chilled during all PCR preparation to prevent nucleic acid degradation and unwanted enzyme activity prior to RT-PCR reaction.
21. Nontemplate control should be added at the master mix preparation area but seal the tube until after all the other samples have been added. The suggested addition order is to minimize cross contamination and reduce the chance of false positives.
22. Avoid direct exposure to UV radiation or use UV-protective glasses during this procedure.
23. Visualization of a correct sized amplicon following flavivirus RT-PCR should be considered presumptive positive. Confirmation by a specific real-time RT-PCR assay, sequencing of the PCR product, and/or virus isolation and characterization is required to rule out false positives resulting from non-specific amplification.
24. Troubleshooting

    (a) The positive controls are negative: (1) RT-PCR should be optimized. Reaction conditions that can be modified for this optimization include annealing temperature (±5°C), reagent concentrations (primers, RNA template and $MgCl_2$). (2) Positive control RNA may be degraded. Prepare small aliquots to avoid several freezing and thawing cycles. (3) Reagents may be degraded. Verify that all reagents are in proper condition (expiration date, handling procedures, etc.) to ensure that enzymes are not inactivated and other reagents not degraded. (4) Working area and reagents may be contaminated with RNases; follow recommended decontamination procedures described above.

(b) The negative controls are positive: (1) Reagents may be contaminated with positive control RNA or positive sample material. Repeat reaction with freshly prepared reagents avoiding cross contamination. (2) Positive control or sample RNA may have been carried over into the negative control well during the preparation of the reaction in the plate or tubes. Change pipette tubes after application of RNA and master mix and avoid splashing or spilling on the plate or tube.

(c) Visualization of different size amplicons than expected: unspecific amplification may be occurring. Use second primer set to confirm results.

(d) Background levels too high or too low in real-time RT-PCR: (1) Probe concentration may be too high or too low. (2) Probe may be degraded. (3) RT-PCR conditions should be optimized.

25. Positions presented on Table 1 correspond to genome positions on those strains that were used for primer design (14). However, among the members of the mosquito-borne flavivirus cluster pairwise nucleotide sequence identity varies from 69% to 95% with a median genome length of 1,035 base pairs. For this reason primer's target sequence may not be on the same position for all viruses.

References

1. CDC: Centers for Disease Control and Prevention (1999) Outbreak of West Nile-like viral encephalitis – New York, 1999. MMWR Morb Mortal Wkly Rep 48:845–859
2. Lanciotti RS, Roehrig JT, Deubel V, Smith J, Parker M, Steele K, Crise B, Volpe KE, Crabtree MB, Scherret JH, Hall RA, MacKenzie JS, Cropp CB, Panigrahy B, Ostlund E, Schmitt B, Malkinson M, Banet C, Weissman J, Komar N, Savage HM, Stone W, McNamara T, Gubler DJ (1999) Origin of the West Nile virus responsible for an outbreak of encephalitis in the northeastern United States. Science 286:2333–2337
3. Nash D, Mostashari F, Fine A, Miller J, O'Leary D, Murray K, Huang A, Rosemberg A, Greenberg A, Sherman M, Wong S, Layton M (2001) The outbreak of West Nile virus infection in the New York City area in 1999. N Engl J Med 344:1807–1814
4. Komar N, Clark GG (2006) West Nile virus activity in Latin America and the Caribbean. Pan Am J Public Health 19:112–117
5. Estrada-Franco JG, Navarro-Lopez R, Beasley DWC, Coffey L, Carrara A-S, Travassos da Rosa A, Clements T, Wang E, Ludwig GV, Campomanes-Cortes A, Paz Ramirez P, Tesh RB, Barrett ADT, Weaver SC (2003) West Nile virus in Mexico: evidence of widespread circulation since July 2002. Emerg Infect Dis 9:1604–1607
6. Morales-Betoulle ME, Morales H, Blitvich BJ, Powers AM, Davis EA, Klein R, Cordón-Rosales C (2006) West Nile virus in horses, Guatemala. Emerg Infect Dis 12:1038–1039

7. Cruz L, Cardenas VM, Abarca M, Rodriguez T, Reyna RF, Serpas MV, Fontaine RE, Beasley DW, Da Rosa AP, Weaver SC, Tesh RB, Powers AM, Suarez-Rangel G (2005) Serological evidence of West Nile virus activity in El Salvador. Am J Trop Med Hyg 72:612–615
8. Dupuis AP II, Marra PP, Reitsma R, Jones MJ, Louie KL, Kramer LD (2005) Serologic evidence for West Nile virus transmission in Puerto Rico and Cuba. Am J Trop Med Hyg 73:474–476
9. Mattar S, Edwards E, Laguado J, González M, Alvarez J, Komar N (2005) West Nile virus antibodies in colombian horses. Emerg Infect Dis 11:1497–1498
10. Diaz LA, Komar N, Visintin A, Dantur Juri MJ, Stein ML, Allende R, Spinsanti L, Konigheim B, Aguilar J, Laurito M, Almirón W, Contigiani M (2008) West Nile virus in birds, Argentina. Emerg Infect Dis 14:689–690
11. Elizondo-Quiroga D, Davis CT, Fernandez-Salas I, Escobar-Lopez R, Velasco Olmos D, Soto Gastalum LC, Aviles Acosta M, Elizondo-Quiroga A, Gonzalez-Rojas JI, Contreras Cordero JF, Guzman H, Travassos da Rosa A, Blitvich BJ, Barrett ADT, Beaty BJ, Tesh RB (2005) West Nile virus isolation in human and mosquitoes, Mexico. Emerg Infect Dis 11:1449–1452
12. Barrera R, Hunsperger E, Muñoz-Jordán JL, Amador M, Diaz A, Smith J, Bessoff K, Beltran M, Vergne E, Verduin M, Lambert A, Sun W (2008) First isolation of West Nile virus in the Caribbean. Am J Trop Med Hyg 78:666–668
13. Morales MA, Barrandeguy M, Fabbri C, Garcia JB, Vissani A, Trono K, Gutierrez G, Pigretti S, Menchaca H, Garrido N, Taylor N, Fernandez F, Levis S, Enría D (2006) West Nile virus isolation from equines in Argentina, 2006. Emerg Infect Dis 12:1559–1561
14. Kuno G, Chang GJJ, Tsuchiya KR, Karabatsos N, Cropp CB (1998) Phylogeny of the genus Flavivirus. J Virol 72:73–83
15. Gubler DJ, Kuno G, Markoff L (2007) Flaviviruses. In: Knipe DM, Howley PM (eds) Field's virology, 5th edn. Lippincott Williams & Wilkins, Philadelphia, pp 1153–1252
16. Cook S, Bennett SN, Holmes EC, De Chesse R, Moureau G, de Lamballerie X (2006) Isolation of a new strain of the flavivirus cell fusing agent virus in a natural mosquito population from Puerto Rico. J Gen Virol 87:735–748
17. Morales-Betoulle ME, Monzón Pineda ML, Sosa SM, Panella N, López BMR, Cordón-Rosales C, Komar N, Johnson BW (2008) A new mosquito flavivirus from Izabal, Guatemala. J Med Entomol 45:1187–1190
18. Mackay IM, Arden KE, Nitsche A (2002) Real-time PCR in virology. Nucleic Acids Res 30:1292–1305
19. Lanciotti RS, Kerst AJ, Nasci RS, Godsey MS, Mitchell CJ, Savage HM, Komar N, Panella NA, Allen BC, Volpe KE, Davis BS, Roehrig JT (2000) Rapid detection of West Nile virus from human clinical specimens, field-collected mosquitoes, and avian samples by a TaqMan reverse transcriptase-PCR assay. J Clin Microbiol 38:4066–4071
20. Hadfield TL, Turell M, Dempsey MP, David J, Park EJ (2001) Detection of West Nile virus in mosquitoes by RT-PCR. Mol Cell Probes 15:147–150
21. Shi P-Y, Kauffman EB, Ren P, Felton A, Tai JH, Dupuis AP, Jones SA, Ngo KA, Nicholas DC, Maffei J, Ebel GD, Bernard KA, Kramer LD (2001) High-throughput detection of West Nile virus RNA. J Clin Microbiol 39:1264–1271
22. Jiménez-Clavero MA, Agüero M, Rojo G, Gómez-Tejedor C (2006) A new fluorogenic real-time RT-PCR assay for detection of lineage 1 and lineage 2. West Nile viruses. J Vet Diagn Invest 18:459–462
23. Busch MP, Tobler LH, Saldanha J, Caglioti S, Shyamala V, Linnen JM, Gallarda J, Phelps B, Smith RI, Drebot M, Kleinman SH (2005) Analytical and clinical sensitivity of West Nile virus RNA screening and supplemental assays available in 2003. Transfusion 45(4): 492–499
24. Linnen JM, Deras ML, Cline J, Wu W, Broulik AS, Cory RE, Knight JL, Cass MM, Collins CS, Giachetti C (2007) Performance evaluation of the PROCLEIX West Nile virus assay on semi-automated and automated systems. J Med Virol 79(9): 1422–1430
25. Reiter P (1983) A portable, battery-powered trap for collecting gravid *Culex* mosquitoes. Mosq News 43:496–498
26. Hengen PN (1995) Methods and reagents – Wayward PCR primers. Trends Biochem Sci 20:42–44
27. Cady NC, Stelick S, Kunnavakkam MV, Batt CA (2005) Real-time PCR detection of *Listeria monocytogenes* using an integrated microfluidics platform. Sens Actuators B Chem 107:332–341
28. Wang J, Chen Z, Corstjens PLAM, Mauka MG, Bau HH (2006) A disposable microfluidic cassette for DNA amplification and detection. Lab Chip 6:46–53
29. Liu H-B, Gong H-Q, Ramalingam N, Jiang Y, Dai C-C, Hui KM (2007) Micro air bubble formation and its control during polymerase chain reaction (PCR) in polydimethylsiloxane (PDMS) microreactors. J Micromech Microeng 17:2055–2064

Chapter 9

Detection of Antisense RNA Transcripts by Strand-Specific RT-PCR

Eric C.H. Ho, Michael E. Donaldson, and Barry J. Saville

Abstract

Comprehensive genome annotation requires extensive cDNA analysis. This analysis has identified natural antisense transcripts (NATs), which are distinct from the microRNAs, siRNAs, and piRNAs, in a number of diverse eukaryotes. This wide conservation supports the possibility of an important role for NATs in regulating cellular processes. Investigating their roles requires the confirmation of expressed sequence tag (EST) data and the detection of antisense transcripts in distinct cellular backgrounds. This chapter describes the use of a reverse transcription polymerase chain reaction (RT-PCR) method for the detection of antisense transcripts. The protocol was designed to reduce the number of first strand synthesis reactions during screening for antisense transcripts through the utilization of antisense directed primers and oligo dT to prime first strand synthesis. These results are further confirmed using sense and antisense directed primers in first strand synthesis. Results indicate that optimization of the screens requires proper controls to confirm removal of gDNA contamination and to rule out self-priming as a source of first strand products.

Key words: RT-PCR, Strand-specific, Antisense transcripts, Gene expression, RNA self-priming in reverse transcription

1. Introduction

It has become established that a diverse group of eukaryotes expresses natural antisense transcripts (NATs) that are significantly longer than the microRNAs (1–6). In contrast to the well characterized RNA interference and silencing roles for microRNAs first discovered by Fire and Mello (7), the molecular roles for longer NATs are just beginning to be uncovered. These now include a mode of silencing (8) and transcriptional interference (9, 10). The physiological functions of NATs are so far limited to altering cell fate decisions (9) and development (10), but further roles will undoubtedly be uncovered (11). Determining the function of NATs predicted by transcript sequencing screens requires the

Nicola King (ed.), *RT-PCR Protocols: Second Edition*, Methods in Molecular Biology, vol. 630,
DOI 10.1007/978-1-60761-629-0_9, © Springer Science+Business Media, LLC 2010

validation of their predicted transcription and the determination of their expression pattern. Detecting antisense RNA by northern hybridization using strand-specific probes generally requires relatively large amount of RNA and is difficult to scale up. In contrast, we use reverse transcription polymerase chain reaction (RT-PCR) for NAT detection. It is preferable owing to its high sensitivity and the ability to scale up for the simultaneous analysis of multiple transcripts. Detection of NATs by RT-PCR requires the production of strand-specific cDNA template. The protocol below describes a screen for antisense transcripts with the simultaneous detection of sense transcripts. This is accomplished using two separate first strand synthesis reactions, one with gene and strand-specific primers (Fig. 1) and the other with oligo dT, followed by PCR amplification (Fig. 2). This protocol leads to separate pools of first strand synthesis reaction products, one in which only the desired antisense-specific template is present and the other, a control, in which all polyadenylated antisense and sense transcripts are represented. Following this initial screen, the relative levels of sense and antisense transcript can be estimated using strand-specific first strand synthesis followed by PCR. Since the amount of cDNA template produced during first strand synthesis is proportional to the amount of starting RNA, the results of the PCR amplification can provide an estimate of NAT quantity. This, in turn, allows NAT expression profiles to be determined for different cell types or under distinct physiological conditions. While quantitative RT-PCR can be used for more accurate NAT quantification, the focus of this chapter will be the initial screens for NAT detection and cell-type specific expression.

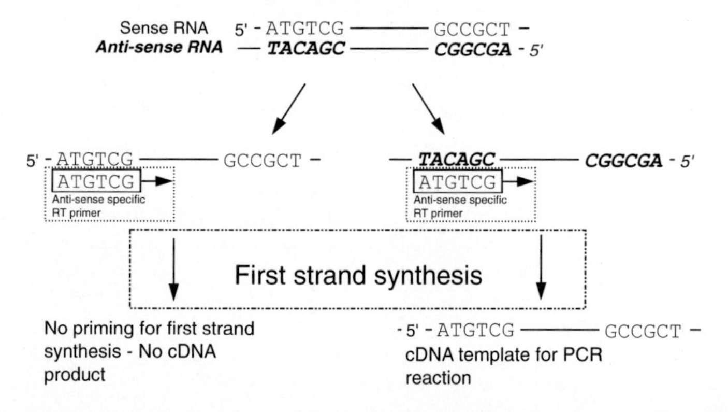

Fig. 1. Antisense strand-specific priming. This figure illustrates the outcome of first strand synthesis using antisense-specific primers. The absence of priming to the sense strand, and the presence of priming to the antisense strand, leads only to cDNA synthesis from the antisense transcript

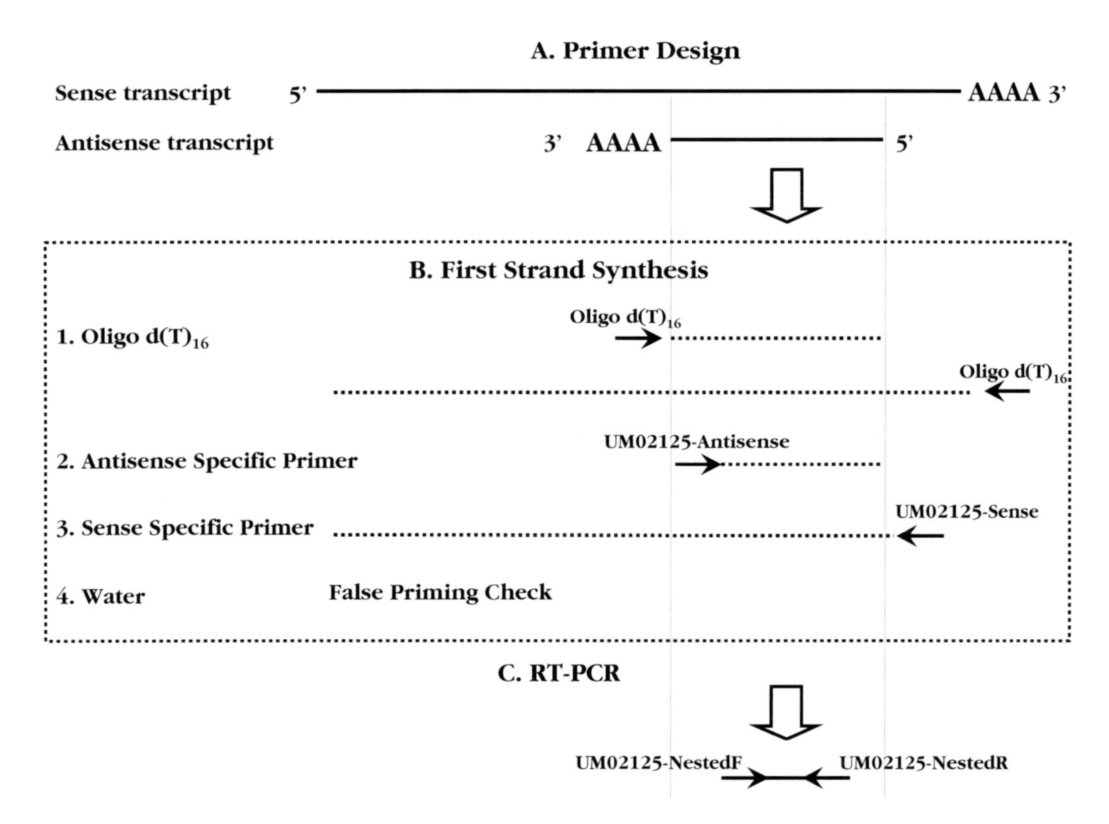

Fig. 2. Strand-specific RT-PCR. This diagram illustrates the process of strand-specific RT-PCR including: (**a**) Primer design, (**b**) First strand synthesis, and (**c**). RT-PCR. As an example, primers used during first strand synthesis and RT-PCR are indicated for strand-specific RT-PCR of UM02125

In this chapter, we outline a method for detecting antisense transcripts in the model pathogenic basidiomycete fungus *Ustilago maydis* (12). The method reflects modifications from the original protocol (3) to allow cost effective analysis of a large number of transcripts. The genes investigated were selected from those identified in *U. maydis* EST libraries (3, 13, 14). They include UM00133 (related to acetylhydrolase) for which antisense ESTs were detected in libraries created from haploid cells grown in rich media (HCM), haploid cells grown in nitrogen limited media (HMN), haploids cells grown in carbon limited media (HMC), and diploid filamentous cells (D12), as well as UM02125 (related to glutamic acid decarboxylase) for which antisense transcripts were only detected in the dormant diploid teliospore (TDO). Herein we describe RNA isolation from haploid cells, and dikaryotic and diploid colonies stimulated to grow as filaments. After the *U. maydis* directed RNA isolation, the techniques described can be generalized to RNA originating from the organism of interest.

2. Materials

2.1. Designing Strand-Specific Primers

1. Primer Designing Software (e.g., Primer 3).
2. Basic Local Alignment Search Tool (e.g., NCBI BLAST).

2.2. U.maydis RNA Isolation (see Note 1)

1. Media:
   (a) Complete Medium: 0.25% (w/v) casamino acids, 0.15% (w/v) anhydrous ammonium acetate, 1.0% (w/v) yeast extract, 1.0% (w/v) sucrose, 6.25% (v/v) salt solution (see Subheading 2.2, step 1c), adjust pH to 7.0 before adjusting final volume. Autoclave to sterilize.
   (b) Minimum Medium: 0.3% (w/v) potassium acetate (anhydrous), 6.25% (v/v) salt solution (see Subheading 2.2, step 1c), adjust pH to 7.0 before adjusting final volume. Autoclave to sterilize. 1.0% (w/v) glucose (add sterily after autoclaving solution).
   (c) Salt Solution: 1.6% (w/v) K_2HPO_4, 0.4% (w/v) Na_2SO_4, 0.8% (w/v) KCl, 0.2% (w/v) $MgSO_4{\cdot}7H_2O$, 0.1% (w/v) $CaCl_2{\cdot}2H_2O$, 0.8% (v/v) trace elements solution (see Subheading 2.2, step 1d).
   (d) Trace Elements Solution: 0.006% (w/v) H_3BO_3, 0.014% (w/v) $MnCl_2$, 0.04% (w/v) $ZnCl_2$, 0.004% mg (w/v) Na_2Mo, 0.01% (w/v) $FeCl_3$, 0.04% (w/v) $CuSO_4$.
2. Liquid nitrogen.
3. Pestle and Mortar.
4. BD Falcon 14 ml snap-cap test tubes.
5. Diethylpyrocarbonate (DEPC)-treated water or other water treated to be RNase free.
6. TRIZOL® regent.
7. Chloroform.
8. 75% Ethanol.
9. Heating block.
10. 10× BPTE Buffer: 100 mM PIPES, 300 mM Bis–Tris, and 10 mM EDTA, no pH adjustment required.
11. Glyoxal reaction mixture: 6 ml DMSO, 2 ml deionized glyoxal, 1.2 ml of 10× BPTE electrophoresis buffer, 0.6 ml of 80% glycerol, and 0.2 ml of ethidium bromide (10 mg/ml in H_2O).
12. RNA size marker (e.g., 0.5–10 Kb RNA Ladder).
13. Spectrophotometer, while any cuvette-based spectrophotometer would suffice, for our work we use NanoDrop 8000 which does not require cuvettes and can measure RNA concentration in a 1 µl volume.

2.3. DNase I Treatment and Cleanup of Total RNA

1. Total RNA (10 μg).
2. Amplification Grade DNase I, includes: 10× buffer and EDTA.
3. Column-based RNA isolation kit, we have used the Qiagen RNeasy Mini Kit but have also successfully completed the procedure with other kits.
4. Spectrophotometer.

2.4. Testing RNA for Genomic DNA Contamination

1. Thermocycler: while any thermocycler can be used, please note that these protocols were developed using the Applied Biosystems 96-Well GeneAmp® PCR System 9700.
2. Hot Start Taq polymerase: we prefer Applied Biosystems AmpliTaq Gold® DNA Polymerase which comes with: 10× PCR Gold Buffer, and 25 mM $MgCl_2$ solution.
3. 10 mM dNTP mix: 2.5 mM each dNTP; we prefer the Applied Biosystems GeneAmp® dNTP blend, but others also work.
4. Test Primers (e.g., GAPDH-Forward, GAPDH-Reverse).
5. DNA Template (e.g., 10–100 ng/μl genomic DNA (gDNA) sample).
6. 50× TAE Buffer: 242 g Tris–base, 57.1 ml of glacial acetic acid, 100 ml of 0.5 M EDTA (pH 8.0), volume adjusted to 1 L with H_2O.
7. DNA Molecular Size Marker (e.g., 100 bp DNA Ladder).
8. 6× gel loading buffer: 0.25% bromophenol blue (w/v), 0.25% xylene cyanol FF (w/v), 30% glycerol (v/v) in H_2O. The loading buffer stock is made up in a final volume of 10 ml.
9. UV transilluminator.

2.5. First Strand cDNA Synthesis

1. TaqMan Gold RT-PCR kit: 10× Taqman RT Buffer, 25 mM $MgCl_2$, 2.5 mM dNTP Mixture, oligo $(dT)_{16}$ primer, RNase Inhibitor, and Multiscribe Reverse Transcriptase (50 U/μl).
2. Strand-specific primers (e.g., UM02125-Antisense, and UM02125-Sense).
3. DEPC-treated water.

2.6. Additional Materials for Sense- and Antisense-Specific RT-PCR

1. Test Primers (e.g., UM02125-NestedF, and UM02125-NestedR).

3. Methods

3.1. Primer Design for RT-PCR

1. Primer design is a critical element of successful first strand synthesis using antisense transcripts as a template (Fig. 1). Using the antisense transcript sequence, design primers that are complementary to a region as close to the 3′ end of the predicted antisense transcript as possible (for example, refer to UM02125-Antisense primer in Fig. 2). This will produce the longest possible cDNA during the reverse transcriptase reaction, providing sufficient template region for PCR primer design (see Subheading 3.1, step 3). Select suitable primers using free open source primer design programs (e.g., Primer3, http://primer3.sourceforge.net/). Set the melting temperature at 55–60°C to allow synthesis by reverse transcriptase at an elevated temperature. We use 50°C to minimize RNA secondary structure interference with first strand synthesis.
2. Similarly, to detect the presence of sense transcripts using RT-PCR, design primers that are complementary to a region close to the 3′ end of the predicted sense transcript (for example, refer to UM02125-Sense primer in Fig. 2).
3. PCR Primers, for amplification from cDNA following first strand synthesis, should be "nested," or inside the 5′ ends of the antisense- and sense- specific primers (for example, refer to UM02125-NestedF and UM02125-NestedR primers in Fig. 9.2). In order to use the same primer pair for the detection of antisense and sense transcripts, it is necessary for both primers to fall within the region of overlap between the antisense and sense transcripts (see Fig. 2 and Note 1).

3.2. Isolation of Total RNA (see Note 2)

1. Culture *U. maydis* haploid cells in complete medium or in nitrogen starvation medium overnight at 30°C with gentle shaking. Alternatively, grow *U. maydis* dikaryon and diploid cultures at 28°C for 4–5 days for optimum filamentous growth.
2. Centrifuge the haploid culture and remove the supernatant. Filamentous cultures are scraped from plates and collected in 14 ml snap-cap tubes.
3. Freeze the resulting pellet or filamentous culture quickly in liquid nitrogen and homogenize with a mortar and pestle.
4. Extract total RNA using TRIZOL® reagent according to the manufacturer's protocol (see Note 3).
5. Assess RNA quality by electrophoresis. Load 1 µl of RNA (up to 10 µg) on a glyoxal denaturing gel (see Note 4).

(a) Add 10 μl of glyoxal reaction mixture to 1 μl of RNA in a microcentifuge tube. Prepare 1 μl of RNA ladder in a similar manner.

(b) Heat the tubes at 65°C for 10 min, chill on ice and briefly centrifuge to collect the contents.

(c) Separate the glyoxalated RNA using electrophoresis on an agarose gel prepared in 1× BPTE solution, at 100 V for approximately 1 h.

6. Visualize the gel on a UV transilluminator to inspect for degradation (see Note 4).
7. Determine the quantity of the RNA using a spectrophotometer (see Note 4).

3.3. DNase I Treatment and Cleanup of Total RNA

1. Use amplification (amp) grade DNase I to treat 10 μg of total RNA according to the manufacturer's instructions.
2. Prepare a 100 μl reaction: 10 μg of total RNA, 10 μl (10 U) of amp grade DNase I, and 10 μl of DNase I Reaction Buffer (1× final concentration), and use DEPC-treated water to bring the reaction volume to 100 μl.
3. Incubate samples at room temperature for 15 min and inactivate DNase I reaction by adding 10 μl of 25 mM RNase-free EDTA (final concentration of EDTA must be greater than 2 mM) and heat the samples at 65°C for 10 min. Place the RNA samples on ice for 1 min and centrifuge briefly to collect the RNA (see Note 6).
4. Purify DNase I treated RNA using the QIAGEN RNeasy Mini Kit according to the manufacturer's RNA cleanup protocol. Use 30 μl of Elution Buffer (supplied in the kit) to elute RNA from the spin-column.
5. Prepare a 1:10 dilution (2 μl of RNA and 18 μl of DEPC-treated water) of the purified, DNase I treated RNA.
6. Determine RNA concentration (see Note 5). Dilute the stock solution of DNase I treated RNA to a final concentration of 100 ng/μl.

3.4. Testing RNA for Genomic DNA Contamination (see Note 7)

1. Perform PCR reactions using a hot start DNA polymerase following the manufacturer's suggested protocol and using a primer pair designed to amplify GAPDH (or another housekeeping gene, see Note 8). Use the DNase I treated RNA as template without prior reverse transcription to determine if it is contaminated with gDNA (see Note 9).
2. Make up a mixture for all components of the PCR reaction except the template (a master mix). We generally prepare mixes for eight or more reactions in which the relative concentrations for a 25 μl reaction are: 2.5 μl of 10× PCR

Gold Buffer (1× final concentration), 3.0 μl of 25 mM $MgCl_2$ (3.0 mM final concentration), 2.0 μl of 2.5 mM dNTPs (200 μM final concentration), 1.0 μl each of 5 mM GAPDH-forward and reverse primers (0.2 μM final concentration each), 0.125 μl of 5 U/μl Amplitaq Gold (1 μl per eight reactions or 0.625 U per 25 μl reaction), and 13.375 μl of water (91 μl per eight reactions). For each reaction, add 23 μl master mix to 2 μl of total RNA (200 ng of total RNA, from Subheading 3.3, step 6). In addition, use 2 μl of water, and 2 μl of gDNA (10–100 ng/μl) as templates in separate PCRs for negative and positive controls.

3. Carry out the PCR reactions in a 96-Well thermocycler using the program: 95°C for 10 min, 35 cycles of 95°C for 30 s, 65°C for 15 s, and 72°C for 1 min, followed by 72°C for 10 min.
4. After the PCR amplification, add 5 μl of 6× gel loading dye to each reaction and load 6 μl of this reaction mix for electrophoresis at 90 V for approximately 60 min using an ethidium bromide-stained agarose gel prepared in 1× TAE buffer. Include a DNA size ladder on the gel.
5. Using a UV transilluminator, visualize and photograph the gel. (see Note 9)
6. If the RNA sample was devoid of gDNA contamination, perform strand-specific RT-PCR. If gDNA contamination was detected in the total RNA sample, as indicated by the presence of a PCR product, it is necessary to repeat Subheading 3.3, step 2, using the previously treated RNA sample (from Subheading 3.3, step 6).

3.5. Strand-Specific First Strand Synthesis

1. For each sample of RNA successfully treated with DNase I, perform first strand synthesis reactions separately, using different primers (see Note 10, see results following PCR in Fig. 3):
   (a) Oligo $(dT)_{16}$ primer
   (b) Water (see Note 11, see results following PCR in Fig. 4)
   (c) Antisense-specific primer (see results following PCR in Figs. 3 and 5)
   (d) Sense-specific primer (see results following PCR in Fig. 5)
2. Reverse transcribe 200 ng of total RNA (see Note 12) using the TaqMan Gold RT-PCR kit, according to the manufacturer's protocol. Add 200 ng of total RNA, and a final concentration of 2.5 μM Oligo $(dT)_{16}$ primer or 200 nM of strand-specific primer separately to each reaction. For each 10 μl reaction, add total RNA and primer, so that their combined volume is equal to 4 μl. To the RNA and primer, add 6 μl of a master mix (see Note 13) that includes for each 10 μl reaction: 1 μl

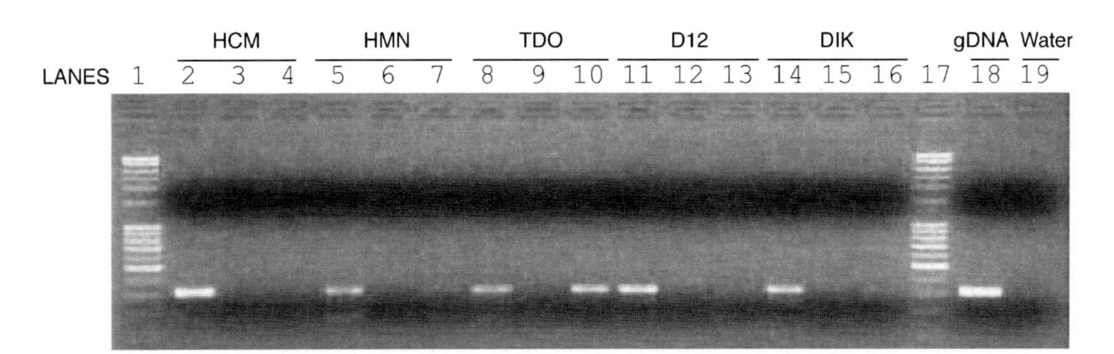

Fig. 3. Antisense-specific RT-PCR screen for UM02125. Antisense-specific RT-PCR detected high levels of antisense transcript in TDO relative to the other cell types. *Lanes 1* and *17* Full Ranger Ladder (Norgen Biotek). The cellular origin of the cDNA templates are: *Lanes 2–4* HCM; *Lanes 5–7* HMN; *Lanes 8–10* TDO; *Lanes 11–13* D12; *Lanes 14–16* DIK. The primers used in first strand synthesis were: *Lanes 2, 5, 8, 11, 14* Oligo $(dT)_{16}$ primer; *Lanes 3, 6, 9, 12, 15* no Primer; *Lanes 4, 7, 10, 13, 16* Antisense-specific primer. The following controls were included: *Lane 18* gDNA PCR control; *Lane 19* Water PCR control. *DIK* dikaryotic filamentous mycelia; *HCM* haploid cells grown in complete media; *HMN* haploid cells grown in nitrogen limited media; *TDO* dormant diploid teliospore

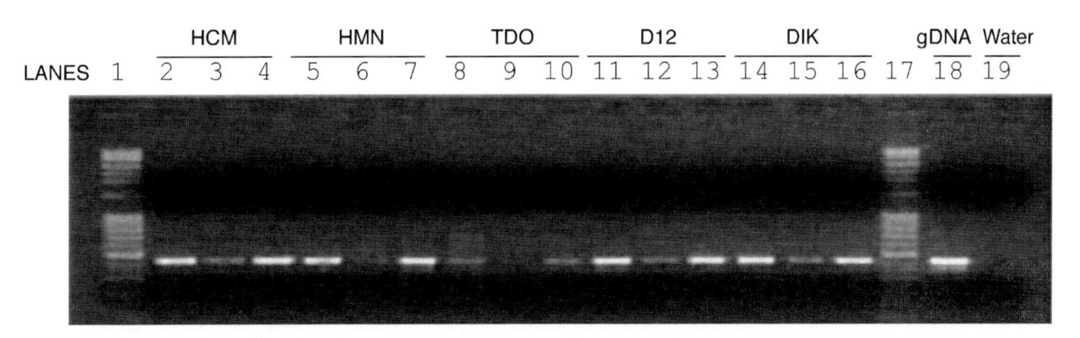

Fig. 4. False-priming in RT-PCR, as seen in an antisense-specific RT-PCR screen for UM00133. High background can be seen in HCM, HMN, D12, and DIK when no primer is used during first strand synthesis. This makes interpretations of the presence, or absence of antisense transcripts difficult using RT-PCR. *Lanes 1 and 17* Full Ranger Ladder. The cellular origin of the cDNA templates are: *Lanes 2–4* HCM; *Lanes 5–7* HMN; *Lanes 8–10* TDO; *Lanes 11–13* D12; *Lanes 14–16* DIK. The primers used in first strand synthesis were: *Lanes 2, 5, 8, 11, 14* Oligo $(dT)_{16}$ primer; *Lanes 3, 6, 9, 12, 15* Water Primer; *Lanes 4, 7, 10, 13, 16* Antisense-specific primer. The following controls were included: *Lane 18* gDNA PCR control; *Lane 19* Water PCR control. *DIK* dikaryotic filamentous mycelia; *HCM* haploid cells grown in complete media; *HMN* haploid cells grown in nitrogen limited media; *TDO* dormant diploid teliospores

of Taqman RT Buffer (1× final concentration), 2.2 μl of 25 mM $MgCl_2$ (5.5 mM final concentration), 2.0 μl of 2.5 mM dNTPs (500 μM final concentration), 0.2 μl of 20 U/μl RNase Inhibitor (0.4 U/μl final concentration), 0.25 μl of 50 U/μl Multiscribe Reverse Transcriptase (1.25 U/μl final concentration) with the remaining volume made up with DEPC-treated water.

3. Gently mix the reactions by flicking the tube with your index finger and centrifuge briefly to collect the reaction contents. Reactions containing no reverse transcriptase do not need to

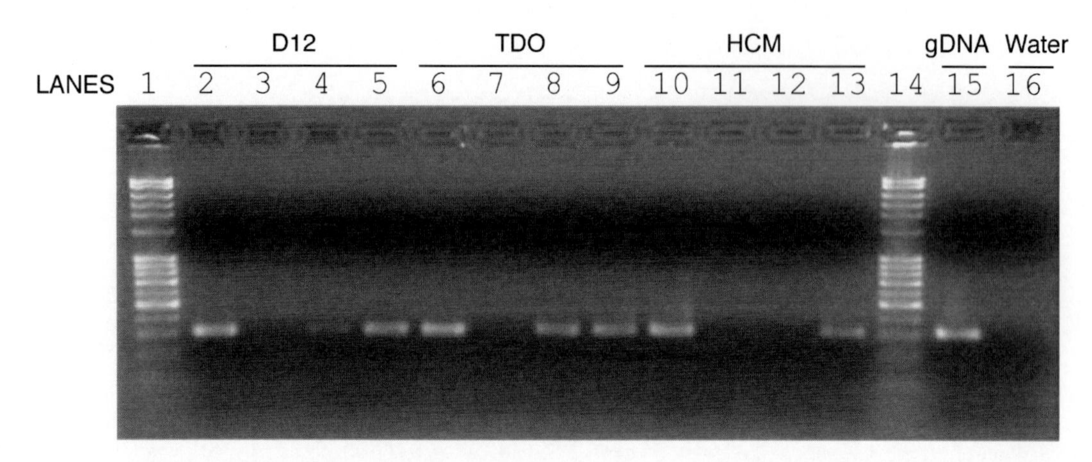

Fig. 5. Cell-type specific sense and antisense RT-PCR for UM02125. Strand-specific RT-PCR detected high levels of sense transcripts for all three cell types. Low levels of antisense transcript in D12 relative to high levels in TDO were detected using strand-specific RT-PCR. Antisense-specific transcripts were not detected in HCM. *Lanes 1* and *14* Full Ranger Ladder; The cellular origin of the cDNA templates are: *Lanes 2–5* D12; *Lanes 6–9* TDO; *Lanes 10–13* HCM. The primers used in first strand synthesis were: *Lanes 2, 6* and *10* Oligo (dT)16 primer; *Lanes 3, 7* and *11* no primer; *Lanes 4, 8* and *12* Antisense-specific primer; *Lanes 5, 9 13* Sense-specific primer; The following controls were included: *Lanes 15* gDNA PCR control; *16* Water PCR control. *D12* diploid filamentous cells; *HCM* haploid cells grown in complete media; *TDO* dormant diploid teliospores

be performed because the total RNA has already been screened for the presence of gDNA (Subheading 3.4, step 6).

4. Place the first strand synthesis reactions containing Oligo $(dT)_{16}$ primer in a 96-Well thermocycler preset with the following program: 25°C for 10 min, 50°C for 30 min, and 95°C for 5 min. Place the first strand synthesis reactions containing water primer, an antisense-specific primer, and a sense-specific primer in a 96-Well thermocycler preset with the following program: 50°C for 30 min, and 95°C for 5 min.
5. At the end of the reverse transcription reaction, add 30 μl DEPC-treated water to each tube to dilute the first strand synthesis products. Gently mix each dilution and briefly centrifuge each tube to collect the contents.

3.6. Polymerase Chain Reaction

1. Following the procedure outlined under Subheading 3.4, step 2, setup 25 μl reactions using 2 μl of template from Subheading 3.5, step 5, and the forward and reverse primers designed in Subheading 3.1, step 3. Once again, use water and gDNA as template for negative and positive controls. At the end of the PCR, resuspend samples in gel loading buffer as outlined under Subheading 3.4, step 4 and visualize the strand-specific RT-PCR results as outlined in Subheading 3.4, step 5 (see Figs. 3–5).

4. Notes

1. In this procedure, strand-specific first strand synthesis using reverse transcriptase, requires the use of strand-specific primers, and not oligo dT. The use of strand-specific primers means that the region of the gene synthesized by reverse transcriptase using the sense strand-specific primer differs from that synthesized using the antisense stand-specific primer. To streamline the PCR screening process, we designed these sense and antisense specific primers so that they produce cDNA that overlaps in sequence (see Fig. 2). This allows the use of a single primer pair to amplify both cDNAs.
2. Strict RNA handling procedures must be used to avoid RNase contamination. Handle all reagents with fresh gloves (just removed from the box). All glassware should either be baked at a high-temperature (300°C) or DEPC-treated. All solutions and plastic ware, with some exceptions such as Tris based solutions, should be DEPC-treated to remove RNase.
   (a) DEPC-treatment of solution: dilute DEPC in ethanol prior to its addition to aqueous solutions. For example, to treat 1 l of water; add 1 ml of DEPC to 9 ml of 100% ethanol, mix well and add to 990 ml water (0.1% DEPC (v/v) final concentration). Incubate the solution overnight at room temperature (or 37°C for an hour) and autoclave.
   (b) DEPC-treatment of plastic ware: soak plastic ware in DEPC-treated water (0.1% v/v). Cap the top of the container, or cover the top with Parafilm, and incubate overnight at room temperature (or 37°C for an hour).
   (c) Additionally, treat nonautoclavable labware (such as electrophoresis devices) with DEPC as follows:
      - Wash with detergent solution.
      - Rinse in autoclaved distilled water.
      - Dry with ethanol.
      - Fill with 3% solution of hydrogen peroxide and incubate for 10 min at room temperature.
      - Rinse with DEPC-treated water (0.1% v/v).
      - Isolate teliospore RNA following Zahiri et al. 2005 (15).
3. Perform an extra chloroform extraction step if RNA purity appears to be a problem.
4. Good quality eukaryotic RNA will have sharp, clear 28 S and 18 S rRNA bands. The 28 S rRNA bands should be approximately twice as intense as the 18 S rRNA bands.

5. In determining the concentration of RNA using absorbance, make three replicate measurements of O.D. using a spectrophotometer or, in our case a NanoDrop 8000. To calculate the concentration of RNA from absorbance, use a modification of the Beer–Lambert equation: $c = (Ae)/b$. Where c is the nucleic acid concentration in ng/μl, A is the absorbance in AU, e is the wavelength-dependent extinction coefficient in ng-cm/μl, and b is the path length in cm. The generally accepted extinction coefficient for RNA is 40 ng-cm/μl. For the NanoDrop 8000 Spectrophotometer, path lengths of 1.0 mm and 0.2 mm are used compared to a standard spectrophotometer using a 10.0 mm path. Therefore, in our calculation $c = (A40$ ng-cm/μl)/0.1 cm or (RNA) = $A400$ ng/μl. Whereas with most spectrophotometers, the calculation is (RNA) = $A40$ ng/μl because the path length is 1 cm.
6. For DNase I treatment of RNA:
   (a) The entire reaction can be scaled up or down linearly relative to the amount of input RNA.
   (b) Do not exceed the manufacturer's recommended incubation time, or increase the reaction temperature, as doing so may lead to hydrolysis of the RNA.
   (c) The addition of EDTA is paramount for the heat inactivation step, as unchelated magnesium ions will cause RNA hydrolysis.
7. It is important to ensure that the RNA is free of gDNA before RT-PCR. A PCR may be used to screen the RNA sample prior to first strand synthesis. This step can save both time and money. Since RNA is not a template for polymerases used in PCR under standard amplification conditions, a polymerase chain reaction will only amplify from a DNA template. PCR of an RNA sample with water and gDNA amplification as controls allows one to determine if there is any DNA contaminating the RNA sample.
8. Quantifying the level of a given transcript is often carried out relative to that of a gene whose transcription is not expected to vary under the conditions of the experiment. Such genes are considered to be expressed at similar levels in many cell types and under varying physiological conditions. They are so considered because they are critical to normal cell physiology and, as such, are termed "house keeping" genes. An example of such a gene is glyceraldehyde-3-phosphate dehydrogenase (GAPDH). It is noteworthy that most researchers do not confirm by independent means, that the expression level of these genes does not vary between cell types, although doing so is preferred.
9. The purpose of this screen was to detect even the slightest amount of gDNA that may contaminate the purified,

DNase I treated, RNA sample (from Subheading 3.3, step 6). Therefore, it is recommended that the picture of the gel be taken using a long exposure and that the gel image be studied to ensure that there are no faint bands that would indicate gDNA contamination.

10. For rapid screening of antisense transcripts, or candidate gene selection; oligo dT, water, and antisense-specific primers should be used to prime the first strand synthesis reaction. In the case where candidate screening has yielded results warranting further investigation, such as cell-type specific antisense expression (Fig. 3), design a sense-specific primer to study the relative amounts of sense and antisense transcripts in a given cell-type (Fig. 5).
11. Nonspecific- or "false"-priming may occur during first strand synthesis due to reverse transcriptase reactions primed by hairpin structures of the target RNA (Fig. 4). In addition, antisense and sense transcripts can act to prime their complementary strand during first strand synthesis (16). Therefore, it is necessary to include a no-primer first strand synthesis reaction in order to account for false-priming background. Place the "no-primer" reaction in the same cycle conditions as the strand-specific primer reactions.
12. RT-PCR may be semiquantitative if equal amounts of total RNA are used in each first strand synthesis reaction. Examine the manufacturer's protocol to determine the optimal starting amount of RNA for a given reverse transcriptase. For example, for a 10 μl reaction, the TaqMan Gold RT-PCR kit will convert 2 pg–200 ng of total RNA to cDNA. In our experiments, the upper limit (200 ng of total RNA) was selected for strand-specific RT-PCR in order to maximize the likelihood of detecting antisense-specific transcripts.
13. Master mixes are used here and in other areas of molecular biology to ensure that the same concentration of components is added to each tube in an experiment. Thus, the control reactions have the identical primer, enzyme buffer etc., concentrations as the experimental reactions. To make a master mix, calculate the concentrations of each component in the volume of the reaction you will be using and then multiply this by the number of reactions you will be carrying out, plus one or two. The extra volume is included to account for pipetting error which may occur during the dispensing of aliquots. This technique also enables researchers to add quantities of reactants per reaction that are not readily measured such as fractions of a microliter. For example: when a master mix for eight reactions is made up with 0.125 μl of enzyme per reaction, the actual enzyme volume measured and added to the master mix is 1 μl.

Acknowledgments

Funding provided to BJS by NSERC of Canada.

References

1. Werner A, Berdal A (2005) Natural antisense transcripts: sound or silence? Physiol Genomics 23:125–131
2. Steigele S, Nieselt K (2005) Open reading frames provide a rich pool of potential natural antisense transcripts in fungal genomes. Nucleic Acids Res 33:5034–5044
3. Ho EC, Cahill MJ, Saville BJ (2007) Gene discovery and transcript analyses in the corn smut pathogen Ustilago maydis: expressed sequence tag and genome sequence comparison. BMC Genomics 8:334
4. Osato N, Yamada H, Satoh K, Ooka H, Yamamoto M, Suzuki K, Kawai J, Carninci P, Ohtomo Y, Murakami K, Matsubara K, Kikuchi S, Hayashizaki Y (2003) Antisense transcripts with rice full-length cDNAs. Genome Biol 5:R5
5. Chen J, Sun M, Kent WJ, Huang X, Xie H, Wang W, Zhou G, Shi RZ, Rowley JD (2004) Over 20% of human transcripts might form sense-antisense pairs. Nucleic Acids Res 32:4812–4820
6. Beiter T, Reich E, Williams RW, Simon P (2009) Antisense transcription: a critical look in both directions. Cell Mol Life Sci 66:94–112
7. Fire A, Xu S, Montgomery MK, Kostas SA, Driver SE, Mello CC (1998) Potent and specific genetic interference by double-stranded RNA in Caenorhabditis elegans. Nature 391:806–811
8. Bonoli M, Graziola M, Poggi V, Hochkoeppler A (2006) RNA complementary to the 5' UTR of mRNA triggers effective silencing in Saccharomyces cerevisiae. Biochem Biophys Res Commun 339:1224–1231
9. Hongay CF, Grisafi PL, Galitski T, Fink GR (2006) Antisense transcription controls cell fate in Saccharomyces cerevisiae. Cell 127:735–745
10. Petruk S, Sedkov Y, Riley KM, Hodgson J, Schweisguth F, Hirose S, Jaynes JB, Brock HW, Mazo A (2006) Transcription of bxd noncoding RNAs promoted by trithorax represses Ubx in cis by transcriptional interference. Cell 127:1209–1221
11. Lapidot M, Pilpel Y (2006) Genome-wide natural antisense transcription: coupling its regulation to its different regulatory mechanisms. EMBO Rep 7:1216–1222
12. Kamper J, Kahmann R, Bolker M, Ma LJ, Brefort T, Saville BJ, Banuett F, Kronstad JW, Gold SE, Muller O, Perlin MH, Wosten HA, de Vries R, Ruiz-Herrera J, Reynaga-Pena CG, Snetselaar K, McCann M, Perez-Martin J, Feldbrugge M, Basse CW, Steinberg G, Ibeas JI, Holloman W, Guzman P, Farman M, Stajich JE, Sentandreu R, Gonzalez-Prieto JM, Kennell JC, Molina L, Schirawski J, Mendoza-Mendoza A, Greilinger D, Munch K, Rossel N, Scherer M, Vranes M, Ladendorf O, Vincon V, Fuchs U, Sandrock B, Meng S, Ho EC, Cahill MJ, Boyce KJ, Klose J, Klosterman SJ, Deelstra HJ, Ortiz-Castellanos L, Li W, Sanchez-Alonso P, Schreier PH, Hauser-Hahn I, Vaupel M, Koopmann E, Friedrich G, Voss H, Schluter T, Margolis J, Platt D, Swimmer C, Gnirke A, Chen F, Vysotskaia V, Mannhaupt G, Guldener U, Munsterkotter M, Haase D, Oesterheld M, Mewes HW, Mauceli EW, DeCaprio D, Wade CM, Butler J, Young S, Jaffe DB, Calvo S, Nusbaum C, Galagan J, Birren BW (2006) Insights from the genome of the biotrophic fungal plant pathogen Ustilago maydis. Nature 444:97–101
13. Nugent KG, Choffe K, Saville BJ (2004) Gene expression during Ustilago maydis diploid filamentous growth: EST library creation and analyses. Fungal Genet Biol 41:349–360
14. Sacadura NT, Saville BJ (2003) Gene expression and EST analyses of Ustilago maydis germinating teliospores. Fungal Genet Biol 40:47–64
15. Zahiri AR, Babu MR, Saville BJ (2005) Differential gene expression during teliospore germination in Ustilago maydis. Mol Genet Genomics 273:394–403
16. Beiter T, Reich E, Weigert C, Niess AM, Simon P (2007) Sense or antisense? False priming reverse transcription controls are required for determining sequence orientation by reverse transcription-PCR. Anal Biochem 369:258–261

Chapter 10

RT-PCR Amplification and Cloning of Large Viral Sequences

Xiaofeng Fan and Adrian M. Di Bisceglie

Abstract

A long reverse transcription polymerase chain reaction (LRP) protocol is described for the amplification of large RNA sequences. The amplification of near full-length hepatitis C virus (HCV) genome from serum samples is used as an example to detail each step in LRP procedure, including primer design, RNA extraction, reverse transcription, and PCR. The protocol for efficient cloning of such large amplicons is also presented. Since HCV represents a difficult template in terms of its near full-length amplification due to extensive secondary structures and low titers in clinical samples, methods described in this chapter should be applicable for other RNA viruses and cellular RNA templates.

Key words: RT-PCR, Long RT-PCR, Hepatitis C virus, Cloning

1. Introduction

Polymerase chain reaction (PCR) is an indispensable technique in biomedical research. With known primer sequences, it can easily amplify a DNA target less than 3 kb, but it has diminished power when the target is larger than 3 kb. Barnes et al. first hypothesized that the inability to amplify large DNA fragments was due to the misincorporation of nucleotides by most thermostable DNA polymerases, which resulted in premature termination of PCR (1). Based on this hypothesis, mixed polymerases, one of which has 3′–5′ exonuclease "proofreading" activity to correct the misincorporation, have successfully amplified DNA targets up to 42 kb (2). However, there has been limited success in applying this concept to the amplification of large RNA genomes that require the reverse transcription (RT) step prior to PCR amplification. The inability to amplify large RNA template by long RT-PCR (LRP) may be attributed to multiple factors, such as incompatibility between RT and PCR buffers, inhibitive role of

Nicola King (ed.), *RT-PCR Protocols: Second Edition*, Methods in Molecular Biology, vol. 630, DOI 10.1007/978-1-60761-629-0_10, © Springer Science+Business Media, LLC 2010

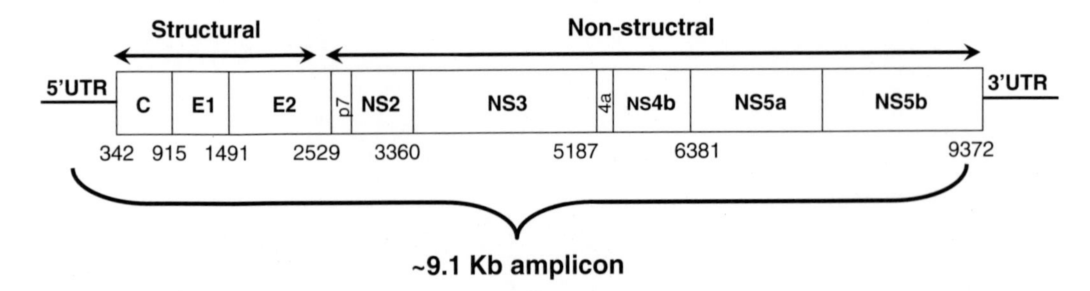

Fig. 1. Organization of the HCV genome. The 5′ untranslated region (UTR) and 3′UTR are shown as *lines*. Between them is an open reading frame encoding a single polyprotein processed into structural (Core, E1, and E2) or nonstructural proteins (p7, NS2, NS3, NS4a, NS4b, NS5a, and NS5b) as shown in *boxes*. Nucleotide numbering is according to HCV H77 strain, GenBank accession number AF009606. The target domain for the long RT-PCR, about 9.1 kb, is indicated

RT enzymes on thermostable DNA polymerases (3), and retention of RNA template secondary structure.

In this chapter, we present a robust and simple protocol for amplification of the near full-length genome of hepatitis C virus (HCV), a positive sense single-strand RNA virus in the family of *flaviviridae* (Fig. 1). HCV is difficult to culture in vitro, and the clinical samples in which HCV has a low titer are the only source of its RNA template for LRP. Such template limitations together with its extensive secondary structure throughout the entire genome (about 9.6 kb) (4–6) makes HCV represent one of the most challenging targets for LRP amplification. Thus, the method described in this chapter should be readily adapted to other RNA viruses as well as cellular RNA templates.

2. Materials

2.1. LRP

2.1.1. RT-PCR

1. SuperScript III reverse transcriptase (Invitrogen, Carlsbad, CA).
2. dNTPs (see Note 1).
3. Avian Myeloblastosis Virus reverse transcriptase (AMV).
4. rTth DNA polymerase, XL (Applied Biosystems, Foster City, CA).
5. $Mg(OAc)_2$, 25 mM.
6. Dithiothreitol (DTT) (100 mM).
7. Trnc-21 (see Note 2).
8. Nuclease-free water.
9. Mineral oil.

2.1.2. Electrophoresis

1. Seakem GTG agarose gel.
2. 1 kb DNA ladder.

3. Ethidium bromide (EB), 10 mg/ml.
4. 50× TAE buffer: To make 1 liter of 50× TAE buffer, add 242 g Tris, 100 ml of 0.5 M EDTA (pH 8.0) and 57.1 ml glacial acetic acid into enough H_2O, dissolve solids and adjust pH with HCl to 7.6–7.8, then bring up to final volume of 1,000 ml.
5. Electrophoresis equipment.
6. Benchtop UV transilluminator (302/365 nm).

2.1.3. Others

1. –80°C Freezer.
2. EuGene version 1.01 (Daniben Systems Inc., Cincinnati, OH), a software for primer design.
3. QIAamp Viral RNA Mini Kit.
4. DNA Thermal Cycler 480.
5. PCR box, 23×23×34 in., a clean plastic box made by Saint Louis University.

2.2. Cloning

2.2.1. Restriction Digestion and DNA Purification

1. Spectrophotometer NanoDrop 1000.
2. QIAquick PCR Purification Kit.
3. Restriction enzymes Pac I and Fse I.
4. Bovine Serum Albumin (BSA) (10 mg/ml).
5. Razor blade.
6. QIAEX II Gel Extraction Kit.

2.2.2. Ligation and Transformation

1. T4 DNA ligase.
2. pClone vector (see Note 14).
3. ElectroMAX DH10B cells.
4. MicroPulser Electroporation Apparatus.
5. Eppendorf electroporation cuvette, 1 mm gap, 100 µl volume.
6. Kanamycin solution, 50 mg/ml.
7. LB plates containing 30 µg/ml kanamycin, freshly made and stored at 4°C no more than 1 month.
8. LB medium.
9. SOC medium.
10. Falcon 2059 Polypropylene Tubes.
11. 37°C incubator.
12. Shaking incubator.

2.2.3. Plasmid Purification

Using QIAprep Spin Miniprep Kit.

3. Methods

3.1. LRP Procedure

3.1.1. Primer Design

1. The selection of primers plays a critical role in successful LRP. Because HCV has a highly variable RNA genome, locate all LRP primers within conserved domains close to either 5′ or 3′ extremes.
2. Conduct primer design with the software EuGene version 1.01 (see Note 3).
3. Set the melting temperatures (T_m) for the reverse primers for the RT step to approximately 60°C (see Note 4).
4. Set the T_m for the first and second round PCR to around 61 and 68°C, respectively (see Note 5).
5. Primers used for HCV LRP (genotype 1a) are shown in Table 1.

3.1.2. RNA Extraction

1. Serum samples should be stored in a –80°C freezer (see Notes 6 and 7).
2. QIAamp Viral RNA Mini Kit is used to extract viral RNA from serum. According to the manual, RNA extracted from 280 μl serum is finally eluted into 60 μl of Tris buffer (provided in the kit). Vigorous vortexing should be avoided to prevent shearing of long RNA template (7) (see Note 8).
3. Extracted RNA is used for subsequent PCR or stored at –80°C until use.

3.1.3. RT

1. Based on 20 μl of reaction volume, RT matrix, containing all reaction components except for the template, is prepared to contain 1× SuperScript III buffer, 10 mM DTT, 1 μM QR2

Table 1
LRP primers of HCV genotype 1a. Stars indicate that primers contain restriction sites in their 5′ ends. Primer numbering is according to HCV H77 strain (GenBank accession no. AF009606). All primers were designed with software EuGene version 1.01 based on 9 full-length HCV genotype 1a isolates

Primer	Polarity	Sequence (5′–3′)	Position	T_m (°C)
QR2	Antisense	Tagccagccgtgaaccag	9,248–9,265	61
WF33	Sense	Acgcagaaagcgtctagccat	66–86	63.5
WF5*	Sense	ggatctgacgttaattaacatagtggtctgcggaaccggt	139–160	67.5
WR55*	Antisense	atagctgggtggccggccatggcagccctacctcctctgg	9,139–9,160	67.9

(reverse primer), 2 mM dNTPs, 20 U of Rnasein Ribonuclease Inhibitor, 200 U of SuperScript III and 5 U of AMV.

2. Vortex briefly and aliquot 9.4 μl into each 0.5-ml microfuge tube.
3. Add 10. 6 μl of extracted RNA into tubes, followed by the addition of two drops of mineral oil.
4. Vortex for seconds and centrifuge tubes briefly.
5. Incubate the tubes at 50°C for 75 min, followed by heating at 70°C for 15 min (see Note 9).
6. Store tubes at −20°C if not used for subsequent PCR immediately.

3.1.4. PCR

1. Prepare PCR matrix that contains reaction components except for templates. The matrix includes 1.25 mM $Mg(OAc)_2$, 1× XL PCR buffer, 0.8 mM of dNTPs, 0.4 μM of Trnc-21, 0.4 μM of each primer (WF33 and QR2), and 2 U of rTth XL DNA polymerase.
2. Aliquot 45 μl of PCR matrix into 0.5-ml microfuge tubes.
3. Add 5 μl of RT reaction into tubes.
4. Add two drops of mineral oil.
5. Vortex for 10 s and centrifuge tubes briefly.
6. Load tubes onto PCR machine and start programmed cycles: 94°C for 1 min followed by the first 10 cycles of 94°C for 30 s and 65°C for 9 min and final 20 cycles, in which the annealing/elongation temperature is reduced to 60°C for 9 min with a 3-s autoextension at each cycle. The reaction is ended with 10-min incubation at 72°C (see Note 10).
7. Perform second round PCR by repeating steps 1–4 with changes as follows:
   (a) In PCR matrix, primers WF33 and QR2 are replaced with WF5 and WR55.
   (b) 2 μl of first round PCR product is added into 48 μl of PCR matrix.
   (c) Cycle parameters are the same as the first round PCR except the annealing/elongation temperature is changed to 72°C for the first 10 cycles and 68°C for the last 20 cycles, respectively (see Note 11).
8. Prepare 0.8% of Seakem GTG agarose gel containing 0.5 mg/ml ethidium bromide.
9. 10 μl of LRP product is electrophoresed in 1× TAE buffer at 95 V for 1 h.
10. The gel is visualized on a UV transilluminator and photographed. The expected size of the LRP product is 9,095 bp (Fig. 2).

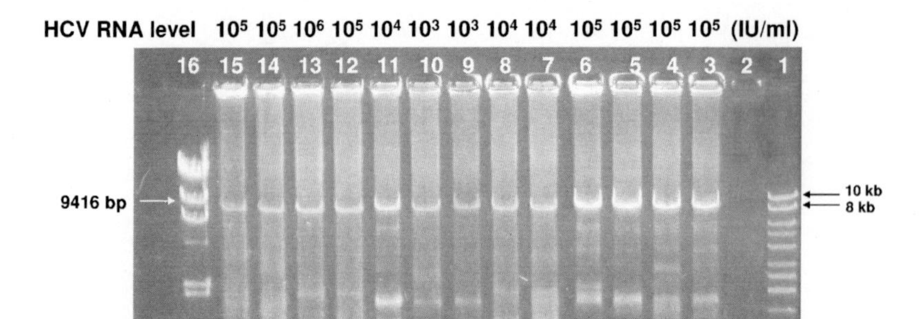

Fig. 2. Amplification of a 9.1 kb fragment of the HCV genome from human serum samples from patients infected with HCV containing various HCV RNA levels. The PCR product was electrophoresed on a 0.8% Seakem GTG agarose gel. *Lane 1* 1 kb DNA ladder; *Lane 2* negative control; *Lane 16* Lambda DNA/Hind III markers. Other lanes represent different serum samples

11. A bench LRP protocol used daily in our lab is presented in Table 2.

3.2. Cloning the LRP Product

1. Purification of LRP product with QIAquick PCR Purification Kit. LRP product is finally eluted into 30 µl of Tris buffer (pH 8.0, provided in the kit) (see Note 12).
2. Digest purified LRP product with restriction enzymes PacI and FseI.

Purified LRP product	15 µl
PacI	1.0 µl (10 units)
FseI	5.0 µl (10 units)
10× NEBuffer 4	2.5 µl (This buffer is provided with enzymes)
BSA (10 mg/ml)	0.25 µl
H_2O	1.25 µl
Total	25 µl

3. Incubate the tubes at 37°C for 2 h, followed by the inactivation of enzyme activity at 65°C for 20 min.
4. After a brief centrifuge, load digestion product onto the gel. Run the gel at 95 V for 1 h.
5. Guided by UV light, cut correct DNA bands with a clean blade (see Note 12).
6. Purify DNA by using QIAEX II Gel Extraction Kit according to the manufacturer's instruction.
7. Take 1 µl of purified DNA to check the concentration and purity with NanoDrop1000.

Table 2
A bench LRP protocol for HCV genotype 1a

Components	1	5	10	20
RT: 9.4 μl RT matrix plus 10.6 μl of RNA template[a]				
5× SuperScript III buffer	4.0	20	40	80
Reverse primer (QR2, 50 μM)	0.4	2.0	4.0	8.0
DTT (100 mM) (Invitrogen)	2.0	10	20	40
dNTPs (40 mM) Invitrogen	1.0	5.0	10	20
Rnasein (40 U/μl) (Promega)	0.5	2.5	5.0	10
SuperScript III (200 U/μl) (Invitrogen)	1.0	5.0	10	20
AMV (10 U/μl) (Promega)	0.5	2.5	5.0	10
First round: 45 μl of PCR matrix + 5 μl of RT reaction[b]				
Sterile DI H_2O	24.2	121	242	484
$Mg(OAc)_2$ (25 mM)	2.5	12.5	25	50
3.3× XL PCR buffer	15.1	75.5	151	302
dNTPs (40 mM) Invitrogen	1.0	5.0	10	20
Trnc-21 (50 μM)	0.4	2.0	4.0	8.0
Forward primer (WF33, 50 μM)	0.4	2.0	4.0	8.0
Reverse primer (QR2, 50 μM)	0.4	2.0	4.0	8.0
rTth polymerase (2U/μl) (ABI)	1.0	5.0	10	20
Second round: 48 μl of PCR matrix + 2 μl of first round of PCR product[c]				
Sterile DI H_2O	27.3	136.5	273	546
$Mg(OAc)_2$ (25 mM)	2.4	12	24	48
3.3× XL PCR buffer	15.1	75.5	151	302
dNTPs (40 mM) Invitrogen	1.0	5.0	10	20
Trnc-21 (50 μM)	0.4	2.0	4.0	8.0
Forward primer (WF5, 50 μM)	0.4	2.0	4.0	8.0
Reverse primer (WR55, 50 μM)	0.4	2.0	4.0	8.0
rTth polymerase (2 U/μl)	1.0	5.0	10	20

[a]50°C 75 min – 70°C 15 min – 4°C hold (File 61 – File 39)
[b]94°C 1 min (File 62); 94°C 30 s – 65°C 30 s – 68°C 9 min × 10 cycles (File 63); 94°C 30 s – 60°C 30 s – 68°C 9 min × 20 cycles (File 64) Auto extension 3 s every cycle; 72°C 10 min – 4°C hold (File 73 – File 39)
[c]94°C 1 min (File 65); 94°C 30 s – 72°C 9 min × 10 cycles (File 66); 94°C 30 s – 68°C 9 min × 20 cycles (File 67) Auto extension 3 s every cycle; 72°C 10 min – 4°C hold (File 73 – File 39)

Digested pClone vector	1.0 μl (~25 ng)
LRP product	*x* μl (227.5 ng)
10× reaction buffer	1.0 μl
T4 DNA ligase	1.0 μl (1 ~ 3 units)
H_2O	*y* μl ($x+y=7.0$ μl)
Total	10 μl

8. Digestion and purification of pClone vector:
   (a) Digest pClone vector with restriction enzymes PacI and FseI, similar to the procedures described in the step 2 Subheading 3.2 (see Note 14).

(b) After a brief centrifuge, load digestion product onto the gel. Run the gel at 95 V for 1 h.

(c) Guided by UV light, cut correct pClone DNA bands with a clean blade (see Note 13).

(d) Purify pClone DNA by using QIAEX II Gel Extraction Kit according to the manufacturer's instruction.

(e) Take 1 μl of purified pClone DNA to check the concentration and purity with NanoDrop1000.

(f) pClone DNA concentration should be adjusted to about 25 ng/μl.

9. Set up ligation reaction:
10. Gently mix by pipetting and incubate at 14°C overnight.
11. Save the ligation product at −20°C if not used for transformation immediately.
12. Transformation

(a) Chill electroporation cuvettes, 1.5-ml microfuge tubes and safety chamber slide from electroporator on ice.

(b) Dissolve DH10B cells on ice.

(c) Add 25 μl of DH10B cells into 1.5-ml microfuge tubes.

(d) Add 2.0 μl of ligation prod uct into the tubes and gently mix using a pipette (see Note 16).

(e) Set controls on the electroporator as follows for 1 mm cuvettes:

Pulser capacitance:	25 μF
Pulse controller:	200
Set volts:	1.65 V

(f) Leave controls set on Time Constant when electroporating.

(g) Transfer mixture of cells and DNA from ligation into a cold electroporation cuvette.

(h) Place the cuvette in a chilled safety chamber slide. Push the slide into the chamber until the cuvette is sealed.

(i) Press both red buttons on electroporator simultaneously until the buzz sounds.

(j) Remove the cuvette from the chamber immediately and add 1 ml of SOC medium to the cuvette and quickly resuspend the cells by pipetting.

(k) Transfer the cell suspension to a Falcon 2059 tube and place in a shaking incubator at 37°C for 1 h.

(l) Cast cell suspension to LB plates containing 30 μg/ml kanamycin (see Note 17).

(m) Incubate LB plates at 37°C overnight.

4. Inoculate clones from LB plates into Falcon 2059 tubes containing 2 ml of LB medium with 30 μg/ml kanamycin.
5. Place Falcon 2059 tubes in a shaking incubator and culture at 37°C overnight.
6. Purify recombinant plasmids using QIAprep Spin Miniprep Kit according to manufacturer's instruction.
7. Check the DNA purity and concentration with NanoDrop 1000. Typically, a 2 ml miniculture yields approximately 5 μg of recombinant plasmid,
8. Confirm the existence of ~9.1 kb insert by restriction digestion with PacI and FseI, followed by electrophoresis (see Note 18).
9. Plasmids are ready for a downstream experiment, such as sequencing.

4. Notes

1. There is no clear definition with regard to the concentration of mixed dNTPs. For example, 2.5 mM of dNTPs from Applied Biosystems means individual concentration of each dNTP while in dNTPs from Invitrogen, it indicates accumulated values of all four dNTPs. In our experience, we obtain the best results with dNTPs from Invitrogen, in which dNTPs are dissolved in 0.6 mM Tris–HCl (pH 7.5) rather than water.
2. Trnc-21 is a 51-bp oligonucleotide (5′- TGG CGG AGC GAT CAT CTC AGA GCA TTC TTA GCG TTT TGT TCT TGT GTA TGA -3′), which is reported to specially inhibit DNA polymerase isolated from *Thermus thermophilus* (*Tth pol*) at low temperature (8, 9). The addition of Trnc-21 allows the PCR setup to be performed at room temperature and a hot-start procedure becomes unnecessary.
3. Other primer design software should also work well.
4. Reverse primers for full-length cDNA synthesis are not easily determined. Besides an appropriate T_m value (60°C), primers which are located entirely within potential stem-loop regions should be avoided if template RNA genomes are highly structured, such as in HCV. It is often required to test several reverse primers. In highly structured RNA templates, self-priming events generate cDNA with variable lengths. Thus, PCR amplification of multiple small regions through entire

genomes becomes necessary when monitoring the synthesis of full-length cDNA.

5. Unlike primers for RT, the requirement for primers in long PCR is less stringent as long as the T_m values of primers are approximately 61 and 68°C, respectively for the first and second PCR.
6. Serum samples should be stored in a –80°C freezer and avoid frequent thaw-freezing cycles, which may result in potential degradation of viral RNA. Amplification in HCV LRP is successful in archived serum samples of up to 12-year storage.
7. It is best to quantitate HCV RNA level in serum samples. In our experience, LRP is successful in samples with various HCV RNA levels, ranging from 10^3 to 10^6 IU/ml (Cobas Amplicor HCV Monitor v2.0, Roche Molecular Systems, Pleasanton, CA), indicating that our LRP protocol is robust and sensitive. For those samples with HCV RNA levels lower than 10^3 IU/ml, LRP amplification is improved by extracting total RNA from 560 μl of serum instead of 280 μl of serum.
8. Besides QIAamp Viral RNA Mini Kit, RNA extracted by using TRIzol LS reagent gives similar LRP results. However, in the latter procedure, the addition of tRNA or glycogen in the RNA precipitation step, even at a low concentration, has resulted in the failure of subsequent PCR, suggesting that these carriers have a detrimental role on rTth XL activity. TRIzol LS reagent is also appropriate for the extraction of RNA from liver tissue.
9. Using an incubation of more than 75 min does not improve the PCR result. Inactivation at 94°C for 5 min will result in breakage of cDNA templates.
10. We use a touchdown cycle protocol that improves LRP results. Cycle parameters are specially designed for DNA Thermal Cycler 480, which is an old-fashioned style of PCR machine. When using other types of PCR machines, the time lengths at each step should be revised according to the manufacturer's instruction.
11. It should be emphasized that high annealing/elongation temperature is a critical factor for successful LRP. Thus, T_m values of primers should be around 68°C.
12. When the DNA band is weak or a large amount of DNA is required for downstream applications, such as pyrosequencing, the necessary level of LRP product may be generated by repeating the second round of PCR.
13. In this step, use of long-wave UV and minimization of UV exposure are required to reduce potential damage to DNA templates.

14. We constructed pClone vector by replacing PacI-Bam HI fragment of pAdTrack-CMV, a pBR322 derived vector (10), with a ~120 bp fragment that was assembled to include rare restriction enzymes not found in the HCV genome based on an analysis of 13 full-length HCV genotype 1a isolates. Two combinations of restriction enzymes were designed to cover the cloning of LRP product of HCV genotype 1a isolates, PacI/FseI, and AsiSI/XbaI.

15. A common formula to calculate the amount of LRP product is useful. The ratio of vector/insert should be set at 3 to 1.

$$\text{LRP product(ng)} = \frac{25\,\text{ng(pClone amount)} \times 9.1\,\text{kb(insert length)}}{3.0\,\text{kb pClone length}} \times \frac{3}{1}$$

$$= 227.5\,\text{ng}$$

16. The addition of ligation product over 10% cell volume results in frequent arcs that destroy the electroporation.

17. In the initial experiment, try different casting volumes of cell suspension to obtain reasonable clone density on LB plates.

18. In different samples, the positive rate for recombinant clones ranges from 30 to 100%.

References

1. Barnes WM (1994) PCR amplification of up to 35-kb DNA with high fidelity and high yield from lambda bacteriophage templates. Proc Natl Acad Sci U S A 91:2216–2220
2. Cheng S, Fockler C, Barnes WM, Higuchi R (1994) Effective amplification of long targets from cloned inserts and human genomic DNA. Proc Natl Acad Sci U S A 91:5695–5699
3. Chumakov KM (1994) Reverse transcriptase can inhibit PCR and stimulate primer-dimer formation. PCR Methods Appl 4:62–64
4. Tuplin A, Wood J, Evans DJ, Patel AH, Simmonds P (2002) Thermodynamic and phylogenetic prediction of RNA secondary structures in the coding region of hepatitis C virus. RNA 8:824–841
5. Tuplin A, Evans DJ, Simmonds P (2004) Detailed mapping of RNA secondary structures in core and NS5B-encoding region sequences of hepatitis C virus by RNase cleavage and novel bioinformatic prediction methods. J Gen Virol 85:3037–3047
6. Simmonds P, Tuplin A, Evans DJ (2004) Detection of genome-scale ordered RNA structure (GORS) in genomes of positive-stranded RNA viruses: Implications for virus evolution and host persistence. RNA 10:1337–1351
7. Lu L, Nakano T, Smallwood GA, Heffron TG, Robertson BH, Hagedorn CH (2005) A refined long RT-PCR technique to amplify complete viral RNA genome sequences from clinical samples: application to a novel hepatitis C virus variant of genotype 6. J Virol Methods 126:139–148
8. Dang C, Jayasena SD (1996) Oligonucleotide inhibitors of Taq DNA polymerase facilitate detection of low copy number targets by PCR. J Mol Biol 64:268–278
9. Lin T, Jayasena SD (1997) Inhibition of multiple thermostable DNA polymerases by a heterodimeric aptamer. J Mol Biol 271:100–111
10. He TC, Zhou S, da Costa LT, Yu J, Kinzler KW, Vogelstein BA (1998) A simplified system for generating recombinant adenoviruses. Proc Natl Acad Sci U S A 95:2509–2514

Part II

The RT-PCR Mathematician: Assessing Gene and RNA Expression

Chapter 11

One-Step RT-LATE-PCR for mRNA and Viral RNA Detection and Quantification

Cristina Hartshorn and Lawrence J. Wangh

Abstract

LATE-PCR is an optimized form of asymmetric PCR that efficiently generates high levels of single-stranded DNA amplicons. Single-stranded amplicons are advantageous because, as shown in this chapter, they can be probed at low temperature(s) with one or more probes. Based on its properties, LATE-PCR is also useful for constructing multiplex assays.

Viral RNA or RNA present in cells can be detected and quantified using LATE-PCR preceded by a reverse transcription (RT) step that converts RNA into cDNA. According to the conventional two-step approach, RT is primed using nonspecific random oligonucleotides prior to performing PCR amplification in a separate step. Recently, however, the so-called "one-step" RT-PCR strategy has gained increasing popularity because all reagents needed for both reactions are added together in a single mix, thus reducing the possibility of contamination, a consideration particularly relevant to viral samples analysis. We describe below two protocols using RT-LATE-PCR and provide specific guidelines for the general application of this technique. The first protocol is devised to quantify mRNA from mouse embryos by real-time PCR; the second protocol employs end-point analysis to detect a number of different viral-like sequences in a multiplex reaction.

Key words: LATE-PCR, One-step RT-PCR, qRT-PCR, RNA virus assay, mRNA assay, End-point quantification, Single-stranded amplicon, Low-temperature probe, Multiplex PCR

1. Introduction

1.1. The Logic of LATE-PCR

Symmetric PCR is currently the predominant method used to amplify relatively short segments of DNA. Each amplicon is generated using two primers having equal melting temperatures (T_ms) and equal concentrations (T_m is the temperature at which half of the primer molecules are hybridized to their targets). The resulting reactions are reproducible during the phase of exponential amplification, until a point when they become inefficient and

Nicola King (ed.), *RT-PCR Protocols: Second Edition*, Methods in Molecular Biology, vol. 630,
DOI 10.1007/978-1-60761-629-0_11, © Springer Science+Business Media, LLC 2010

eventually plateau. Thus, real-time analysis (fluorescent measurement of total product after each cycle) of these reactions reveals a characteristic, sigmoidal curve and shows that the levels of double-stranded amplicon accumulated at the end are not correlated to the number of targets present at the beginning of the reaction. In addition, exit from exponential amplification in symmetric PCR is stochastic and generates highly variable levels of double-stranded DNA among replicates. For the above reasons, symmetric PCR reactions must be analyzed in real-time to quantify the starting numbers of target molecules using measurements of the threshold cycle or C_T. The C_T is the cycle at which the fluorescence resulting from the accumulated product reaches a predefined threshold and is generally close to the end of exponential amplification.

Asymmetric PCR potentially allows to correlate end-point fluorescence with template numbers and to solve the problem of scatter at end-point. This is achieved by replacing equimolar primers with an excess primer (XP) at higher concentration and a limiting primer (LP) at lower concentration. As a result, amplification switches from exponential production of double-stranded DNA (ds DNA) to linear production of single-stranded DNA (ss DNA) when the limiting primer is used up (1). However, the timing of the transition from ds DNA to ss DNA synthesis in traditional asymmetric PCR is primer-pair specific, making it difficult to construct multiplex reactions using this method.

Our laboratory has discovered that simply diluting one member of a pair of symmetric primers does not take into account the fact that lowering a primer's concentration also lowers its T_m, resulting in less efficient amplification. We found that the transition from ds DNA to ss DNA can be made predictable and uniform for all asymmetric reactions by designing primers with appropriate length and composition to insure that the actual melting temperature of each LP at its initial concentration, $T_{m\,[0]}^{L}$, is as high or higher than that of the excess primer at its respective initial concentration, $T_{m\,[0]}^{X}$. Thus, the primary axiom of LATE-PCR (see Note 1) is $T_{m\,[0]}^{L} - T_{m\,[0]}^{X} \geq 0$ (2, 3). As a result, LATE-PCR assays carry out efficient exponential amplification during Phase I and then reliably switch to linear amplification in Phase II, when the LP is exhausted (Fig. 1). In real-time, probe-based LATE PCR assays, the LP runs out shortly before single-strand production is first detected, i.e., at the C_T. Single-stranded XP-strand molecules accumulate linearly thereafter. As a result, LATE-PCR fluorescence end-points with sequence-specific probes are reproducible for a given amount of target. If LATE-PCR reactions are stopped after 20–30 cycles of linear amplification, the amount of ss DNA at end-point quantitatively reflects the amount of target present at the start of the reaction just as the C_T values do (see Fig. 2, right panel). However, LATE-PCR reactions keep on generating more

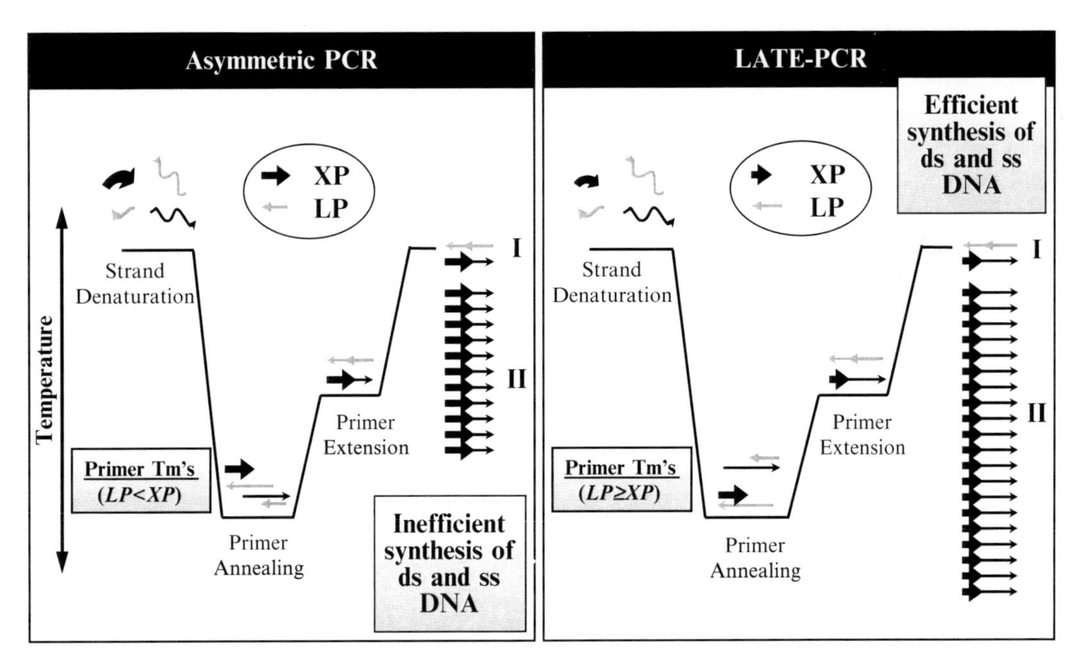

Fig. 1. Comparison of asymmetric PCR and LATE-PCR. In asymmetric PCR, the primers are designed to have equal T_ms, but the T_m of the LP is actually lower than that of the XP because its concentration is lower. This leads to inefficient priming. In LATE-PCR, this problem is corrected by designing the LP to have a concentration-corrected T_m equal or higher than that of the XP. Using a 50–100 nM LP concentration in LATE-PCR assays, real-time reactions switch from exponential to linear at the C_T, which can be used for template quantification

and more single-stranded molecules for at least 100 cycles. Thus, even reactions initiated with very few molecules (including single molecules) eventually generate detectable levels of ss DNA. Moreover, very abundant and very scarce templates can be amplified in the same LATE-PCR reaction (see Fig. 3) because unlike double-stranded amplicons, accumulating ss DNA does not bind the DNA polymerase (4, 5). Phase I ds-DNA amplicons of a LATE-PCR can be detected in real-time by using nonspecific intercalating fluorescent dyes such as SYBR Green, while detection of Phase II ss-DNA amplicons requires the use of fluorescently tagged hybridization probes (Fig. 2). Both types of LATE-PCR products can also be analyzed by agarose gel electrophoresis. Double-stranded amplicons are strongly stained by ethidium bromide, while single-stranded amplicons stain less intensely because ethidium bromide binds only to hairpin regions in which the ss DNA is folded on itself (GC-rich regions, see Fig. 4). In addition, single-stranded amplicons migrate more slowly than the corresponding double-stranded products due to lower linear charge density (6).

1.2. Properties of One-Step RT-LATE-PCR

In one-step RT-PCR, the same set of primers drive both RT and PCR, which actually take place sequentially in the same closed tube (Fig. 5). RT is primed by the primer antisense to the RNA

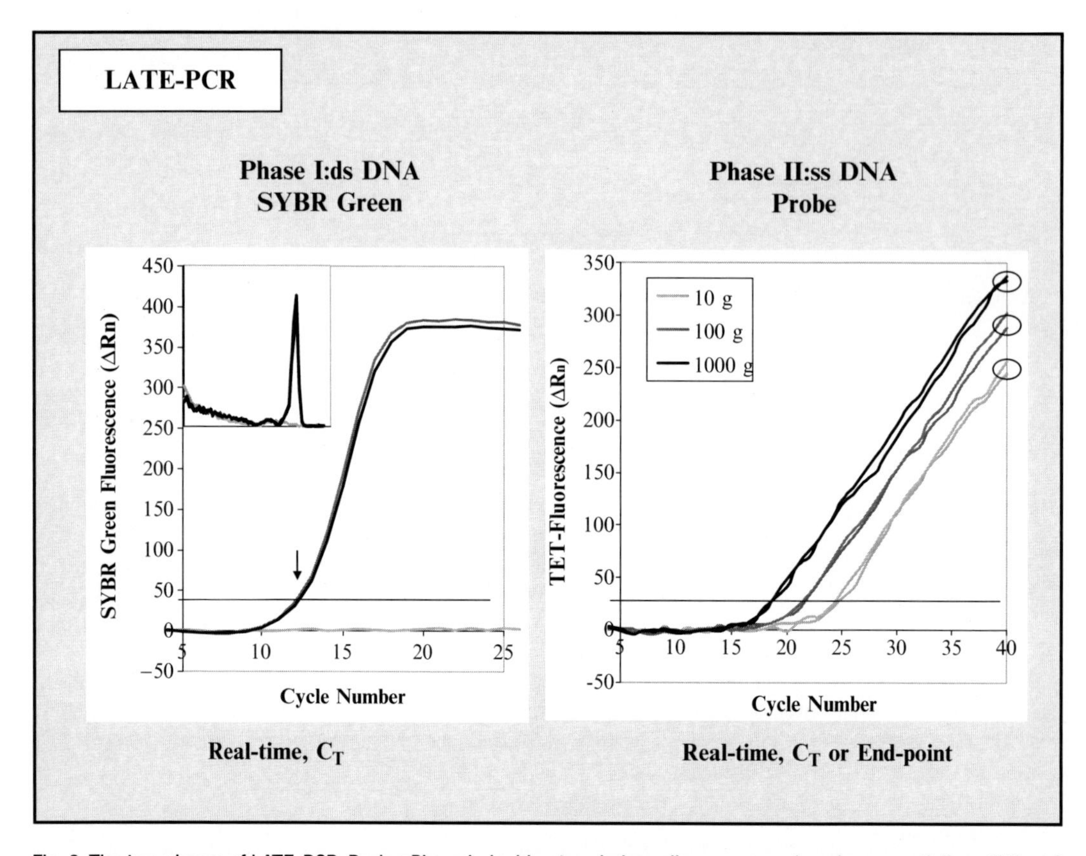

Fig. 2. The two phases of LATE-PCR. During Phase I, double-stranded amplicons are produced exponentially until the LP is exhausted. Real-time LATE-PCR in the presence of nonspecific intercalating dyes such as SYBR Green generates sigmoidal curves that visualize this first phase of the reaction. Template quantification can be achieved using the C_T values from these curves but not using end-point fluorescence because all curves reach a plateau. The purity of the product can be monitored by melt peak analysis (inset of left panel). During Phase II of LATE-PCR, only the XP is still available and is elongated into single-stranded product, thus requiring sequence-specific probes for visualization. The reaction switches to linear and, within a certain number of cycles, end-point fluorescence is proportional to template number and can be used for quantification as much as the C_T values

target (which is in almost all cases single-stranded), and PCR is primed by both primers. In RT-LATE-PCR, either the XP or the LP can function as the RT primer. In both protocols presented in this chapter, the XP was chosen to prime RT in one-step RT-LATE-PCR reactions. Also in both protocols, we took advantage of the fact that LATE-PCR amplicons can be probed at low temperature because they are single-stranded, and the probe is not out-competed by an amplicon's complementary strand. Low-temperature probing was achieved by adding data acquisition steps at the end of PCR amplification, either at every cycle in real-time (first protocol, Subheading 3.1.2) or after all cycles at end-point (second protocol, Subheading 3.2.4).

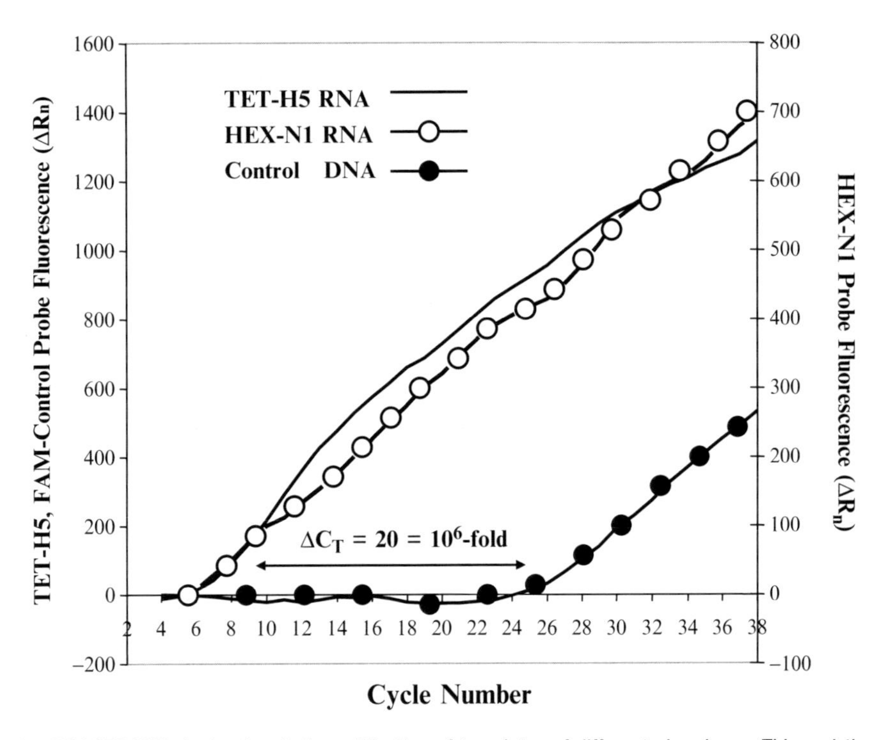

Fig. 3. Triplex RT-LATE-PCR single-stranded amplification of templates of different abundance. This real-time reaction contained 10^3 copies of a control DNA and two viral-like RNAs at 10^6-fold higher concentration. Three probes with different fluors were used to monitor individual amplicon's generation in an ABI 7700 thermocycler

2. Materials

All instrumentations and reagents listed in this chapter can be replaced by different products having equivalent characteristics. Specific identification of the products used in our laboratory is provided below for reference purpose only, in consideration of the wide variety of instruments and RT-PCR reagents commercially available and having different properties.

2.1. Instrumentation and Materials

1. Set of micropipettes
2. Sterile filter pipette tips
3. Vortex mixer
4. Microcentrifuge (capable of reaching at least 14,000 × *g*)
5. Optical 200-µl PCR tubes with optical caps
6. Sterile 0.5-mL Eppendorf tubes.
7. Thermocyclers: iQ™5 Real-Time PCR Detection System (BioRad), or instrument with equivalent capabilities, and (optional) ABI Prism 7700 Sequence Detector (Applied Biosystems) or equivalent (see Note 2)

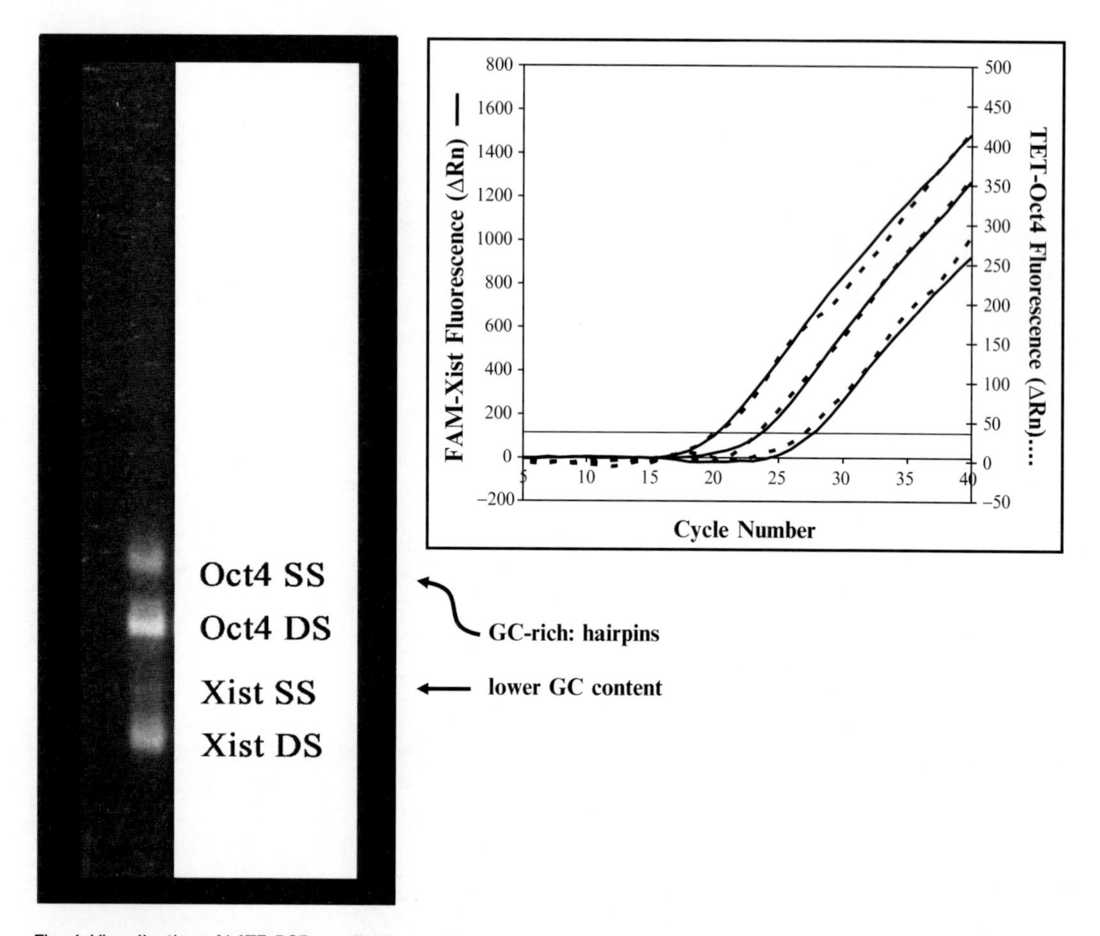

Fig. 4. Visualization of LATE-PCR amplicons by gel electrophoresis. The products of a duplex LATE-PCR assay were analyzed on agarose gel and stained with ethidium bromide (*left panel*). Both templates (sequences from the Oct4 and Xist mouse genes) generated ds DNA amplicons during the first phase of LATE-PCR, as shown by the presence of strongly stained bands of the expected size. In contrast, the single-stranded amplicons of LATE-PCR's second phase show differential staining, much stronger for the GC-rich Oct4 than for the AT-rich Xist. However, the real-time curves of this assay demonstrate that both Oct4 and Xist single-stranded amplicons were efficiently produced at template concentrations ranging from 10 to 1,000 copies (*right panel*)

8. Freezers (–20°C and –70°C)
9. Heating block or water bath
10. Sterile disposable gloves
11. Bleach solution (10%) for decontamination

2.2. Oct4 RT-PCR

All reagents need to be DNase- and RNase-free, molecular biology grade.

1. LATE-PCR primers (see Note 3)
2. FAM-conjugated probe with Black Hole Quencher 1 (see Note 4)

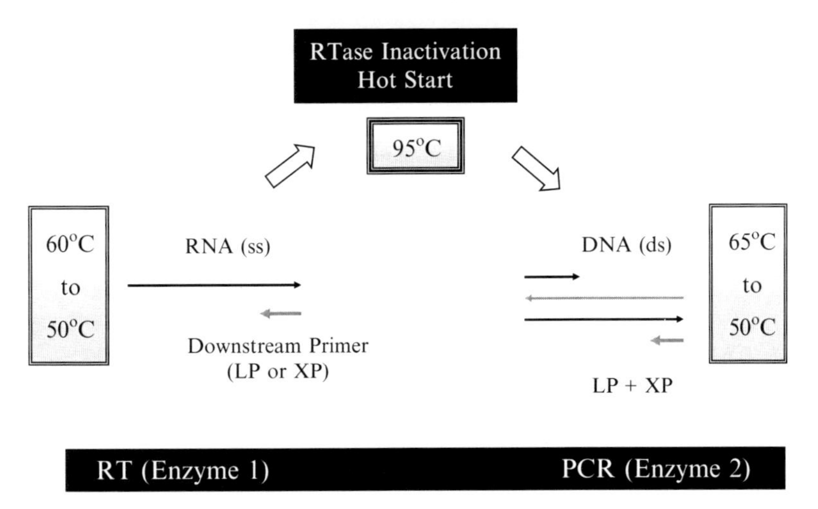

Fig. 5. Diagram of the two reactions within one-step RT-LATE-PCR. All reagents are present together in one mix. The reaction is started when the primer antisense to the RNA strand anneals to target and primes RT, which is mediated by a reverse transcriptase (*left*). RT temperature usually ranges between 50°C and 60°C. At this stage, PCR is blocked by a DNA polymerase-specific antibody. (The use of a DNA polymerase-specific antibody is highly recommended in one-step RT-PCR to prevent nonspecific PCR activity during RT.) RT is terminated by increasing the temperature to 95°C, which denatures the reverse transcriptase and the antibody, thus freeing the DNA polymerase ("hot start", middle). The polymerase is now able to perform PCR amplification by elongating both primers; annealing temperatures range in most cases between 50°C and 65°C (*right*)

3. PrimeSafe 043 or PrimeSafe 002 (7), available at biodetection@smithsdetection.com, or other mispriming preventing compound (see Note 5)
4. Purified mouse genomic DNA (see Note 6)
5. RNase-degrading product to treat surfaces and instrumentation (see Note 7)
6. SuperScript™ III Platinum® One-Step Quantitative RT-PCR System (see Note 8)
7. Tris–Cl, 10 mM, pH 8.3
8. Platinum® Taq DNA polymerase, also containing 10× PCR Buffer and 50 mM $MgCl_2$ (see Note 8)
9. dNTPs, 10 mM
10. Nuclease-free water
11. RNaseOUT, 40 U/µl (see Note 8) or other ribonuclease inhibitor (optional)

2.3. Viral Multiplex RT-PCR (Additional/Alternative Reagents)

1. LATE-PCR primers (see Note 3)
2. Tris–EDTA Buffer Solution, pH 8.0: 10 mM Tris–Cl, 1 mM disodium EDTA, pH 8.0

3. SYBR® Green (any brand)
4. SuperScript™ III Reverse Transcriptase (see Note 8)
5. Platinum® Tfi Exo(-) DNA polymerase, also containing 5× Platinum® Tfi Reaction Buffer and 50 mM $MgCl_2$ (see Note 8)
6. FAM, CalOrange 560, CalRed 610 and Quasar 670-conjugated probes with the appropriate Black Hole Quencher (see Note 4)
7. PrimeSafe 001 (7) (see Note 5)
8. Armored RNA (Asuragen, (8)) or other RNA control (see Note 9)

2.4. Software for Nucleic Acid Design, Folding and Analysis

Visual OMP (DNA Software, Inc.) (see Note 10)

2.5. In Vitro Transcription (Additional Solutions, Kits, and Reagents)

1. Expression plasmids containing a transcription promoter and the target sequence inserts (see Note 11)
2. *Eco*R V, also containing 10X REACT®2 Buffer (see Notes 8 and 11) or another appropriate restriction enzyme
3. Riboprobe® in vitro Transcription System (Promega) or other appropriate IVT kit (see Note 12)
4. Citrate-saturated phenol (pH 4.7):chloroform:isoamyl alcohol (125:24:1)
5. Chloroform:isoamyl alcohol (24:1)
6. Ammonium acetate solution for molecular biology, 7.5 M
7. EtOH, absolute, 200 proof, for molecular biology
8. Yeast transfer RNA or equivalent nucleic acid co-precipitant (see Note 13)

3. Methods

Important: Perform all assays in a dedicated, "clean" lab space. Treat all surfaces, micropipettes, and small equipment with an RNase-degrading product; rinse well prior to use. Use only DNase- and RNase-free reagents and plastic ware. Wear gloves at all times and change them frequently. At the end of the PCR, seal the tubes in a disposable container and discard them in a bin outside the lab. Do not reopen them and do not autoclave them (tubes may pop open during this process), as this could lead to amplicon contamination of the working space (see Note 14).

3.1. Quantification of Oct4 mRNA from Mouse Embryos by RT-Real-Time-LATE-PCR

3.1.1. Primer and Probe Design and Analysis for RT-LATE-PCR

1. Design one LP and one XP according to LATE-PCR guidelines (see Introduction under Subheading 1.1 and (2, 3)). Design primers within an exon of the gene of interest, so that cDNA templates derived from mRNA after RT are identical to the corresponding genomic sequence (5, 9). This strategy allows the use of commercially available mouse DNA to build quantification standards having the same amplification efficiency as the cDNA from the embryos (see Subheading 3.1.2.) (see Note 15).
2. Establish which primer will prime RT (primer antisense to mRNA).
3. Design a probe. The probe is comprised by a linear portion complementary to the target plus a "2-nucleotide (nt) stem." This stem is formed by adding two nucleotides at the 5′ end and two complementary nucleotides at the 3′ end. In the example below, the FAM fluor is conjugated at the 5′ end and the quencher (Black Hole Quencher 1) at the 3′ end.

   RT-LATE-PCR primer and probe sequences for Oct4 mRNA quantification:

LP:	5′-TGGCTGGACACCTGGCTTC AGACT-3′	$T_m = 71°C$[a]
XP (RT primer):	5′-CAACTTGGGGGACTAGGC-3′	$T_m = 67°C$[a]
Probe:	5′-FAM-CTGGGATGGCATACTG TGAG-BHQ1-3′	$T_m = 60°C$[a]

[a]T_ms are calculated for each primer/probe at its starting concentration, at annealing temperature and in the presence of a ds DNA target

4. Analyze the interactions and T_ms of your primers and probe in silico under both RT and PCR conditions, as illustrated in Fig. 6. Visual OMP analysis calculates "Effective T_m" values that take into account all the reaction's components and template folding. As shown in Fig. 6, the Effective T_m of Oct4 XP during the annealing step of PCR is only slightly lower than its calculated T_m (66°C rather than 67°C). Prior to PCR, however, this primer also needs to prime RT. When the template is ss RNA (rather than ds DNA), the Effective T_m of Oct4 XP drops to 53°C, although the RT temperature was set to be the same as the PCR annealing temperature (55°C). At 55°C, about half of the Oct4 XP molecules bind to the RNA target, providing efficient priming (the XP is present at high concentration). Above this temperature, however, priming would become inefficient. Lowering the RT temperature below 55°C leads to an increase in ss RNA secondary structure, further decreasing Oct4 XP's Effective T_m (see Note 16).

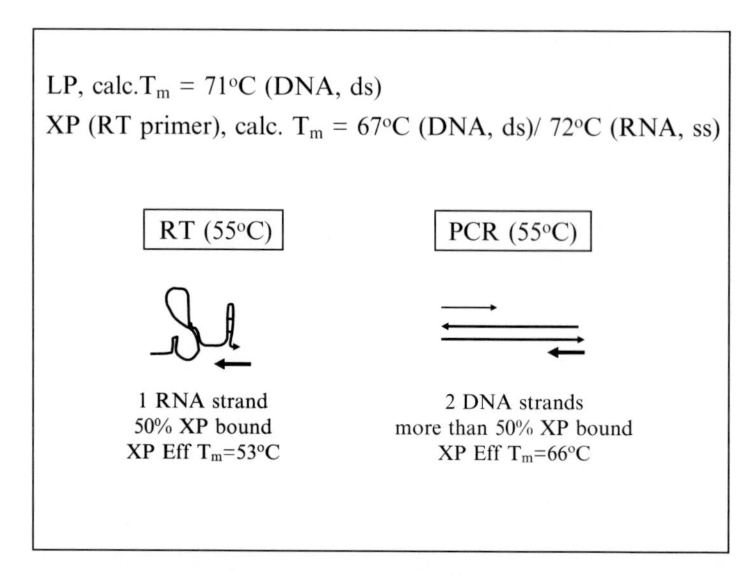

Fig. 6. Visual OMP analysis of the Oct4 RT-LATE-PCR primers. The T_ms of Oct4 LP and Oct4 XP, calculated based on the primer and target composition and salt concentration, are shown at the top for DNA and RNA targets. Oct4 XP functions as a primer during both RT and PCR. Its Effective T_ms, which also take in account the secondary structure of the target at different temperatures, are shown at the bottom of the figure. The secondary structure of ss RNA results in decreased binding efficiency of Oct4 XP during RT when compared to the PCR annealing step, although both reactions are simulated to take place at 55°C

5. Based on the above analysis, determine the best RT temperature. For this protocol, we chose an RT temperature of 55°C because Oct4 XP primes RT very poorly above or below 55°C due to its effective T_m in the presence of ss RNA (Fig. 7).
6. Choose a reverse transcriptase highly active and thermostable at the set RT temperature. Keep in mind that SuperScript™ III, used in this protocol, works well at 55°C but has a half-life of only 2.5 min at 60°C ((10) and Fig. 7).

3.1.2. One-Step RT-LATE-PCR Protocol

Prepare three sets of assays, with each containing in replicate: "+RT", "No RT" and No Template Controls (NTCs). Also prepare a DNA standard (see step 5 below).

1. For "±RT" and "No RT" samples: Preincubate the RT primer with the RNA template for 5 min at 65°C in the following mix (volumes given per reaction):

Tris–Cl, 10 mM, pH 8.3	(to 10 μl)
Oct4 XP[a], 100 μM (RT primer)	0.5 μl
RNA from mouse embryo (see Note 17)	approx. 6 μl
	10 μl

[a]2 μM in the final RT-PCR assay, 5 μM during preincubation

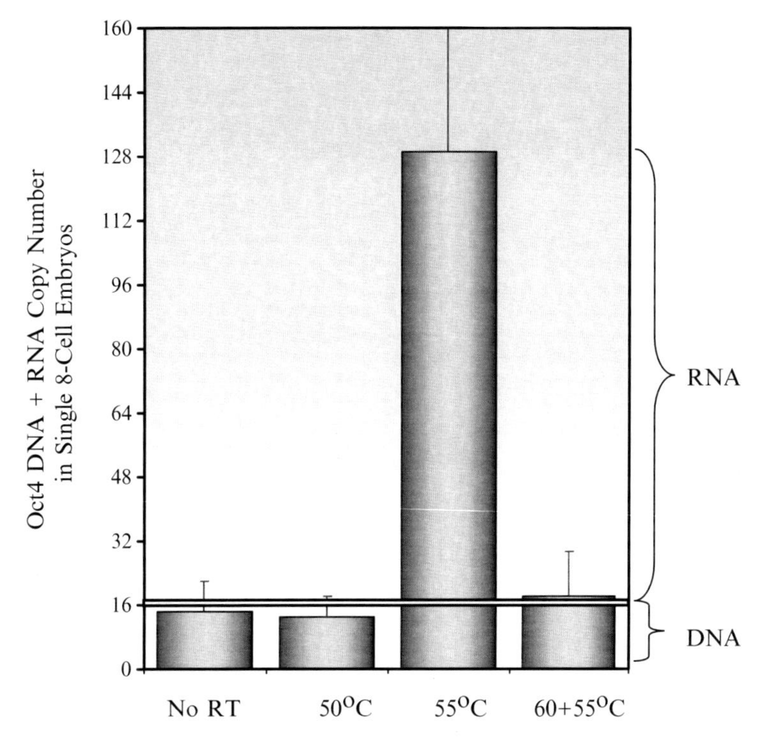

Fig. 7. Thermo-dependency of RT efficiency in one-step RT-LATE-PCR. Single mouse embryos at the 8-cell stage were processed with the PurAmp procedure, so that samples contained both genomic DNA and mRNA (5, 9). The Oct4 RNA+DNA content of each embryo was measured with one-step RT-LATE-PCR, varying the RT temperature as shown on the abscissa; Oct4 DNA copy numbers were obtained from "No RT" samples and amounted to about 16 copies/embryo as expected (two copies of the Oct4 gene/cell). Columns are the average of several embryos; the portion of each column above the horizontal bar gives a measure of Oct4 RNA in the samples. It is evident that Oct4 RNA (present in all embryos) is primed efficiently during RT only when the reaction is carried out at 55°C. When the 55°C incubation is preceded by 10 min at 60°C (column on the far right), RT fails because the enzyme is denatured

2. Cool down and add to the RT-PCR Mix below (10 µl from preincubation + 15 µl of RT-PCR Mix = *25 µl final volume of RT-PCR assay*) or to the "No RT" PCR Mix described below (step 3):

   RT-PCR Mix (per reaction):

	Final concentration	
10× PCR Buffer[a] (see Note 18)	1×	2.5 µl
$MgCl_2$, 50 mM	3 mM	1.5 µl
dNTPs, 10 mM	0.25 mM	0.625 µl
Oct4 LP, 10 µM	*100 nM*	0.25 µl
FAM-Oct4 probe, 100 µM (see Note 19)	500 nM	0.125 µl

	Final concentration	
PrimeSafe 043, 10 μM (see Note 20)	300 nM	0.75 μl
RNaseOUT, 40 U/μl (optional)	0.8 U/μl	1 μl
SuperScript™ III RT/Platinum® Taq Mix[b]	(1 U Taq)	0.5 μl
Nuclease-free H_2O		7.75 μl
		15 μl

[a]From Platinum® Taq kit: 200 mM Tris–HCl, pH 8.4, and 500 mM KCl
[b]From SuperScript™ III Platinum® One-Step Quantitative RT-PCR System

3. Prepare a "No RT" PCR Mix by substituting the SuperScript™ III RT/ Platinum® Taq Mix with 1 U of Platinum® Taq DNA polymerase.
4. For NTCs: Follow the same procedure as for the "+RT" samples, but do not add RNA to the preincubation mix.
5. Prepare a DNA standard using genomic DNA (5, 9):

   Prepare serial dilutions of mouse genomic DNA in "No RT" PCR Mix, so that tubes (in triplicate) will contain 10, 100, and 1,000 genomes. (The size of a diploid mouse genome is 6 pg (11)).
6. Run all assays ("+RT", "No RT", NTCs, genomic standard dilutions) with the following thermal profile (optimized for Prism 7700 Sequence Detector; it can also be used on iQ5 thermocycler, see Note 19).

55°C, 30 min (RT) →	95°C, 5 min→	95°C, 10 s→	95°C, 15 s→
		63°C, 20 s	55°C, 25 s
		72°C, 30 s	72°C, 35 s
			45°C, 30 s[a]
1×	1×	15×	40×

[a]Data acquisition (see Note 21)

Plot the data from the genomic standard, C_T values vs template copy number. Keep in mind that each genome contains two copies of the autosomal Oct4 gene. Quantify Oct4 RNA copy numbers in the unknown samples using the standard's regression line (9).

3.1.3. PrimeSafe Titration

Prior to performing the RNA assays, it is recommended to titrate PrimeSafe on the DNA standard, as some batch-to-batch variability is possible (see Note 22). The slope of LATE-PCR reactions, which affects end-point fluorescence, is a very sensitive tool to monitor efficient amplification, in addition to C_T value analysis (see Fig. 8 and (5)).

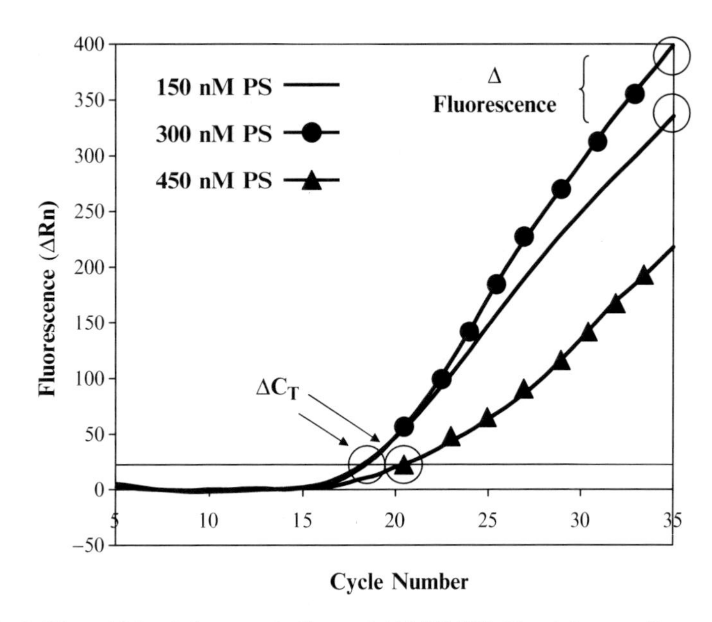

Fig. 8. Effect of PrimeSafe concentration on Oct4 LATE-PCR. All real-time reactions were carried out in the presence of 2,000 copies of genomic DNA and different concentrations of PrimeSafe 043. The LATE-PCR curves show that a concentration of 300 nM PrimeSafe is optimal because 450 nM results in excessive C_T delay and low final fluorescence (overall PCR inhibition), while 150 nM does not improve the C_T value but end-point fluorescence is still lower than with 300 nM. In this second case, a lower-than-optimal end-point fluorescence suggests that nonspecific products are generated as PCR progresses, not "seen" by the probe, because the PrimeSafe concentration is too low

3.1.4. Duplex RT-LATE-PCR and Assay Optimization by Slope Analysis

Two different types of transcript can be simultaneously measured in a sample as small as a single cell using RT-LATE-PCR, independent from their abundance (5, 12). For one-step reactions, use the protocol described above with the following modifications:

1. Design primers and a probe for the second mRNA, all having T_ms as close as possible to those of the Oct4 primers and probes.
2. Select primers so that the XP for the second target will also prime RT, as shown above for Oct4 XP.
3. The probe for the second mRNA needs to have a different fluor. The fluor choice for the two probes depends on the PCR cycler used. We use FAM and TET with a Prism 7700 cycler, and CalOrange 560 and Quasar 670 with an iQ5 cycler.
4. Use the second pair of primers and the second probe at the same concentrations used for the Oct4 primers and probe. (Adjust if necessary.)
5. Increase dNTPs concentration to 0.4 mM.

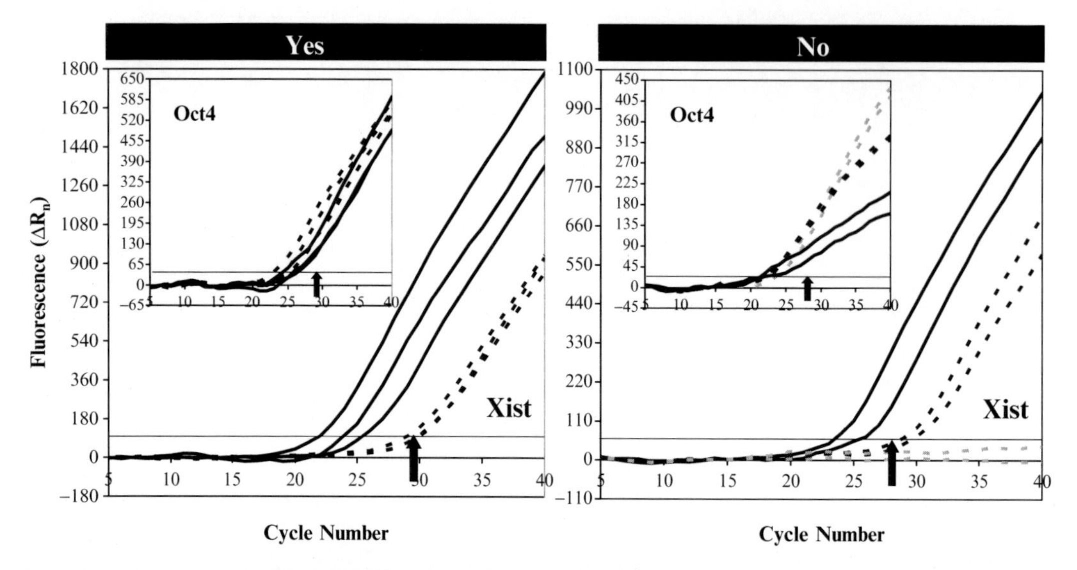

Fig. 9. Optimization of duplex RT-LATE-PCR by slope analysis. The linear phase of LATE-PCR allows assay optimization by slope analysis. Two templates, Xist and Oct4 (RNA + DNA), were simultaneously amplified from single mouse embryonic cells. As shown in the charts, male cells (*broken lines*) contain a single copy of the Xist gene and no Xist RNA while female cells also contain abundant Xist RNA (*solid lines*). Black arrows indicate the C_T value expected for a single template copy based on a genomic DNA standard. Slope analysis reveals that under suboptimal conditions (*right panel*), amplification of the second template, Oct4 (inset), is much less efficient in the female cells that contain high amounts of Xist RNA than in male cells that only contain one Xist template. An optimal Oct4 slope is only obtained when Xist amplification fails (*broken grey lines*). After optimization (*left panel*), all Oct4 slopes are similar, although the Xist RNA + DNA contents of the cells analyzed span from one to several hundred copies

6. Retitrate PrimeSafe. This is very important for a duplex because the two amplicon sequences will always present differences affecting PCR efficiency and, hence, sensitivity to PrimeSafe. Again, use slope analysis to optimize the assay, as shown in Fig. 9.

3.2. End-Point Detection/Quantification of Multiplex Viral Sequences by RT-LATE-PCR

3.2.1 Multiplexing Scheme

Select the desired RNA sequences to be probed in the assay. Sequences can include genes from the same virus and be subtype specific. Additionally, one same, long sequence can contain the targets for two different probes (see Note 23). Sequences from different viruses are amplified in alternative outcomes, so that not all primers and probes are expected to be used during one assay although they are all present together in the reaction mix. The final number of selected probe targets depends on the available PCR cycler. The iQ5 cycler allows detection of a maximum of five different fluors at the same time; Fig. 10 shows our preferred fluor selection for this instrument: FAM, CalOrange 560, CalRed 610, and Quasar 670, the last one being the brightest (see Note 24). LATE-PCR allows probing at low temperature at the end of the reaction (rather than during annealing) because the product is single-stranded. For this reason, at least two sets

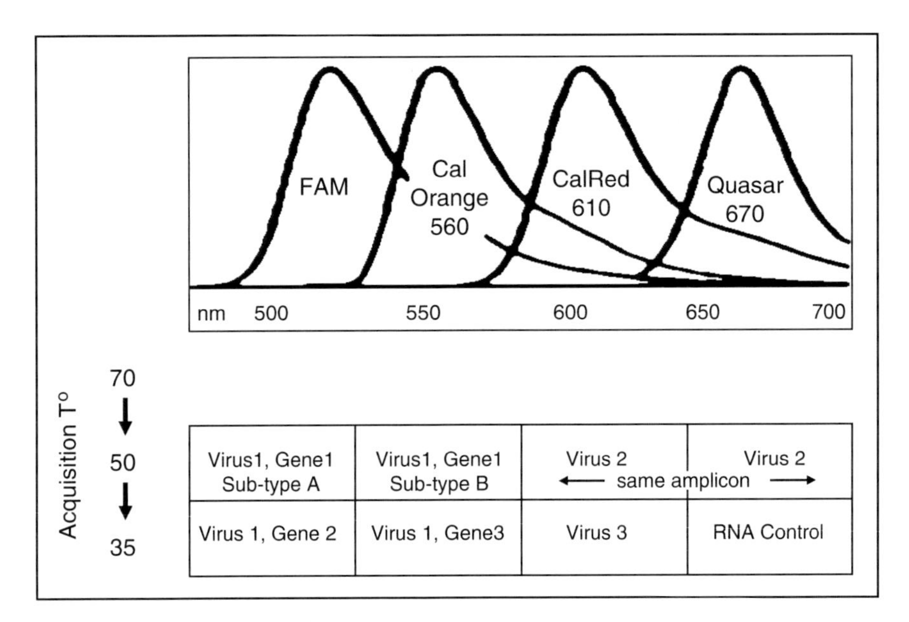

Fig. 10. Eight-target multiplex for virus detection. Four different fluors are chosen with minimally overlapping emission spectra (see Note 11; spectra in the figure are from (15)) and two sets of probes are designed to hybridize to the selected viral targets (see text), so that each fluor is used twice in the multiplex. Both sets of probes have low T_ms compared to traditional probes, but one set has T_ms higher than the other set. Fluorescence intensities are acquired at 70°C (normalization temperature, see Fig. 11), 50°C (higher Tm probes) and 35°C (lower Tm probes). In the scheme shown here, three alternative outcomes are possible: Virus 1 is detected by multiplex amplification, while Virus 3 produces a single amplicon; Virus 2 generates one amplicon probed on two segments. An external RNA control is added to all assays to verify efficient RT-LATE-PCR

of probes can be designed having the same fluor but different T_ms. These probes can be distinguished because they hybridize at different temperatures (35°C and 50°C, see Fig. 10). Thus, a total of eight different probe targets are chosen for this multiplex (see Note 25), one being a designated external RNA control (see Note 26).

Caveat: Probes that bind at 35°C will not be bound at 50°C, but probes that bind at 50°C will still be bound at 35°C (see Fig. 11). Keep this in mind when choosing each target/probe pair. In the example of Fig. 10, the presence of Virus 1 is flagged by FAM and CalOrange 560 probes binding at 35°C. If the virus is of either subtype A or B, fluorescence of either of these two fluors will also increase at 50°C (see Note 27).

3.2.1.1. Development of the Alternative-Outcome Multiplex RT-LATE-PCR Assay

Keep in mind that all assay components have to work under the same conditions during both RT and PCR.

1. Design all LPs with similar T_ms (67–69°C).
2. Design all XPs with similar T_ms (63–65°C).
3. Make sure that each amplicon's T_m is no more than 20°C higher than its XP's T_m (see Note 28).

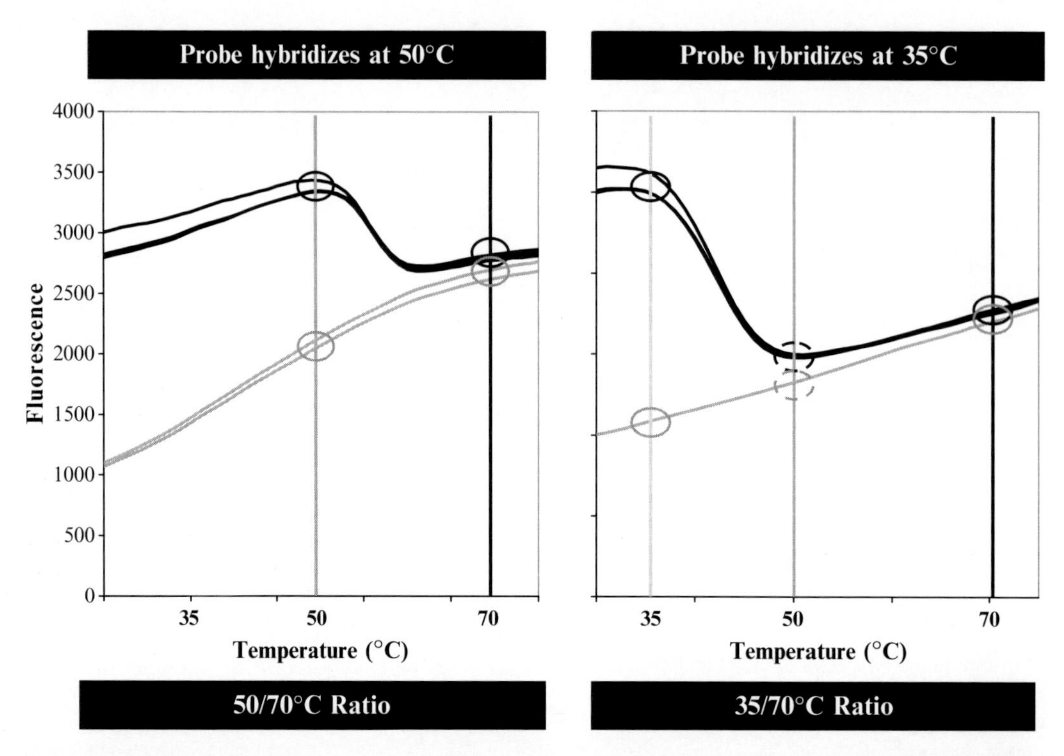

Fig. 11. Temperature-specific low-T_m probes. Two sets of probes are required for a multiplex RT-LATE-PCR viral assay as shown in Fig. 10. The higher-T_m probes are designed to be bound at and below 50°C. The lower-T_m probes are designed to be bound at 35°C but not at 50°C. Probe/target melting curves for two such probes are shown here (*black lines*), together with the respective NTCs melting curves (*grey lines*). Left panel: higher-T_m CalOrange 560-probe. *Right panel*: lower-T_m CalRed 610-probe. As shown by the NTCs curves, each fluor has a characteristic temperature-dependent fluorescence profile; in addition, there are sample-to-sample variations in fluor content (curves do not overlap perfectly at 70°C, when all probes are open). The effect of both these factors is eliminated by normalizing each sample's fluorescence at probe-binding temperature (50°C and 35°C) by that same sample's fluorescence at 70°C

4. Prime RT with the same type of primer (XP) for all targets (see Note 29).
5. Design all probes expected to bind at 50°C with similar T_ms (58–61°C).
6. Design all probes expected to bind at 35°C with similar T_ms (47–50°C).
7. Make sure that the probes that bind at 35°C are not still partially bound at 50°C (see Fig. 11).
8. Analyze all the primer/probe/amplicon sequences with nucleic acid folding software. *Avoid any interaction that could lead to primer elongation or mispriming. Take into account buffer composition and salt concentration when calculating* T_m *values.* In our experience, Visual OMP provides the most rigorous sequence interaction analysis and the most accurate T_m estimates.

3.2.1.2. Low-T_m Probe Design

As for Oct4 mRNA (Subheading 3.1.1.), probes are designed to include a linear portion complementary to the target and having the appropriate T_m. A "2-nt stem" is included in the probe by adding two nucleotides at the 5′ end and two complementary nucleotides at the 3′ end. The fluor of choice is conjugated at the 5′ end and the appropriate Black Hole Quencher at the 3′ end. For FAM probes, the stem must include a GC base pair in order to improve fluor/quencher interaction (see Notes 30 and 31).

1. Design a set of probes hybridizing at 50°C.
2. Design a set of probes hybridizing at 35°C keeping in mind the following: (a) fluor and quencher need to be sufficiently distant once the probe is bound to target (avoid very short probes; the probe should be about 20-nt long including the stem); (b) you can introduce mismatch(es) to reach a low T_m without sacrificing length; alternatively, add a few nucleotides noncomplementary to target between the target-specific sequence and the stem (avoid making the stem longer because it would not open at low temperature); (c) expect the experimentally determined T_m of this type of probes to be several degrees lower than the calculated T_m. Visual OMP estimates are less accurate at this very low temperature.

Figure 12 shows an example of two probes having the same fluor but different T_ms and their use in a viral RNA multiplex assay.

3.2.2. Preparation of Viral RNA Sequences In Vitro (see Note 32)

RNA sequences identical to selected viral gene sequences are generated in the lab from plasmids containing the desired sequence preceded by the T7 RNA polymerase promoter. We use Plasmid pGS-21a, custom-modified in order to eliminate any intervening sequence between the T7 promoter and the desired insert sequence (ClaI/NcoI modification) (see Note 11). With this method we have successfully generated RNA targets over 400 nt-long.

3.2.2.1. Plasmid Linearization Protocol

1. Resuspend plasmid (plasmid miniprep = ~4–5 µg) in 20 µl of nuclease-free water.
2. Heat at 50°C for 15 min before taking an aliquot for linearization.
3. Prepare the *Linearization Mix:*

   8 µl plasmid (=~1.6–2 µg)

   1 µl 10X REACT 2 Buffer

   1 µl *Eco*R V stock

   10 µl total

   Final glycerol concentration = 5% (*Important*: higher glycerol concentrations affect enzyme's specificity)

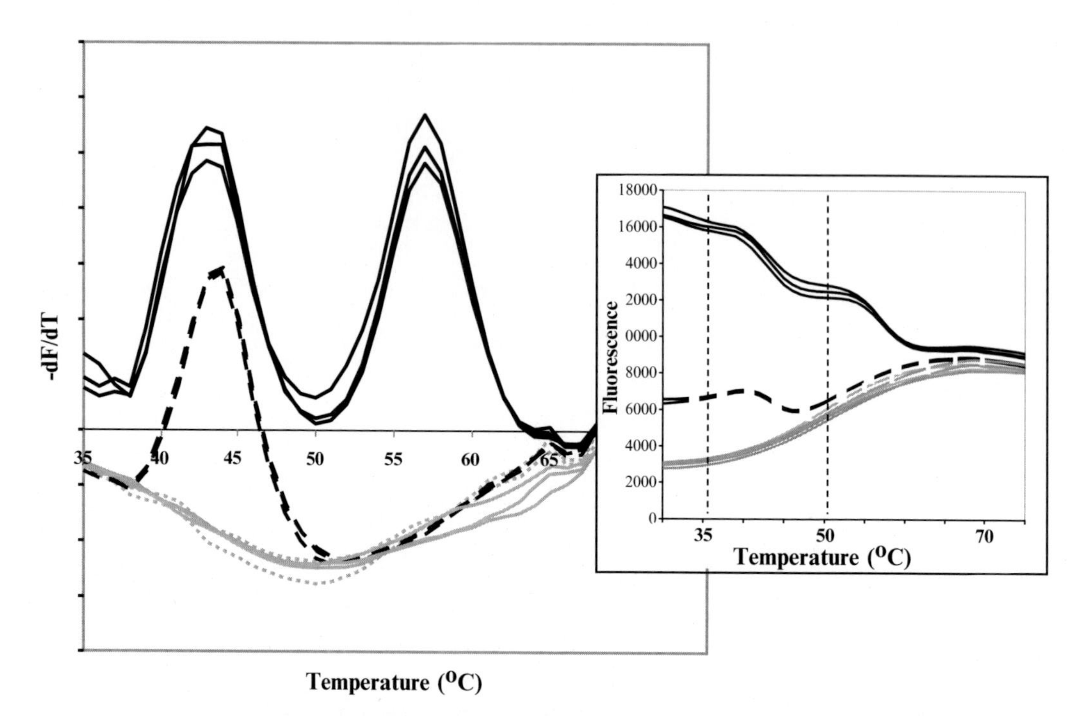

Fig. 12. Probes with the same fluor and different T_ms in a multiplex RT-LATE-PCR assay. Two Quasar 670-probes were designed with different target-specificity and different T_ms, as detailed in Fig. 11.11. The target for the lower-T_m probe is an external RNA control that is added to all assays; the target for the higher-T_m probe is a sequence from the H5N1 avian influenza virus (see also Fig. 11.15). RT is required for target amplification. (Neither probe hybridizes in "No RT" assays, *grey solid lines*; *grey broken lines*=NTCs.) The lower-T_m probe binds to target in all RT-assays, because the RNA control is always present. The higher-T_m probe binds to target only when the viral RNA sequence is added to the assay: in this case, both probes hybridize to target (bimodal *black solid line*; broken *black line*: only RNA control present in the assay). Main chart: negative derivatives of the probe-to-target melting curves (in the inset)

Final *Eco*R V concentration = 10 U/10 μl

4. Incubate at 37°C for 1 h, then inactivate enzyme by heating at 75°C for 10 min.
5. Store at –20°C.

3.2.2.2. In Vitro Transcription (IVT)

Use the Riboprobe in vitro Transcription System.

1. *Have all reagents at room temperature* (except for enzymes and rNTPs) to avoid RNA precipitation in the spermidine-containing buffer.
2. Prepare the IVT Mix:

Transcription Opt. 5× Buffer	4 μl
DTT, 100 mM	2 μl
RNasin, 40 U/μl	1 μl

rNTPs mix, 2.5 mM each[a]	4 µl
T7 RNA polymerase, 20 U/µl	1 µl
Nuclease-free water	4 µl
Linearized plasmid[b]	4 µl
	20 µl total

[a]Freshly prepared by adding 1:1:1:1 each of the 10-mM stock solutions

[b]Added last to avoid contamination; 4 µl = ~0.6–0.8 µg

3. Incubate at 37-to-39°C for 1 h.

3.2.2.3. DNase Digestion and RNA Purification

1. Add 0.8 U DNase from the same IVT kit to the sample (0.8 µl from 1 U/µl stock): ~1 U DNase/µg of DNA.
2. Incubate at 37°C for 15 min.
3. Without delay, perform phenol–chloroform extraction as explained on p.13 of the kit manual.
4. In the last step (ammonium acetate/EtOH), add a coprecipitant: 1 µl (10 µg) of yeast transfer RNA.
5. Allow to precipitate at −70°C O/N (longer is fine)
6. Centrifuge in a microcentrifuge for 1 h at max speed (at least 14,000 × *g*).
7. Discard supernatant and wash pellet with 0.5 ml of 70% EtOH (prepared with nuclease-free water).
8. Centrifuge in a microcentrifuge for 10 min at max speed (at least 14,000 × *g*).
9. Remove supernatant and allow pellet to dry for a few minutes, but not completely (complete drying of the RNA pellet hinders its resuspension).
10. Resuspend pellet in 100 µl of Tris–EDTA buffer.
11. Divide in 10 µl-aliquots and store at −70°C (see Notes 33 and 34).

3.2.3. Test of Selected Primers' Ability to Prime RT on Individual RNA Targets

Prior to multiplexing, test each RNA prep with its own RT primer, in order to establish efficient cDNA generation (see Figs. 6 and 7, Note 16 for possible problems due to RNA secondary structure). Prepare three sets of assays, each in duplicate: "+RT", "No RT" and No Template Controls (NTCs).

1. For "±RT" and "No RT" samples: Preincubate the RT primer with the RNA template for 5 min at 65°C, in the following mix (volumes given per reaction):

Tris–Cl, 10 mM, pH 8.3	8.75 µl
XP[a], 100 µM (RT primer)	0.25 µl
RNA prep diluted 1:500 in Tris–EDTA buffer	1 µl
	10 µl

[a]1 µM in the final RT-PCR assay, 2.5 µM during preincubation

2. Cool down and add to the RT-PCR Mix below (10 µl from preincubation + 15 µl of RT-PCR Mix = *25 µl final volume of RT-PCR assay*) or to the "No RT" PCR Mix described in Step 3 below:

RT-PCR Mix (per reaction):

	Final concentration	
5X Platinum® Tfi reaction buffer (Platinum® Tfi DNA polymerase kit)	1×	5 µl
$MgCl_2$, 50 mM (Platinum® Tfi DNA polymerase kit)	3 mM	1.5 µl
dNTPs, 10 mM	0.25 mM	0.625 µl
LP, 10 µM	100 nM	0.25 µl
SYBR Green, 10X (see Note 35)	0.25×	0.625 µl
PrimeSafe 001, 10 µM	100 nM	0.25 µl
Platinum® Tfi Exo(-) (see Note 36)	2 U	0.4 µl
SuperScript™ III (200U/µl),	200 U	1 µl
Nuclease-free H_2O		5.35 µl
		15 µl

3. Prepare a "No RT" PCR Mix by substituting SuperScript™ III with water.
4. For NTCs: Follow the same procedure as for the "+RT" samples, but do not add RNA in the preincubation mix.
5. Run all assays ("+RT", "No RT", NTCs) with the following thermal profile (optimized for iQ5 cycler; use the Persistent Well Factor option because the Experimental Well Factor option is not available for SYBR Green):

Cycle	Repeats	Step	Dwell time	Setpoint (°C)	Data acquisition
1	1	1	10 min	55	
		2	5 min	95	
2	50	1	5 s	95	
		2	10 s	58	

Cycle	Repeats	Step	Dwell time	Setpoint (°C)	Data acquisition
		3	20 s	68	Real-time
3	121	1	15 sec	35–95	Melt Curve
				(temp. change 0.5°C)	

6. Data analysis. Compare "+RT" (RNA+DNA) and "No RT" (DNA only) amplification curves. "No RT" curves reveal the number of the plasmid-derived DNA templates still present in the RNA preparation after DNase digestion. This number is usually very low, typically 0.1% of total (RNA+DNA) template copies or less, see Fig. 13. A wide delta between "+RT" and "No RT" curves confirms efficient RT priming. Flat NTCs lines are desirable because they indicate that no primer-dimers are formed during PCR. The negative derivative of the melt curves is used to confirm that only one product, with the correct T_m, is generated during RT-PCR (see inset in Fig. 13).

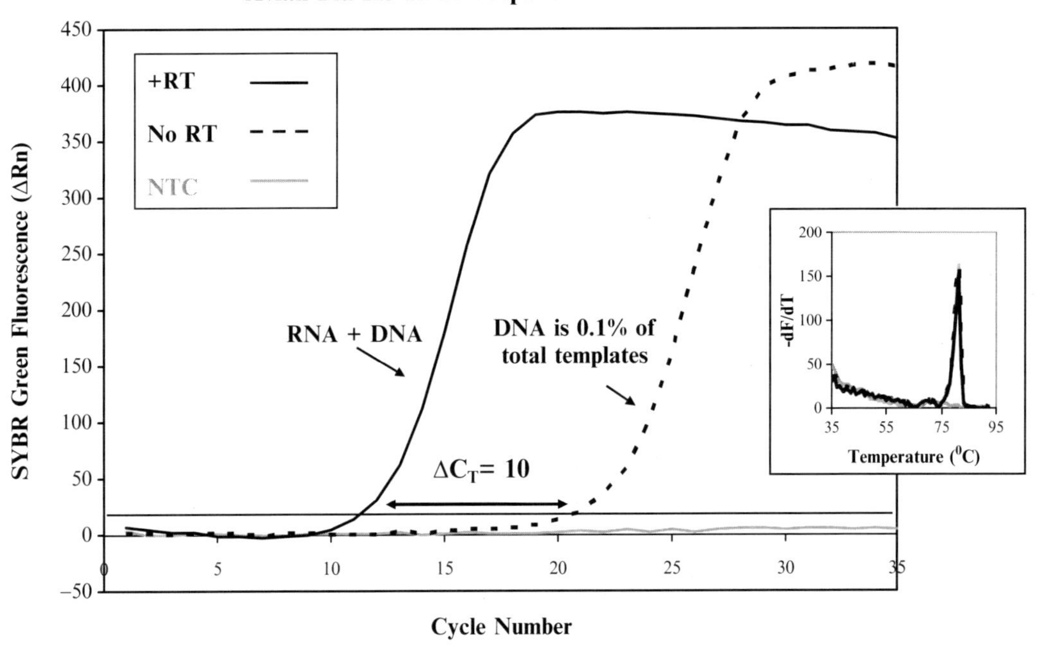

Fig. 13. RNA and DNA content of an IVT-RNA preparation from plasmid. IVT was carried out on a viral-like RNA sequence inserted in a plasmid, as described in the text. The preparation was tested by real-time RT-LATE-PCR in the presence of SYBR Green and the DNA content (carry-over from the plasmid after DNase digestion and RNA purification) was determined based on a set of "No RT" assays. The data show that RNA is 1,000-fold more abundant than DNA in this preparation ($\Delta C_T = 10$ cycles). The RNA and DNA templates are identical and clean, as indicated by the melt peaks in the inset

3.2.4. Multiplex RT-LATE-PCR Mix and Protocol

1. *Templates:* The assay described here is a model for a diagnostic assay. Depending on the presence of different viruses in each diagnostic sample, not all possible targets would be amplified at the same time. (Only the RNA control will be amplified every time.) We mimic these alternative outcomes by grouping sets of IVT-derived RNAs as they would be found in a viral sample. The RNAs in each set are added at equimolar concentration based on the C_T values obtained from real-time qPCR of the individual RNA preps (see Subheading 3.2.3. and Fig. 13). Referring to Fig. 10, the presence of Virus 1, Subtype A is modeled by mixing three RNA preps (probed with FAM at 50°C and 35°C and with CalOrange 560 at 35°C). The presence of Virus 1, Subtype B is also modeled by mixing three RNA preps (two of them also used for Subtype A) and probed with FAM at 35°C and with CalOrange 560 at 50°C and 35°C). The presence of Virus 2 is mimicked by one RNA prep and probed with CalRed 610 and Quasar 670 at 50°C, and the presence of Virus 3 is mimicked by yet another RNA prep and probed with CalRed 610 at 35°C. Fifty-thousand copies of Armored RNA (ARNA, external control) are added in all the assays to ensure that RT-LATE-PCR did not fail.
2. *Armor Lysis:* For control RNA preparation. Heat ARNA (50,000 copies/assay) at 75°C for 3 min, as recommended by the manufacturer (see Note 37).
3. RT-LATE-PCR: The general strategy is similar to the one detailed in Subheading 3.2.3. Begin by preincubating the RT primers (XPs) with the RNA templates for 5 min at 65°C, in the following mix (volumes given per reaction):

Tris–Cl, 10 mM, pH 8.3	(to 10 μl)
XP for each RNA prep[a], 100 μM (RT primer)	0.25 μl each
XP for ARNA, 100 μM (RT primer)	0.25 μl
Lysed ARNA	2 μl
Set of RNA preps diluted in Tris–EDTA buffer	3 μl
	10 μl

[a]1 μM each in the final RT-PCR assay, 2.5 μM during preincubation

4. Cool down and add to the RT-PCR Mix below (10 μl from preincubation + 15 μl of RT-PCR Mix = *25 μl final volume of RT-PCR assay*) or to the "No RT" PCR Mix described in Step 5 below:

RT-PCR Mix (per reaction):

	Final concentration	
5× Platinum® Tfi Reaction Buffer(Platinum® Tfi DNA Polymerase kit)	1×	5 µl
$MgCl_2$, 50 mM(Platinum® Tfi DNA Polymerase kit)	3 mM	1.5 µl
dNTPs, 10 mM,	0.4 mM	1 µl
LP for each RNA prep, 10 µM	100 nM	0.25 µl each
LP for ARNA, 10 µM	50 nM	0.125 µl
FAM probe (50), 100 µM	200 nM	0.05 µl
FAM probe (35), 100 µM	200 nM	0.05 µl
CalOrange560 probe (50), 100 µM	200 nM	0.05 µl
CalOrange560 probe (35), 100 µM	200 nM	0.05 µl
CalRed610 probe (50), 100 µM	200 nM	0.05 µl
CalRed610 probe (35), 100 µM	200 nM	0.05 µl
Quasar670 probe (50), 100 µM	200 nM	0.05 µl
Quasar670-ARA probe (35), 100 µM	200 nM	0.05 µl
PrimeSafe 001, 10 µM	100 nM	0.25 µl
Platinum® Tfi Exo(-)	2 U	0.4 µl
SuperScript™ III (200U/µl),	200 U	1 µl
Nuclease-free H_2O		(to 15 µl)
		15 µl

5. Prepare a "No RT" PCR Mix by substituting SuperScript™ III with water.
6. For NTCs: Follow the same procedure as for the "+RT" samples, but do not add RNA in the preincubation mix.
7. Run all assays ("+RT", "No RT", NTCs) with the following thermal profile (optimized for iQ5 cycler; use the Experimental Well Factor option):

Cycle	Repeats	Step	Dwell Time	Setpoint (°C)	Data Acquisition
1	1	1	10 min	55	
		2	1:30 min	89 (see Note 38)	
2	50	1	5 s	95	
		2	10 s	58	
		3	20 s	68	

Cycle	Repeats	Step	Dwell Time	Setpoint (°C)	Data Acquisition
3	2	1	2 min	71–70 (temp.change –1°C)	Melt C. (see Note 39)
4	2	1	2 min	51–50 (temp. change –1°C)	Melt C.
5	2	1	5 min	36–35 (temp. change –1°C)	Melt C. (see Note 40)
6	1	1	3 min	95	
7	71	1	30 s	95–25 (temp. change –1°C)	Melt C. (see Note 41)

3.2.5. Data Analysis

Positive or negative results (virus present or absent) are scored based on ratios of end-point fluorescence readings (see Note 42). Fluorescence is first acquired at 70°C (all probes are open at this temperature) and these values are used to normalize each assay for its probe content, eliminating assay-to-assay variations, as follows:

1. Divide data acquired at 50°C by the corresponding readings at 70°C (50/70°C ratios).
2. Divide data acquired at 35°C by the corresponding readings at 70°C (35/70°C ratios).
3. The 50/70°C and 35/70°C ratios of the NTCs are the threshold for a "Negative outcome." These ratios are different for different fluors (see Fig. 11 and Note 43).
4. Plot all assays' 50/70°C and 35/70°C ratios versus the NTCs' 50/70°C and 35/70°C ratios and determine a positive/negative result for each of the probes in the assay (see example in Fig. 10).
5. Analyze the melt curves and their derivatives (see Fig. 12) to establish that the probes worked as expected (see Note 44).

3.2.6. End-Point Quantification with Temperature-Specific Probes

A qualitative result is usually sufficient when investigating the presence of viruses in a sample, but quantitative data are also useful. To test the efficiency of end-point quantification for a single LATE-PCR amplicon or in a multiplex, it is necessary to know the amount of target added in the reaction:

1. For target quantification, use the C_T values obtained during real-time amplification of each individual target in the presence of SYBR Green (see Subheading 3.2.3. and Fig. 13).
2. Prepare dilutions of the targets.
3. Run RT-LATE-PCR (if RNA targets are used) or LATE-PCR (if DNA targets are used) assays in the presence of sequence-specific probes for all the targets being tested and different target concentrations.
4. Calculate probe fluorescence ratios (50/70°C and 35/70°C) and plot them as a function of the calculated copy numbers

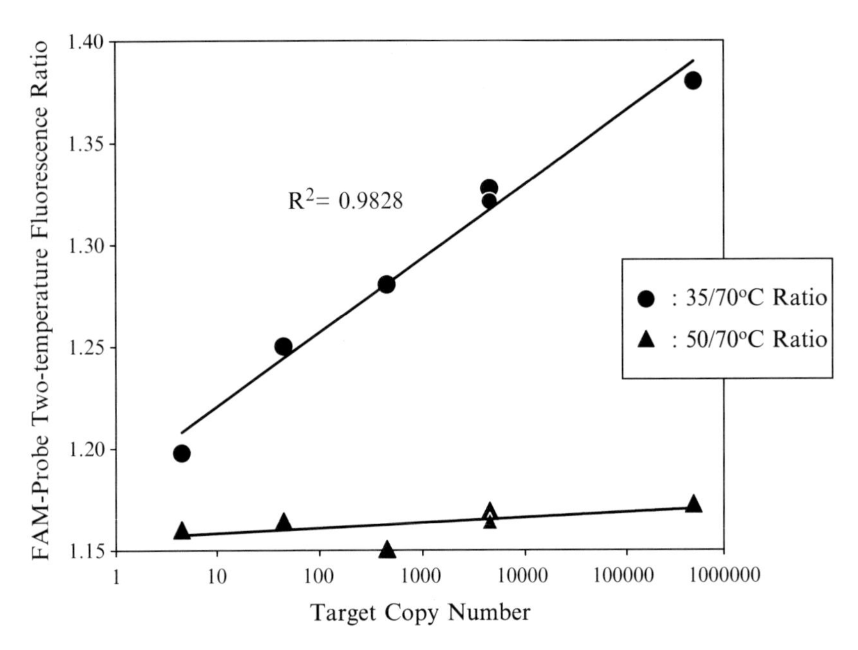

Fig. 14. LATE-PCR end-point quantification with a temperature-specific low-T_m probe. A FAM-conjugated probe was designed with a T_m of 49°C (experimentally-determined T_m = 46°C), so as to be bound to target at 35°C but not at 50°C. DNA oligonucleotides at known concentrations were used as the probe target (*abscissa*) and fluorescence intensity was acquired at 70°C, 50°C, and 35°C. The 35/70°C (*circles*) and 50/70°C (*triangles*) fluorescence ratios (*ordinate*) were calculated and plotted vs the target copy numbers (*abscissa*). The results demonstrate that the probe is temperature-specific as desired and that end-point quantification is linear from 4 to 400,000 target copies

on a log scale (Fig. 14) or of the SYBR Green C_T values on a linear scale (Fig. 15). Figure 14 illustrates the results for a single target probed by a temperature-discriminating probe. Figure 15 shows the results obtained from one of the amplicons in a triplex, probed in two different regions.

4. Notes

1. US Pat No. 7,198,897 issued April 3, 2007, European Patent: 02 795963.4 granted May 7, 2008.
2. Considerable differences exist between thermocyclers regarding the approach used to measure and analyze fluorescence emitted from multiple fluors. In addition, excitation and emission filter sets vary, in some instances, limiting the spectrum of fluors that can be used in the same assay. Finally, not all thermocyclers can reach low temperatures such as 35°C. (This is the case in particular for air-cooled cyclers, which are sensitive to ambient temperatures [see Subheading 2.1].)

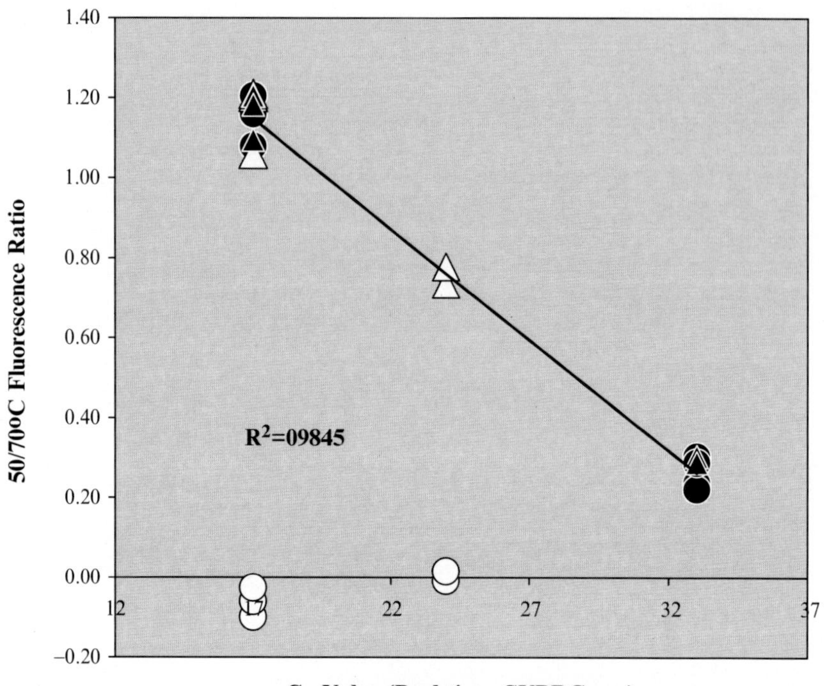

Fig. 15. LATE-PCR end-point quantification using two probes on the same amplicon in a triplex. Three viral-like RNAs were coamplified in sets of RT-LATE-PCR assays mimicking the presence of different substrains of the avian influenza H5N1 virus. This figure summarizes results obtained from the longest RNA in the triplex (~400 nt), which contains the targets for two probes (Quasar 670-probe, *triangles*, and CalOrange 560-probe, *circles*). Both probes bind to target at 50°C and their 50/70°C ratios (*ordinate*) were calculated as a function of target abundance. (A correction was introduced to account for the different fluorescence intensities of the two fluors.) Target abundance was determined using the C_T values of SYBR Green real-time LATE-PCR assays of the individual amplicons comprising the triplex (*abscissa*). The target for the Quasar 670-probe is conserved in both low- and high-pathogenic substrains of H5N1 (*black triangles* and *white triangles*, respectively); the target for the CalOrange 560 probe is found in the low-pathogenic H5N1 substrains (*black circles*) but not in the high-pathogenic substrains (*white circles*). The data demonstrate target-specificity of the two probes and good correlation between end-point quantification and target abundance

All these factors need to be kept in mind when selecting a thermocycler different from those indicated in the Materials section. The commercial names of those thermocyclers are provided to enable perspective users with an accurate term of comparison when selecting an alternative instrument.

3. We have observed that the quality of primers supplied by different manufacturers is not always consistent. It may be worthy to switch supplier if the PCR products are not as expected.

4. A wide array of fluors is available from different manufacturers, with different properties. (Keep in mind brightness, background levels, and fluorescence change at increasing temperatures.) Careful consideration should be given to the

choice of multiple fluors for a single assay, with the aim of minimizing emission spectra overlaps. The choice of the quencher is also important, as quenchers affect background fluorescence levels (when the stem of the probe is closed). When using more than one probe with the same fluor in a single assay, it is essential to lower as much background fluorescence as possible by selecting the best quencher. We had the best results with the the CAL Fluor® dyes conjugated to Black Hole Quenchers®, both proprietary of Biosearch Technologies, Inc., but the final choice of fluors also depends on the thermocycler used for the assay (see Note 2).

5. The PrimeSafe series was developed in our laboratory specifically for use with LATE-PCR. We have not tested other mispriming preventing compounds in our assays.

6. Mouse DNA is commercially available or can be prepared and purified in-house. Attention: if the DNA is sheared in short fragments during the preparation, PCR may fail or may underestimate the number of targets actually present. In fact, the sequences targeted by both primers need to be present on the same DNA segment in order to generate the full PCR product.

7. We used RNase ZAP®, but other similar products are available. Keep in mind that bleach is not sufficient to degrade all RNases.

8. All of these products are available from Invitrogen and contain specific, proprietary enzymes and buffers that determine the experimental conditions and protocols described in this chapter. In addition, the optimal temperature for the selected reverse transcriptase was a factor in the design of the RT primers because the T_m of this type of primer must allow its efficient hybridization to the RNA target during RT. For the above reasons, we believe that it is necessary to specifically identify the enzymes and buffers that we chose after testing numerous other reverse transcriptases and DNA polymerases. Researchers experienced in RT-PCR may make alternative choices keeping in mind that these will potentially require modifications of protocols and experimental methods, including buffers, cations, primer/probe design (affecting their T_ms), etc.

9. Armored RNA consists of an RNA standard sequence enveloped by MS2 bacteriophage coat protein. It is produced in *Escherichia coli* by the induction of an expression plasmid that encodes the coat protein and the RNA standard sequence (8). The packaged RNA is stable and ribonuclease-resistant. The MS2 phage (containing its own native RNA) is a cheaper, if less reliable, alternative for a ribonuclease-resistant RNA standard (13).

10. Other software programs are available for designing primers and probes, but do not offer the same analysis capabilities as the one cited here, particularly regarding T_ms calculations under different conditions. Researchers unable to access Visual OMP may have to assess effective T_m values empirically. (For instance, melt peak analysis can be used to determine probe-to-target T_ms.)
11. We use pGS-21a plasmids (ClaI/NcoI custom modification) from GenScript Corporation. These plasmids contain the T7 promoter directly followed by a DNA sequence which is the template for generation of the desired RNA in vitro. At the end of this sequence, there are restriction sites for an array of restriction enzymes. The *Eco*R V enzyme was selected after restriction sequence analysis owing to its compatibility with all of our targets and because it produces a blunt-ended cut in the DNA template, ensuring better results during IVT.
12. As for Note 8, in order to describe a specific protocol, it is necessary to refer to a specific product, considering the wide range of products and procedure available. Again, an expert researcher will be able to make the necessary modifications to the overall protocol when choosing an alternative IVT kit. Keep in mind that the chosen RNA polymerase needs to bind to the correspondent promoter in the expression plasmid: for instance, we use the T7 RNA polymerase which binds to the T7 promoter in our plasmids.
13. Glycogen solutions are often used as inert coprecipitants of nucleic acid; we, however, had better results with tRNA.
14. Amplicon contamination becomes apparent when the specific PCR product is generated in the absence of any template (see "No Template Controls", described later in this chapter). The best solution to this problem usually consists of cleaning thoroughly surfaces and instrumentation and in replacing all reagent aliquots with new, never used ones.
15. If the sequence between the primers includes a gene's intron or part of it, the PCR product generated from the genomic DNA target will be different from that generated from the corresponding cDNA target, because intron sequences are eliminated from mature mRNA by splicing. (If one of the primers is placed within the intron, no product will actually be generated from the cDNA template.) It cannot be assumed that two amplicons having different sequences and lengths are generated with equal efficiencies during PCR. A standard using genomic DNA templates (mouse DNA) would not, therefore, be absolutely reliable for quantification of the corresponding cDNA templates. If, however, the primer sequences are placed within an exon, the amplicons generated

from genomic DNA and cDNA targets will be exactly identical; hence, quantification of cDNA copy number can be reliably achieved by comparison with a genomic DNA standard (9).

16. RNA targets with high level of secondary structure may require moving the RT primer to a different part of the sequence in order to achieve good results.

17. Mouse embryo RNA can be prepared with any commercial method. Our lab has developed a single-tube RNA/DNA preparation-to-quantification method for small samples (single cell, single embryo) called PurAmp (9), US Pat No. 7,465,562. If using the PurAmp method, maintain an RNA volume of 6 μl, but increase the assay final volume to 50 μl by doubling the volume of all other mix components and adjusting the water volume.

18. We have found this buffer to considerably improve the reaction's efficiency over the buffer provided with the SuperScript™ III Platinum® One-Step RT-PCR Kit.

19. This FAM probe is recommended when using the ABI 7700 thermocycler. The fluor can be substituted by Quasar 670 and the quencher by Black Hole Quencher 2 when using a BioRad iQ5 or a Mx3005P Stratagene cycler. For the ABI assay, we have also successfully used a TET/Dabcyl-Oct4 molecular beacon at 1 μM concentration (5). Traditional molecular beacons with 5-to-6 nt stems (14), however, are harder to design for low temperature annealing due to their longer stem that remains closed when cooling the reaction.

20. PrimeSafe 002 (a more recent version of PrimeSafe) at 200 nM can be substituted for 300 nM PrimeSafe 043 in this assay.

21. Data acquisition starts after the initial 15 cycles in order to limit possible mispriming when the intended amplicon is first generated. Mispriming may take place at the low temperature necessary for probe hybridization.

22. A concentration range between 100 and 300 nM works in most cases. If primers are designed with T_ms lower than those shown in this protocol (see Subheading 3.1.2.), we recommend the use of PrimeSafe 001 as detailed in Subheading 3.2.

23. Unlike symmetric PCR, LATE-PCR allows generation of hundreds-of-nucleotides long amplicons. Two probes on the same sequence are useful to detect mutations that often occur in viral strains. The first probe is located on a conserved region and signals the presence of the virus; the second probe is located on a variable region and only hybridizes when major

mutations have not occurred. (See for instance the insertion of a "cleavage region" responsible for increasing the pathogenicity of influenza strains.)

24. A useful list of fluors with multiplex-compatible absorption/emission spectra can be found at (15). Select your PCR instrument to update the list.
25. Design the assay to avoid the possibility of false positive, or false negative for dangerous agents: build in redundancy for most important targets.
26. If the RNA control is an armored RNA (see protocol), the control is also useful to monitor effective sample preparation. The BioSeeq®-Vet PCR System, currently being developed by Smiths Detection Diagnostics, Watford, UK, includes a field-portable instrument capable of viral sample preparation-to-detection by RT-LATE-PCR in a single, disposable cartridge.
27. An alternative strategy is to detect different viruses with probes hybridizing at different temperatures, see the CalRed 610-probes for Virus 2 and Virus 3 in Fig. 10.
28. A greater difference will decrease amplification efficiency. Ideally, all amplicons should be generated with the same efficiency. If not, there is the risk that the least efficient amplicons will "drop out" if targets are present at low copy number. In practice, however, it is not possible to have all amplicons with the same T_m and length (ideally around 100 nt): in some cases, it is desirable to fit two probes in the same, long amplicon; in other cases, conserved regions (ideal for priming) are separated by long variable regions. Keep also in mind that long sequences are particularly problematic during RT due to the high level of secondary structure of ss RNA.
29. LP can also be used for RT priming, but it may require higher RT temperature because of its higher T_m.
30. Although the fluor/quencher interaction affects the T_m, we found more accurate results by not including this modification when calculating probe T_ms with Visual OMP. If the stem nucleotides are also complementary to the target, they will contribute to the probe's T_m.
31. The positions of fluor and quencher can be switched. When using two probes on the same amplicon, relatively close to each other, it is better to have one quencher on the 3′ and the other on the 5′, in close proximity, rather than having the quencher of the first probe close to the fluor of the second probe. It is also recommended to have the quencher at the 3′ end when using DNA polymerases that, unlike Tfi(-), display exonuclease activity, because Black Hole Quenchers prevent probe cutting.

32. If desired, the multiplex assay can be initially tested on DNA targets. In this case, keep in mind the following caveats: (a) DNA oligos longer than 100–150 nt cannot be synthesized; longer templates will need to be shortened by eliminating some of the intervening sequence and will thus be a poor model for the actual RNA target. In addition, the T_ms of these shorter amplicons will be lower than those of the amplicons generated from the RNA targets, affecting the PCR outcome; (b) DNA oligos do not allow testing for RT efficiency.
33. Storage of RNA at higher temperature greatly affects its stability.
34. You will generate an extremely high number of RNA molecules. Be extremely careful to avoid template contamination when doing RT-PCR after RNA preparation. Work in a different, dedicated area if possible and treat all instrumentations, surfaces, and pipettes used for IVT with bleach.
35. The use of specific probes is not necessary at this point, for single amplicons. These real-time curves are sigmoidal because they visualize generation of ds DNA during phase I of LATE-PCR (see Fig. 2).
36. Platinum® Taq polymerase (1.5 U/reaction) can be used instead of Platinum® Tfi(-), with the appropriate buffer. In this case, use the SuperScript™ III/ Platinum® Taq mix available from the manufacturer as part of a one-step kit (see Subheading 3.1.2.) and switch to PrimeSafe 002 at 300 nM concentration. Also, change annealing temperature to 62°C and extension temperature to 72°C.
37. Avoid product contamination by using a heating block, not the cycler where PCR is carried out.
38. This temperature value is necessary when using "Experimental Well Factor" in an iQ5 cycler because the instrument will collect two readings at 95°C to establish a normalizing value for each well. The readings will be taken before RT, thus denaturing the enzyme, if any of the steps in the first Cycle is above 89°C. Using the settings shown above, two readings at 95°C (1:30 min total) will be taken after RT and before PCR.
39. The iQ5 cycler does not allow data acquisition in more than one cycle, except when using the Melt Curve option. For this reason, we set all end-point collection data (at 70°C, 50°C, and 35°C) as "mini-melts."
40. A long dwell time at 35°C is useful to improve probe binding at this low temperature.
41. This is actually an Anneal Curve, from high to low temperature.

42. The switch from real-time to end-point data gathering is necessary in multiplex LATE-PCR because the probes are designed to hybridize at low T_m. In real-time PCR, data are acquired at every cycle and lowering the temperature at every cycle favors mispriming when multiple pairs of primers are present in the reaction. PrimeSafe can efficiently prevent this problem in duplex real-time PCR (see Subheading 3.1.4.). In triplex real-time reactions (see Fig. 3), we observed scatter at end-point among replicates, indicating the generation of nonspecific products in the background. End-point data acquisition bypasses this problem because the temperature is lowered to allow probe hybridization only at the end of PCR, after amplification.
43. The reason for this difference is that temperature-dependent change in fluorescence is fluor-specific.
44. If a probe binds with T_m lower than expected when the assay is used for tests on real viruses, mutations may have occurred in the viral sequence being probed. LATE-PCR allows direct sequencing of the single-stranded product by simple dilution (7), which would be very informative in this case.

Acknowledgments

This work was funded by Smiths Detection Diagnostics, Watford, UK, with the support of Brandeis University. Arthur Reis, Kenneth Pierce, Jared Ackers, Odelya Hartung, and Bonnie Ronish are acknowledged for contributions to the development of the Oct4 and viral assays.

References

1. Gyllensten UB, Erlich HA (1988) Generation of single-stranded-DNA by the polymerase chain-reaction and its application to direct sequencing of the Hla-Dqa Locus. Proc Natl Acad Sci USA 85:7652–7656
2. Sanchez JA, Pierce KE, Rice JE, Wangh LJ (2004) Linear-after-the-exponential (LATE)-PCR: an advanced method of asymmetric PCR and its uses in quantitative real-time analysis. Proc Natl Acad Sci USA 101:1933–1938
3. Pierce KE, Sanchez JA, Rice JE, Wangh LJ (2005) Linear-After-The-Exponential (LATE)-PCR: primer design criteria for high yields of specific single-stranded DNA and improved real-time detection. Proc Natl Acad Sci USA 102:8609–8614
4. Raja S, El-Hefnawy T, Kelly LA, Chestney ML, Luketich JD, Godfrey TE (2002) Temperature-controlled primer limit for multiplexing of rapid, quantitative reverse transcription-PCR assays: application to intraoperative cancer diagnostics. Clin Chem 48:1329–1337
5. Hartshorn C, Eckert JJ, Hartung O, Wangh LJ (2007) Single-cell duplex RT-LATE-PCR reveals *Oct4* and *Xist* RNA gradients in 8-cell embryos. BMC Biotechnol 7:87
6. Chou CF, Tegenfeldt JO, Bakajin O, Chan SS, Cox EC, Darnton N, Duke T, Austin RH (2002) Electrodeless dielectrophoresis of single- and double-stranded DNA. Biophys J 83:2170–2179

7. Rice JE, Sanchez JA, Pierce KE, Reis AH Jr, Osborne A, Wangh LJ (2007) Monoplex/multiplex linear-after-the-exponential-PCR assays combined with PrimeSafe and Dilute-'N'-Go sequencing. Nat Protoc 2:2429–2438
8. Pasloske BL, Walkerpeach CR, Obermoeller RD, Winkler M, DuBois DB (1998) Armored RNA technology for production of ribonuclease-resistant viral RNA controls and standards. J Clin Microbiol 36:3590–3594
9. Hartshorn C, Anshelevich A, Wangh LJ (2005) Rapid, single-tube method for quantitative preparation and analysis of RNA and DNA in samples as small as one cell. BMC Biotechnol 5:2
10. http://www.invitrogen.com/content/sfs/ProductNotes/F_Superscript%20III%20Enzyme%20RD-MKT-TL-HL0506021.pdf?ProductNoteId=36
11. Laird CD (1971) Chromatid structure: relationship between DNA content and nucleotide sequence diversity. Chromosoma 32:378–406
12. Hartung O, Hartshorn C, Wangh LJ (2007) Cdx2 and Oct4 mRNA in pre-implantation stage mouse embryos and their single cells. Fertil Steril 88:S309–S309
13. Dreier J, Störmer M, Kleesiek K (2005) Use of bacteriophage MS2 as an internal control in viral reverse transcription-PCR assays. J Clin Microbiol 43:4551–4557
14. Tyagi S, Kramer FR (1996) Molecular beacons: probes that fluoresce upon hybridization. Nat Biotechnol 14:303–308
15. http://www.biosearchtech.com/display.aspx?pageid=53

Chapter 12

Changes in Gene Expression of Caveolin-1 After Inflammatory Pain Using Quantitative Real-Time PCR

Fiza Rashid-Doubell

Abstract

This chapter will take the reader through the steps involved in obtaining a gene expression profile using real-time PCR. Real-time PCR is an end-point measure of changes in gene expression that have occurred after a physiological event. It specifically describes the process by which real-time PCR was used to measure caveolin-1 gene expression after an inflammatory pain insult. The method outlines the initial stage of total RNA extraction from the starting material through the main steps involved in producing a gene expression profile including primer design, RNA quantification and cDNA production by reverse transcription and of course the real-time PCR. The caveolin-1 gene measured here is the main component of specialised domains found in the plasma membrane with a primary function described as signal transduction 'hot spots'. The result of this experiment shows that there was a twofold increase in caveolin-1 gene expression in rat dorsal root ganglions 24 h after Complete Freund's Adjuvant administration and a decrease at 48 h.

Key words: Caveolin-1, Complete Freund's Adjuvant, Inflammatory pain, SYBR-green, Real-time PCR

1. Introduction

Real-time PCR is a fluorescence-based detection of amplified gene products using a DNA-binding dye. The method combines amplification and detection into a single step by correlating PCR product formation with fluorescent dye intensity (1). SYBR green is one such fluorescent dye used in real-time PCR (2). It is a sensitive binding dye for double-stranded DNA, which can detect single PCR products if there are no primer/dimers formed. This quantitative method offers two main advantages over conventional methods; tremendous sensitivity (10,000 to 100,000-fold) over RNase protection assay, (3) and no post-amplification processing.

The aim of this experiment was to detect and quantify caveolin-1 mRNA after an inflammatory pain insult. Inflammatory pain

Nicola King (ed.), *RT-PCR Protocols: Second Edition*, Methods in Molecular Biology, vol. 630,
DOI 10.1007/978-1-60761-629-0_12, © Springer Science+Business Media, LLC 2010

is precipitated by damage to the integrity of tissues at a cellular level. Studies have shown that a Complete Freund's Adjuvant (CFA) injection induces both thermal and mechanical pain hypersensitivity, which appears after 2 h, then peaks between 6 and 24 h and is maintained for at least 72 h after the CFA injection (4, 5). Multiple mechanisms mediate these inflammatory processes, including proteins in the plasma membrane; one such protein could be caveolin-1. Caveolin-1 is the protein component of caveolae (6). Caveolae are specialised domains found as invaginations on the plasma membrane (7, 8) with a high affinity for cholesterol (9). The presence of this protein has previously been determined in dorsal root ganglion (DRG) neurons by immunofluorescence (10). This method describes the steps involved to produce a gene expression profile (as outlined in Fig. 1 below) for caveolin-1 in rat DRG neurons over a period of time after an inflammatory pain insult.

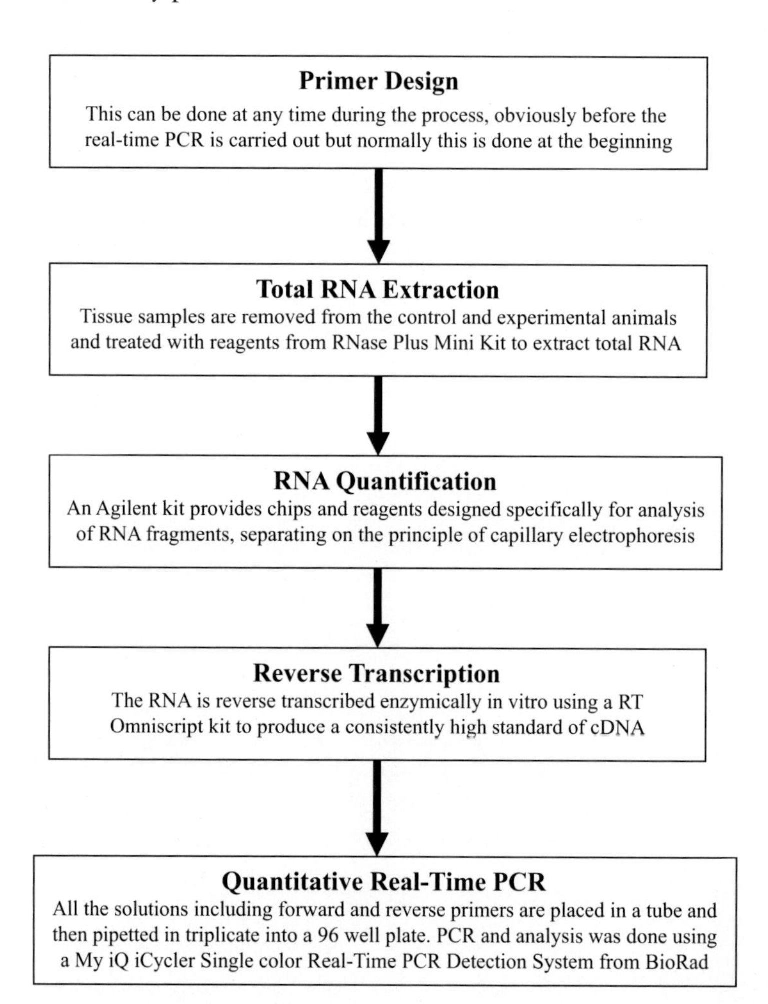

Fig. 1. A diagram outlining the five steps required to produce a gene expression profile for analysis using quantitative real-time PCR

2. Materials

2.1. Primer Design

1. http://www.ncbi.nlm.nih.gov/entrez/query.fcgi
2. http://www-genome.wi.mit.edu/cgi-bin/primer/primer3

2.2. Total RNA Extraction

1. Male Sprague–Dawley rats
2. Bench top microcentrifuge
3. Mini homogeniser
4. Sterile 2 ml round bottom tubes
5. Sterile 15 ml Falcon tubes
6. Sterile 0.5 ml Eppendorf tubes
7. Sterile, RNase-free pipette tips
8. RNeasy Plus Mini Kit for 50 minipreps containing:

   RNeasy Mini Spin Columns, gDNA Eliminator Spin Columns, Collection Tubes, RNase-Free Water and Buffers
9. –80°C Freezer
10. RNA-later
11. DEPC H_2O
12. Liquid Nitrogen
13. 1 M NaOH in DEPC water
14. 70% Ethanol in DEPC water
15. β-mercaptoethanol (β-ME) at 4°C
16. RNase ZAP

2.3. RNA Quantification Using Agilent RNA 6000 Nano Kit

1. Agilent 2100 bioanalyser
2. RNA nano chip and electrode cleaner
3. Syringe and modified syringe holder
4. PCR Eppendorf safe-lock tubes
5. Spin filters
6. IKA vortex mixer
7. 10 μl and 1,000 μl pipettes
8. RNase-free pipette tips
9. RNase-free 0.5 ml microcentrifuge tubes
10. Chip priming station
11. Agilent RNA 6000 ladder
12. RNA Nano dye concentrate
13. Agilent RNA 6000 Nano marker
14. Agilent RNA 6000 Nano gel matrix

15. RNase ZAP
16. RNase-free water

2.4. Reverse Transcription

1. 10 μl and 100 μl pipettes
2. RNase-free pipette tips
3. Water bath
4. Spectrophotometer measuring in UV range
5. Sterile 0.5 ml microcentrifuge tubes
6. Qiagen Omniscript RT (50) kit containing: 200 units Omniscript reverse transcriptase, 10× buffer RT, dNTP mix (5 mM each), RNase-free water
7. Random Primers
8. RNase Inhibitor (40 U/μl)
9. DNase-treated RNA

2.5. Quantitative Real-Time PCR

1. MyiQ iCycler Single color real-time PCR Detection System (BioRad)
2. PCR Plates 96 well semi-skirt Thermo Fast (Axon Lab)
3. Polypropylene 1.5 ml microtubes with screw caps
4. Ultra PCR Tubes 0.65 ml
5. Aerosol Resistant Tips
6. Hood with UV lamp and laminar flow dedicated to preparing real-time PCR samples
7. Microseal adhesive film
8. Hermle 2300 centrifuge
9. BioRad iQ™ SYBR® Green Supermix
10. Water, Molecular Biology Grade
11. RNase ZAP
12. Designed Primers Forward (sense) and Reverse (anti-sense)

3. Methods

3.1. Primer Design

Design the PCR primers by first obtaining the complete gene sequence using the NCBI website (http://www.ncbi.nlm.nih.gov/entrez/query.fcgi). Check the regions of homologous target sequences against other sequences to avoid spurious results. Finally, use the internet software Primer3 on website http://www-genome.wi.mit.edu/cgi-bin/primer/primer3 to design the primer sets (see Note 1).

In order to further optimise the specificity of the primer sets vary both the dilution and annealing temperature.

3.2. Total RNA Extraction

1. Anaesthetise and decapitate male Sprague–Dawley rats that had previously undergone a CFA injection or sham-injection at specific time points after injection.
2. Rapidly remove ipsilateral L4 and L5 dorsal root ganglions and place in RNA-later solution at 4°C to prevent degradation (see Note 2).
3. Put on gloves and clean work space with RNaseZap (see Note 3).
4. Thoroughly clean the homogeniser with DEPC water, then with 1 M NaOH and again with DEPC water between each sample.
5. Carry out the RNA extraction using the RNase Plus Mini Kit, which contains all the required buffers. Place the tissue into a 2 ml round bottom tube containing 600 μl of RLT Plus buffer (see Note 4).
6. Homogenise tissue for less than 1 min. Adapt the homogenising speed accordingly depending on the type of tissue.
7. Centrifuge the lysate for 3 min at ≥11,000 × *g* at room temperature; remove the supernatant and add to the genomic DNA (gDNA) eliminator spin column resting on top of a collecting tube.
8. Centrifuge the gDNA eliminator spin column for 30 s. at ≥8,000 × *g*. Discard the column and retain the supernatant in the collecting tube. (see Note 5).
9. Add an equal volume of 70% ethanol (diluted in DEPC water) to the lysate and triturate, do not centrifuge.
10. Immediately add the lysate (~600 μl) to an RNeasy mini spin column placed over a 2 ml collecting tube then close the column lid.
11. Centrifuge for 15 s ≥8,000 × *g*. Collect the filtrate, pass through the column again and repeat the centrifugation step. Discard the final filtrate. If the lysate volume exceeds 700 μl, then repeat procedure again.
12. Add 700 μl of RW1 buffer to the column and close the lid (see Note 6).
13. Centrifuge for 15 s >11,000 × *g* and then discard the filtrate.
14. Add 500 μl of RPE buffer to the column; close the lid and centrifuge for 15 s ≥8,000 × *g*. Again discard the filtrate.
15. Repeat step 14 but this time the centrifugation should be for 2 min at maximum speed at room temperature.

16. Aspirate any residual liquid on the periphery of the column using a sterile 200 µl yellow pipette tip.
17. Elute RNA off the column by placing the column on to a new collecting tube and adding 30 µl of RNase-free water, on to the silica gel. Close the lid and leave the tube for 5 min at room temperature and then centrifuge for 1 min ≥8,000 × *g*.
18. Repeat step 17 with 20 µl of RNase-free water, centrifuging at ≥8,000 × *g* straightaway.
19. Pool the filtrates from steps 17 and 18 and store at –80°C in aliquots of 1 µg of RNA.

3.3. RNA Quantification Using the Agilent RNA 6000 Nano Kit

This method of total RNA quantification is quick, precise and saves a lot of material. It uses a kit, which provides chips and reagents designed specifically for the analysis of RNA fragments. It uses the principle of capillary electrophoresis to move the RNA fragments, and these are separated by their size as they are driven through a set of interconnected microchannels. This kit is designed for use with the Agilent 2100 bioanalyser.

1. Put on gloves and clean your work space thoroughly with RNaseZap.
2. Take all reagents taken from the freezer and thaw at room temperature for 30 min and then maintain on ice.
3. Prepare the gel by pipetting 550 µl of the Agilent RNA 6000 Nano gel matrix into a spin filter and centrifuging at 1,500 × *g* ±20% for 10 min at room temperature. Then aliquot the filtered gel into 65 µl portions into RNase-free 0.5 ml microcentrifuge tubes (see Note 7).
4. Prepare the gel-dye mix by vortexing the thawed RNA Nano dye concentrate for 10 s and then transfer 1 µl into the 65 µl aliquot of filtered gel (see Note 8). Vortex this solution well and spin at 13,000 × *g* for 10 min at room temperature (see Note 9).
5. Load the gel-dye mix onto the chip by placing the RNA nano chip on to the priming station. Then pipette 9 µl of the gel-dye mix into the well marked G (see Note 10).
6. Depress the plunger until it is held by the clip, wait 30 s then release the clip. Wait another 5 s and then pull the plunger back to the 1 ml position.
7. Open the chip priming station and pipette 9 µl of the gel-dye mix into the wells marked G. Discard the remaining gel-dye mix.
8. Add the Agilent RNA 6000 Nano marker by pipetting 5 µl of the marker into all 12 wells as well as the well marked with the ladder sign.
9. Load the samples and ladder by pipetting 1 µl of the ladder into the well marked with the ladder sign (see Note 11).

10. Add 1 μl of the sample to each of the 12 wells, if there are than 12 samples then pipette 1 μl Agilent RNA 6000 Nano marker into the unused wells.
11. Place the chip horizontally into the IKA vortex mixer and vortex for 1 min at 2,400 rpm (see Note 12).
12. Put the chip into the Agilent 2100 bioanalyser within 5 min of preparation.

3.4. Reverse Transcription

An RT Omniscript kit is used to produce a consistently high standard of cDNA.

1. For one reaction make the mix using the following solutions: 4 μl Buffer RT 10×; 4 μl dNTP; 2 μl Random Primers (hexamers); 0.5 μl RNase Inhibitor; 2 μl RT Omniscript giving a final volume of 12.5 μl.
2. Place x μl of RNA representing 1 μg of material into a 0.5 ml RNase-free microcentrifuge tube.
3. Add an appropriate volume of water to the RNA to make a volume of 27.5 μl.
4. Add 12.5 μl of mix per RNA sample to make a final volume of 40 μl.
5. Incubate for 60 min at 37°C in a water bath.
6. Measure the OD of nucleic acids at 260 nm and proteins at 280 nm using the spectrophotometer.
7. Store cDNA at −20°C in aliquots of 2 μg (see Note 13).

3.5. Quantitative Real-Time PCR

1. Clean the bench, hood, pipettes and every material that will be used for quantitative real-time PCR with RNaseZap to eliminate RNases.
2. Designate two work places, preferably a bench and a hood along with two different set of pipettes (see Note 14).
3. Expose the pipettes, tubes, PCR plate and water to UV in the hood for 20 min to sterilise them prior to use.
4. After sterilisation, prepare cDNA on the bench using one set of pipettes. Prepare the total volume of cDNA necessary for the entire plate in one screw top vial. Dilute the cDNA in molecular biology grade water to the right concentration as determined in advance from the standard curves obtained from the test for primer efficacy (see Note 15).
5. Continue the remainder of the work in the hood using the second set of pipettes.
6. First prepare the primers. Aliquot the primers at concentrations of 100 μM into screw top tubes and store at −20°C. Thaw and dilute the primers to a working concentration of 3 μM in molecular biology grade water.

7. Prepare the master mix containing the SYBR Green. For one well make the master mix by pipetting the following solutions into a 1.5 ml screw top tube, the final volume should be 20 μl per well:

Mix	1 well	Finale concentration
iQ SYBR Green Mix 2×	10 μl	1×
Primer 1 Fwd at 3 μM	2 μl	300 nM
Primer 2 Rev at 3 μM	2 μl	300 nM
Mol Biol. Grade H_2O	1 μl	
Final volume	15 μl	
cDNA at 10 ng/μl	5 μl	50 ng final
Final volume per well	20 μl	

8. Prepare triplicates of the samples in the same screw-top tube. Construct a schematic template in order to determine the quantity of master mix containing SYBR Green Supermix and cDNA required and place this into a small 0.65 ml sterile PCR tube, i.e. 45 μl of master mix SYBR Green + 15 μl of cDNA (10 ng/μl) = 60 μl final volume for three wells (see Note 16).
9. Add 19.5 μl of the reaction mix (mix + cDNA) into each well, then cover the plate with a transparent adhesive film ensuring that the seal is firm around the edges in order to avoid evaporation (see Note 17).
10. Centrifuge the plate in a Hermle Z300 centrifuge at 2,500 × *g* for 5 min at room temperature.
11. Use thermal cycling conditions comprising of 3 min polymerase activation at 95°C, succeeded by 45 cycles of 10 s at 95°C for denaturation and 45 s at 60°C for annealing and extension, followed by a DNA melting curve for the determination of amplicon specificity. Obtain primer efficiencies by the standard curve method with integration for calculation of the relative expression, which is based on real-time PCR threshold values of different transcripts and groups (see Note 18).
12. Determine the specificity of amplification by DNA melting curve analysis using the built-in feature in the software for automatic fluorescence data capturing during gradual temperature increments (0.5°C) from 55 to 95°C.
13. Determine the efficiency of amplification by serial dilution of starting DNA, and construction of standard curves from the respective mean critical threshold value for endogenous control and caveolin targets.
14. Plot results. Figure 2 shows caveolin-1 expression measured over a 48 h period. Individual control samples (sham-injected

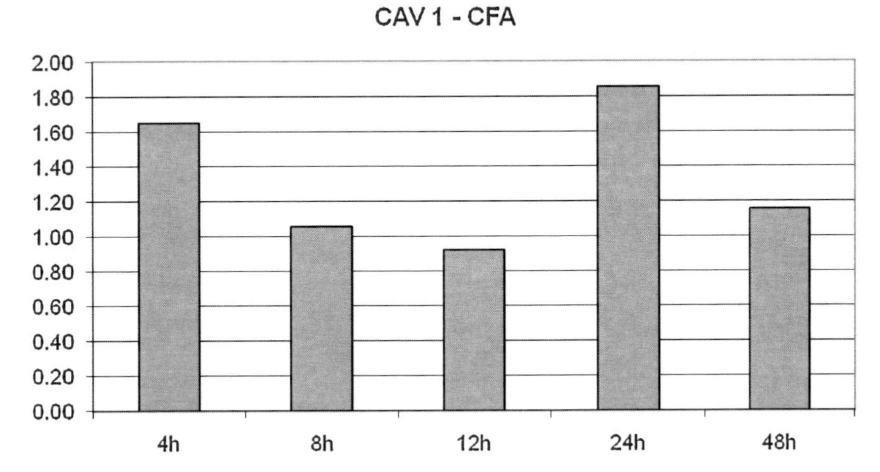

Fig. 2. Effect of inflammatory pain on caveolin-1 expression in rat dorsal root ganglion. These results show that caveolin-1 expression is induced at 4 h after administration of CFA injection and this expression drops after 8 h. The caveolin-1 expression rises again this time almost twofold at 24 h and drops again at 48 h

control group, $n-4$) and treated samples (CFA injected group, $n-4$) were amplified in triplicate. The relative expression of the target gene was calculated based on real-time PCR efficiencies and the threshold value of the unknown sample versus the standard sample (11).

4. Notes

1. Four variables have to be considered when optimising primer design: the primer length (18–27 base pairs), the melting temperature ($Tm = 2(A+T) + 4\ (G+C)$, the PCR product length should be between 75 and 150 base pairs and no greater than 200 and the percentage of primer GC content (40–60 GC%). Finally, check the proposed primers to see if they are on different exons and also BLAST the primers to verify their specificity.
2. Usually, if you place tissue in RNAlater solution, it should stay at least overnight in that solution to protect them. If you do the RNA extraction immediately after dissection, then just freeze the tissue in dry ice or liquid nitrogen.
3. It goes without saying that all equipment and solutions used were free of RNase contamination.
4. The RLT Plus should have had 10 μl β-ME added for every 1 ml of RLT Plus.

5. Always check that there is no liquid remaining on the membrane of the spin column, if there is, centrifuge again until the liquid collects in the tube below.
6. Use the same collecting tube as before.
7. This filtered gel is good for up to 4 weeks.
8. The dye and dye mixtures are light sensitive and should only be removed from the dark when in use and replaced back immediately, otherwise a reduction in signal intensity may occur.
9. Use the prepared gel-dye mix within 1 day of preparation.
10. Ensure that the plunger of the syringe is placed at 1 ml then the chip priming station can be closed.
11. The RNA samples and the ladder should be heated in order to denature the sample before use.
12. It is not possible to give a g value for this centrifuge, as it is not a common centrifuge. But a special vortex for the chip, with a flat platform especially designed for the chip and the graduation for speed are in rpm.
13. Of course, it is a common practice to carry out more than one reaction in which case it is usual to produce a master mix where the reaction is made in bulk.
14. Having these two separate workspaces avoids contamination of mix, primers or water with cDNA or genomic DNA.
15. In our hands, 10 ng/μl was often a good starting point (50 ng/well).
16. The polymerase contained in the SYBR® Green Supermix does not work at room temperature, so you can work at room temperature without problems. The polymerase need to be activated for 2 min at 95°C to be functionally active at 60°C.
17. Avoid touching the film with gloves or fingers, use the edge of the plastic spatula that is provided.
18. The following PCR programme was used.

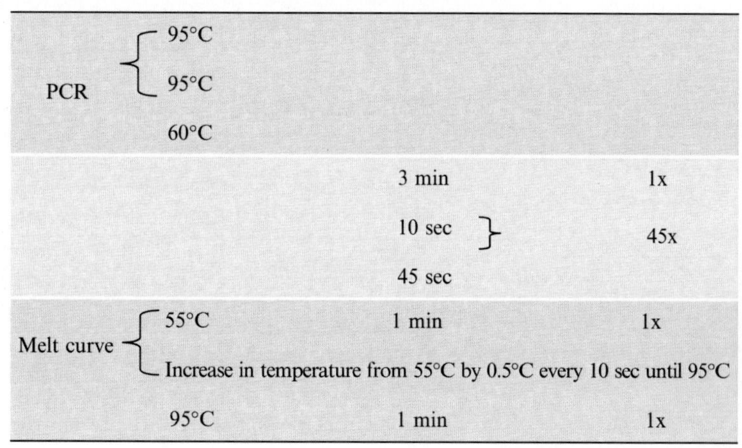

PCR	95°C	3 min	1x
	95°C	10 sec	45x
	60°C	45 sec	
Melt curve	55°C	1 min	1x
	Increase in temperature from 55°C by 0.5°C every 10 sec until 95°C		
	95°C	1 min	1x

Acknowledgements

This work was done in the laboratory of Dr. I. Decosterd and was technically assisted by T. Berta and M. Pertin. The work was funded by a University of Brunei Darussalam research grant. Finally, I would like to offer my sincere thanks to M. Pertin, S. Fredericks and N. King for critical reading of the manuscript.

References

1. Higuchi R, Fockler C, Dollinger G, Watson R (1993) Kinetic PCR analysis: real-time monitoring of DNA amplification reactions. Biotechnology (NY) 11:1026–1030
2. Malinen E, Kassinen A, Rinttila T, Palva A (2003) Comparison of real-time PCR with SYBR green I or 5′-nuclease assays and dot-blot hybridization with rRNA-targeted oligonucleotide probes in the quantification of selected faecal bacteria. Microbiology 149: 269–277
3. Wang T, Brown MJ (1999) mRNA quantification by real-time Taqman polymerase chain reaction: validation and comparison with RNase protection. Anal Biochem 269:198–201
4. Abbadie C, Lindia JA, Cumiskey AM, Peterson LB, Mudgett JS, Bayne EK, DeMartino JA, MacIntyre DE, Forest MJ (2003) Impaired neuropathic pain responses in mice lacking the chemokine receptor CCR2. Proc Natl Acad Sci USA 100:7947–7952
5. Tao Y-X, Rumbaugh G, Wang GD, Petralia RS, Zhao C, Kauer FW, Tao F, Zhou M, Wenthold RJ, Raja SN, Haganir RL, Bredt DS, Johns RA (2003) Impaired NMDA receptor-mediated postsynaptic function and blunted NMDA receptor-dependent persistent pain in mice lacking postsynaptic density-93 protein. J Neurosci 98:201–206
6. Rothberg KG, Heuser JE, Donzell WC, Ying YS, Glenney JR, Anderson RG (1992) Caveolin, a protein component of caveolae membrane coats. Cell 68:673–682
7. Palade GE (1953) Fine structure of blood capillaries. J Appl Phys 24:1424
8. Yamada E (1955) The fine structures of the gall bladder epithelium of the mouse. Biochem Biophys Acta 1746:334–348
9. Murata M, Peranen J, Schreiner R, Weiland F, Kurzchalia TV, Simons K (1995) VIP21/caveolin is a cholesterol-binding protein. Proc Natl Acad Sci USA 92:10339–10343
10. Galbiati F, Volonte D, Gil O, Zanazzi G, Salzer JL, Sargiacomo M, Scherer PE, Engelman JA, Schlegel A, Parenti M, Okamoto T, Lisanati MP (1998) Expression of caveolin-1 and -2 in differentiating PC12 cells and dorsal rot ganglion neurons: caveolin-2 is up-regulated in response to cell injury. Proc Natl Acad Sci USA 95:10257–10262
11. Pertin M, Ru-Rong J, Berta T, Powell AJ, Karchewski L, Tate SN, Isom LL, Woolf CJ, Gilliard N, Spahn DR, Decosterd I (2005) Upregulation of the voltage-gated sodium channel b2 subunit in neuropathic pain models: characterization of expression in injured and non-injured primary sensory neurons. J Neurosci 25(47):10970–10980

Chapter 13

Real-Time Quantitative Reverse Transcriptase Polymerase Chain Reaction

Hongxin Fan and Ryan S. Robetorye

Abstract

The real-time quantitative reverse transcriptase polymerase chain reaction (RQ-PCR) has become the method of choice for the quantification of specific mRNAs. This method is fast, extremely sensitive, and accurate, requires only very small amounts of input RNA, and is relatively simple to perform. These characteristics have made it the method of choice for minimal residual disease monitoring such as in chronic myelogenous leukemia (CML). CML comprises approximately 20% of all leukemias and is characterized by a balanced (9;22) chromosomal translocation that results in the formation of a chimeric gene comprised of the *BCR* (breakpoint cluster region) gene and the *ABL* oncogene (*BCR-ABL* fusion gene). The chimeric gene encodes a fusion protein with constitutively increased tyrosine kinase activity, resulting in growth factor-independent proliferation. This kinase is the target for current CML therapy, and *BCR-ABL* fusion gene levels are monitored to determine the effectiveness of this therapy. This chapter uses *BCR-ABL* transcript detection to illustrate an example for the RQ-PCR and describes a RQ-PCR method to detect the most common form of the *BCR-ABL* fusion transcript in CML, known as p210 *BCR-ABL*.

Key words: Real-time quantitative PCR, p210 *BCR-ABL* fusion gene expression, Applied Biosystems 7900 HT real-time PCR instrument, TaqMan

1. Introduction

The real-time quantitative reverse transcriptase polymerase chain reaction (RQ-PCR) is a powerful tool for accurate quantitation of gene expression. This relative quantitation method measures changes in gene expression in test samples relative to another reference sample (calibrator), and therefore, can be used to compare RNA levels in different tissues, to monitor treatment efficacy, or to measure minimal residual disease (MRD).

The RQ-PCR assay utilizes a well-established three-step protocol that involves (1) a reverse transcription step that converts RNA into a complementary DNA copy (cDNA),

Nicola King (ed.), *RT-PCR Protocols: Second Edition*, Methods in Molecular Biology, vol. 630,
DOI 10.1007/978-1-60761-629-0_13, © Springer Science+Business Media, LLC 2010

(2) multiple rounds of amplification of the cDNA using a heat stable DNA polymerase known as *Taq* polymerase, and (3) the detection and quantification of the amplified products in real-time. Here, we describe RQ-PCR procedures using a quantitative p210 *BCR-ABL* assay as an example. This assay is used to monitor levels of the p210 *BCR-ABL* fusion gene transcript created by the presence of the (9;22) translocation in CML patients. The assay protocol has been adapted from the Europe Against Cancer Program protocol described by Gabert et al. (1) and Beillard et al. (2). Briefly, total RNA is extracted from peripheral blood or bone marrow using TRIzol reagent. The RNA is then converted to single-stranded cDNA using reverse transcriptase. The cDNA then serves as the template for subsequent amplification cycles. The real-time PCR reaction is very similar to conventional PCR, but is carried out in a real-time thermal cycler (ABI 7900HT Fast Real-Time PCR System; Applied Biosystems) that allows measurement of fluorescent molecules in the reaction using TaqMan technology.

TaqMan real-time PCR (3) involves using a dual-labelled fluorogenic probe called a TaqMan probe to measure the accumulation of PCR product (by measuring the amount of the fluorophore) during the exponential stages of the PCR, rather than at the end point as in conventional PCR. The TaqMan probe is a single-stranded oligonucleotide complementary to a segment of 20–60 nucleotides within the cDNA template and is located between the two PCR primers. Fluorescent reporter dyes, such as the fluorophores 6-carboxyfluorescein (FAM) or tetrachloro-fluorescein (TET), and fluorescence quenching dyes, such as tetramethylrhodamine (TAMRA) or dihydrocyclopyrroloindole tripeptide minor groove binder (MGB) (4), are covalently attached to the 5′- and 3′-ends of the probe, respectively. The close proximity of the fluorophore and quencher attached to the intact probe inhibits the fluorescence of the fluorophore by the process of fluorescence resonance energy transfer (FRET). During each PCR extension cycle, the 5′ to 3′ exonuclease activity of *Taq* polymerase degrades the portion of the probe that has annealed to the template. Degradation of the 5′ portion of the probe releases the fluorophore, thus relieving the quenching effect of FRET and allowing the fluorescence of the fluorophore to be detected. The fluorescence detected is directly proportional to the amount of the fluorophore released, and therefore, proportional to the amount of cDNA template present in the PCR. The exponential increase of PCR product is used to determine the threshold cycle, or C_T, which is the number of PCR cycles at which a significant increase in fluorescence is detected. The C_T is directly proportional to the number of copies of cDNA template present in the reaction.

In the methods described in this chapter, TaqMan probes and primers specific for p210 *BCR-ABL* are employed for RQ-PCR.

In a separate tube, a segment of the *ABL* gene is amplified as an endogenous control to normalize for sample variation. Relative quantification of *BCR-ABL* mRNA levels can be accomplished by using either a standard curve method or a comparative C_T method (also known as the $\Delta\Delta C_T$ method). The standard curve method is currently used by most clinical laboratories to calculate *BCR-ABL* levels and requires serially diluted standards to be run in each reaction plate, and unknown sample quantitative values are interpolated from the resulting standard curve. The comparative C_T method uses an arithmetic formula to determine the amount of *BCR-ABL* transcript in an unknown sample, and standard curves are not required for quantification. The comparative C_T method produces essentially equivalent results (5) to the standard curve method, and both these quantification methods will be illustrated in this chapter.

2. Materials

2.1. Samples

1. Peripheral blood or bone marrow aspirate specimens for p210 *BCR-ABL* analysis (see Note 1).
2. K562 CML cell line (see Note 2).
3. SU-DHL-16 non-Hodgkin lymphoma cell line (see Note 3).

2.2. Sample Preparation

1. RBC Lysis Solution (Qiagen).
2. Trizol reagent (Invitrogen), store at 4°C.
3. Chloroform.
4. Isopropanol.
5. 75% ethanol.
6. Nuclease-free H_2O.
7. General laboratory equipment, including vortex mixer, desktop centrifuge, microcentrifuge, and spectrophotometer.
8. PCR thermal cycler.
9. For reverse transcription, we use the Gene Amp RNA PCR Kit (Applied Biosystems), which includes:
   (a) GeneAmp 25 mM $MgCl_2$.
   (b) GeneAmp 10× PCR Buffer II (500 mM KCl; 100 mM Tris-HCl, pH 8.3).
   (c) GeneAmp 10 mM dATP, dTTP, dCTP, dGTP.
   (d) ABI RNase Inhibitor.
   (e) ABI Random Hexamers.
   (f) ABI MuLV Reverse Transcriptase.

2.3. Quantitative Real-Time RT-PCR

1. ABI 7900HT Fast Real-Time PCR System with standard 96-well block module (Applied Biosystems).
2. MicroAmp™ Optical 96-Well Reaction Plate (Applied Biosystems).
3. Optical Adhesive Film (Applied Biosystems).
4. TaqMan Universal PCR Master Mix (Applied Biosystems).
5. Nuclease-free H_2O.
6. Forward and reverse primers specific for target gene transcripts (see Note 4).
7. TaqMan probe specific for target gene transcripts (see Notes 4 and 6).
8. Forward and reverse primers specific for control gene transcripts (see Note 5).
9. TaqMan probe specific for control gene transcripts (see Notes 5 and 6).

3. Methods

Special precautions should be used when handling RNA samples to avoid RNA degradation such as using gloves at all times and using nuclease-free reagents and plasticware.

3.1. Sample Preparation

1. Collect peripheral blood (3–5 ml) and/or bone marrow aspirate (1–3 ml) samples in EDTA (purple top) tubes, and process these samples within 8 h of collection (see Note 7).
2. Lyse red blood cells (RBC) in peripheral blood and bone marrow aspirate samples prior to RNA isolation. This is accomplished using Qiagen RBC Lysis Solution according to the manufacturer's instructions. Extract total RNA from the blood or bone marrow specimens using TRIzol reagent according to the manufacturer's recommendations (see Note 8). After extraction, the RNA samples can be temporarily stored at –20°C or may be placed into longer-term storage at –80°C until ready to proceed. Subsequent DNase treatment of the RNA to remove possible DNA contamination of the sample is optional (see Note 9).
3. Measure RNA concentration using a spectrophotometer (see Note 10), and adjust RNA concentration in nuclease-free H_2O to a concentration of 0.2 μg/μl in preparation for subsequent cDNA synthesis.

3.1.1. Standards

1. Prepare RNA standards for standard curve construction. We use the K562 CML cell line to prepare RNA for both the

p210 *BCR-ABL* and *ABL* standards for standard curve construction. RNA should be isolated from K562 cells using TRIzol reagent and the same protocol as for patient samples (see Subheading 3.1, step 2), and stored in nuclease-free H_2O. For convenience, K562 cell line mRNA may be obtained directly from commercial resources if cell culture facilities are not available. Alternatively, cDNA or plasmid constructs may also be used to prepare standard curves (see Note 11).

2. Determine K562 cell line RNA quantity using a spectrophotometer and aliquot high concentration (>1 μg/μl) K562 cell line RNA stock solution for storage at −80°C). After RNA aliquots are made, repeated freeze-thaw cycles should be avoided (see Note 12). For preparation of standard curves (see following step), one aliquot of K562 RNA stock solution should be adjusted to a concentration of 0.2 μg/μl in nuclease-free H_2O. Aliquots of K562 RNA working solution should be stored at −80°C.
3. Serially dilute 5 μl of the K562 cell line RNA working solution (0.2 μg/μl) tenfold for standard curve determination. Dilute the K562 RNA in nuclease-free H_2O containing 20 ng/μl yeast RNA (see Note 13), and prepare seven standards containing from 1×10^{-6} μg up to 1 μg of K562 RNA to produce standard curves for quantitation of both the p210 *BCR-ABL* and *ABL* genes. Five microliters of each of these RNA dilutions should then be used in a reverse transcription reaction (see Subheading 3.2.1, steps 1 and 2). Both the p210 *BCR-ABL* and *ABL* standard dilutions derived from K562 RNA should be run in each plate if the standard curve quantitation method is used (see Subheading 3.2.2, steps 2 and 3).

3.1.2. Controls

1. Use an endogenous control gene (*ABL*) to standardize the amount of sample RNA added to the PCR reaction. The endogenous control gene should be amplified along with the p210 *BCR-ABL* target gene (see Note 14).
2. Blank or no RNA template controls, consisting of nuclease-free H_2O, should also be included in the reverse transcription reaction and the RQ-PCR to monitor for contamination. Two no template controls are recommended for the RQ-PCR in each reaction plate.
3. A negative control that does not contain the target gene RNA of interest (e.g., *BCR-ABL* negative cell line RNA) should also be included in each RQ-PCR.
4. A positive sensitivity control, consisting of a 10^{-4} dilution (this dilution can vary depending on the sensitivity requirements of the individual assay) of K562 cell line RNA in a background of *BCR-ABL* negative cell line RNA, should also be included in the RQ-PCR. The *BCR-ABL* negative cell line

RNA can be prepared from any cell line that does not contain the *BCR-ABL* fusion gene (we use RNA isolated from the SU-DHL-16 non-Hodgkin lymphoma cell line). K562 cell line RNA and SU-DHL-16 cell line RNA samples should be adjusted to a concentration of 0.2 μg/μl using nuclease-free H_2O. Then, the 10^{-4} dilution is obtained by a series of tenfold serial dilutions of one volume of K562 RNA solution to nine volumes of the 0.2 μg/μl SU-DHL-16 RNA solution. The sensitivity control is used to monitor the sensitivity of the assay and mimics a clinical specimen in which p210 *BCR-ABL* positive tumor cells are present in a background of normal cells.

5. If the RQ-PCR is used for minimal residual disease monitoring, high and low positive controls should also be included in the assay. In this example, p210 *BCR-ABL* positive samples that contain high and low levels of p210 *BCR-ABL* transcripts are included in each RQ-PCR to monitor assay performance parameters such as maintenance of linearity, stability of standards, and the success of each of the RT and RQ-PCR steps.

3.1.3. Calibrators

1. A calibrator sample should be used for comparison of gene expression in different tissues or in the same tissue under different conditions.
2. For our quantitative p210 *BCR-ABL* assay example, comparison of the p210 *BCR-ABL* and *ABL* transcript levels in the same patient samples at diagnosis and after subsequent initiation of treatment is the most appropriate calibrator (see Note 15). The calibrator should be tested in the same RQ-PCR plate as the patient samples to reduce test run variation.

3.2. Real-Time Quantitative RT-PCR

3.2.1. Reverse Transcription

1. Reverse transcribe RNA samples into complementary DNA (cDNA) using the GeneAmp® RNA PCR Kit according to the manufacturer's instructions (see Note 16).
2. For the reverse transcription (RT) reaction, add 1 μg of RNA to the 20 μl total volume RT reaction tube. After the RT reaction, use 5 μl of cDNA (equivalent to approximately 0.25 μg of RNA) for subsequent RQ-PCR analysis (see Subheading 3.2.2, below). If samples are analyzed in duplicate or triplicate, it is recommended that replicate samples be obtained from different RT reaction tubes to monitor assay reproducibility.

3.2.2. Real-Time Quantitative RT-PCR Set up

1. Keep all thawed reagents and patient samples cool on wet ice prior to use. TaqMan probe aliquots should also be protected from light and should not be thawed until immediately prior to use.

2. Prepare two PCR master mix tubes – one for p210 *BCR-ABL* RQ-PCR and one for *ABL* RQ-PCR. The final volume of each master mix tube is based on the total number of reactions required for each RQ-PCR. Each reaction should contain 6.2 μl of nuclease-free H_2O, 12.5 μl of 2× Taqman Universal PCR Master Mix, 0.15 μl of each forward and reverse primers (the final concentration of each primer should be 300 nM), and 1 μl of TaqMan probe (the final concentration of probe should be 200 nM). Appropriate primers and probe should be added to each of the p210 *BCR-ABL* and *ABL* RQ-PCR master mixes.
3. Set up the RQ-PCR in a 96-well plate with 20 μl of master mix per well along with 5 μl of cDNA (0.25 μg RNA equivalent) in appropriate wells. The top four rows (48 wells) of the 96-well plate may be used for the p210 *BCR-ABL* RQ-PCR (target gene), and the bottom four rows for the *ABL* RQ-PCR (endogenous control gene), or vice versa. Each plate should include standards, blank and negative controls, sensitivity controls, and test samples in duplicate or triplicate. When finished pipetting the samples in the 96-well reaction plate, seal the plate with an optical adhesive film.
4. Centrifuge the reaction plate at low speed for 2 min, and then load the plate onto an Applied Biosystems 7900HT real-time PCR instrument. Use Applied Biosystems Sequence Detection Software (SDS) to document and run the assay (see Subheading 3.2.3, steps 1 and 2).
5. Run the RQ-PCR using modified universal thermal cycling conditions: 50°C for 2 min, 95°C for 10 min, and 45 cycles of 95°C for 15 s and 60°C for 1 min.

3.2.3. Program and Run Applied Biosystems 7900HT Instrument

1. Operate the Applied Biosystems 7900HT instrument and SDS software according to the manufacturer's recommendations (6–9).
2. Briefly, launch the SDS software and open a new plate template window to denote well locations on the 96-well plate for the standard dilutions, controls, and testing samples. Record standard values (e.g., 1, 0.1, 0.001 μg of RNA, etc.) for wells containing sample dilutions for the p210 *BCR-ABL* and *ABL* standard curves and replicate samples for duplicate or triplicate testing samples. Save the template window with the recorded data as a SDS run file.
3. Select the connect button, and then load the 96-well plate into the instrument. Select the start button to begin the actual run.

3.2.4. Data Collection and Analysis

1. After the run is completed, remove the plate from the instrument and save the results of the run. When the analysis button

is selected in the SDS software, the results will be analyzed automatically if the standards and testing sample information were recorded as noted above (see Subheading 3.2.3, step 2).

2. View the standard curve results in the amplification plot and standard curve plot views in the SDS software. Select crossing thresholds (C_T) for the standard curves for both the p210 *BCR-ABL* and *ABL* RQ-PCRs in the exponential phase of the reactions, either manually or automatically, to achieve the best slopes and correlation coefficients (R^2) (see Note 17). Examples of standard curves for p210 *BCR-ABL* and *ABL* RQ-PCR standard curves are illustrated in Figs. 1 and 2, respectively.
3. If the standard curves are deemed to have acceptable slopes and R^2 values (see Note 17), export the results from the SDS software as a text file. Import the text file into Microsoft Excel software for subsequent results calculations (see Subheading 3.2.5, steps 3–5, and Subheading 3.2.6, steps 4–6).

3.2.5. Relative Quantitation Calculations Using a Standard Curve

1. Include both the p210 *BCR-ABL* and the *ABL* RQ-PCR results for each sample in the Excel file created for results calculations.

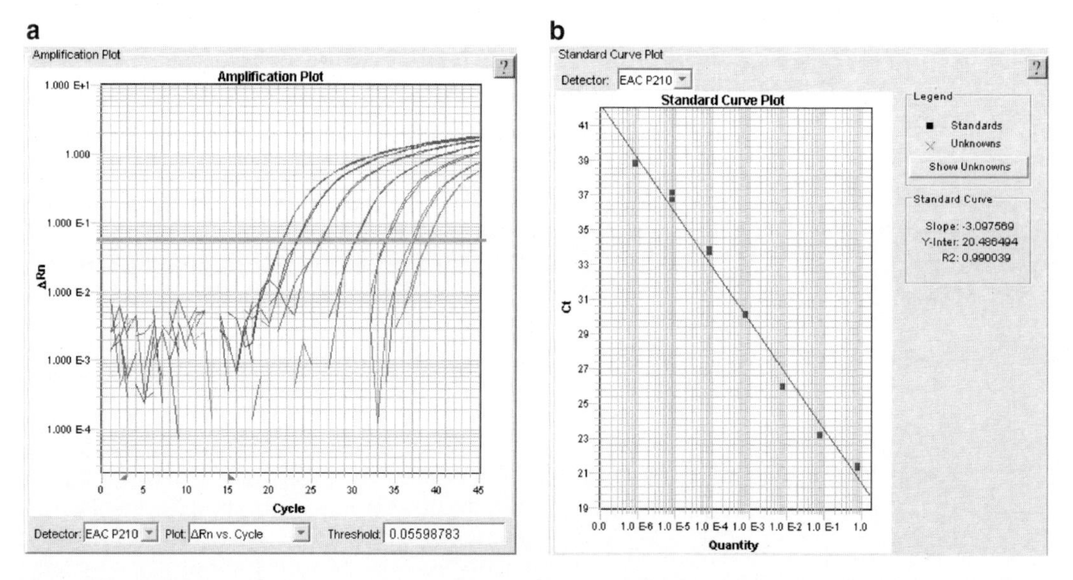

Fig. 1. p210 *BCR-ABL* amplification and standard curve plots. (**a**) Amplification plot for duplicates of seven standards containing from 0.25×10^{-6} µg up to 0.25 µg RNA equivalent of K562 RNA. The *y*-axis is a measure of change in fluorescence versus PCR cycle number on the *x*-axis. The thick horizontal grey bar represents the crossing threshold in the exponential phase of amplification reaction in which the threshold cycle (C_T) is determined for each standard dilution. (**b**) Standard curve plot constructed based on the C_T values determined for each standard from the amplification plot. The standard curve is constructed based on C_T values on the *y*-axis and quantity of the standard on the *x*-axis. This curve is used to determine the amount of p210 *BCR-ABL* transcript in an unknown sample

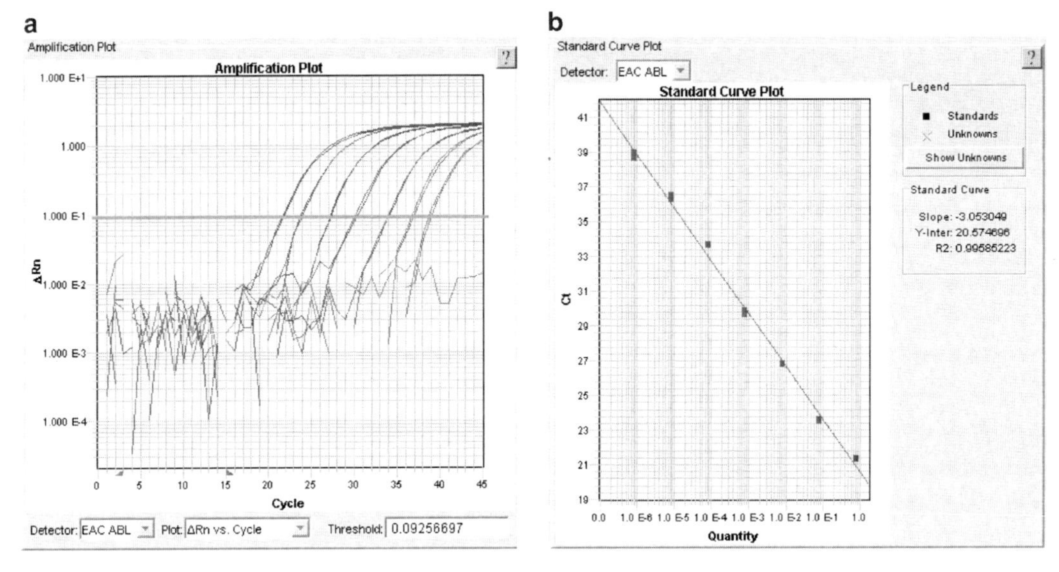

Fig. 2. *ABL* amplification and standard curve plots. (**a**) Amplification plot for duplicates of seven standards containing from 0.25×10^{-6} μg up to 0.25 μg RNA equivalent of K562 RNA. The *y*-axis is a measure of change in fluorescence versus PCR cycle number on the *x*-axis. The thick horizontal grey bar represents the crossing threshold in the exponential phase of amplification reaction in which the threshold cycle (C_T) is determined for each standard dilution. (**b**) Standard curve plot constructed based on the C_T values determined for each standard from the amplification plot. The standard curve is constructed based on C_T values on the *y*-axis and quantity of the standard on the *x*-axis. This curve is used to determine the amount of *ABL* transcript in an unknown sample

2. Required calculations for this method include the following: (1) calculation of averages, if replicate samples are tested (see Note 18); (2) normalization of the target gene results (p210 *BCR-ABL*) to the results obtained for the control gene (*ABL*); (3) normalization of the target gene results to those obtained from the calibrator samples to determine the p210 *BCR-ABL* relative quantitative value (see Note 19).
3. Replicate average example calculations (see Table 1).

Table 1
Example replicate average calculations

p210 *BCR-ABL*	p210 *BCR-ABL* C_T	p210 *BCR-ABL* quantity	*ABL*	*ABL* C_T	*ABL* quantity
Replicate 1	25.90	0.0091	Replicate 1	25.26	0.0092
Replicate 2	25.97	0.0086	Replicate 2	25.55	0.0075
Replicate 3	25.99	0.0085	Replicate 3	25.57	0.0074
Average	*25.95*	*0.0087*	*Average*	*25.46*	*0.0081*

4. Example p210 *BCR-ABL* and *ABL* calibrator C_T and quantity results for subsequent calculations:

   p210 *BCR-ABL* calibrator C_T = 23.23

   p210 *BCR-ABL* calibrator quantity = 0.0568

   ABL calibrator C_T = 22.68

   ABL calibrator quantity = 0.0574

5. Normalization example calculations:

   Test sample p210 *BCR-ABL* normalization = average p210 *BCR-ABL* quantity ÷ average *ABL* quantity = 0.0087 ÷ 0.0081 = 1.074

   Calibrator sample p210 *BCR-ABL* normalization = p210 *BCR-ABL* calibrator sample quantity ÷ *ABL* calibrator sample quantity = 0.0568 ÷ 0.0574 = 0.99

   Final test sample p210 *BCR-ABL* results (p210 *BCR-ABL* relative quantitative value) = normalized test sample p210 *BCR-ABL/ABL* ratio ÷ normalized calibrator sample p210 *BCR-ABL/ABL* ratio = 1.074 ÷ 0.99 = 1.09 (see Note 20).

3.2.6. Relative Quantitation Using the Comparative C_T Method

1. Verify that the amplification efficiencies of the RQ-PCRs for the p210 *BCR-ABL* target gene and *ABL* endogenous reference gene are similar and close to 100% before using the comparative C_T method (see Note 21).
2. Perform the RQ-PCR assay in the same manner as noted above (see Subheadings 3.2.1–3.2.4), except that no standard curves dilutions should be included in the reaction plate. This method involves a calculation method known as the DeltaDeltaC_T ($\Delta\Delta C_T$) method (5) and is based on a C_T number comparison between the p210 *BCR-ABL* target gene and the *ABL* endogenous reference gene relative to a calibrator. The formula for this calculation is: $2^{-\Delta\Delta C_T}$.
3. Required calculations for this method include the following: (1) calculation of the ΔC_T for the p210 *BCR-ABL* test sample relative to the *ABL* endogenous control and the p210 *BCR-ABL* calibrator relative to the *ABL* calibrator; (2 $2^{-\Delta\Delta C_T}$ calculation; (3) $2^{-\Delta\Delta C_T}$ calculation to obtain relative quantitative value.
4. ΔC_T calculations:

   ΔC_T for test samples = average C_T of the p210 *BCR-ABL* test sample – average C_T of the *ABL* endogenous control = 25.95 – 25.46 = 0.49

   ΔC_T for calibrator samples = C_T of the p210 *BCR-ABL* calibrator sample – C_T of the *ABL* calibrator sample = 23.23 – 22.68 = 0.55

5. $\Delta\Delta C_T$ calculation:

   $\Delta\Delta C_T = \Delta C_T$ for test samples $- \Delta C_T$ for calibrator samples $= 0.49 - 0.55 = -0.06$

6. $2^{-\Delta\Delta C_T}$ calculation:

   Relative quantitative value $= 2^{-\Delta\Delta C_T} = 2^{-(-0.06)} = 2^{0.06} = 1.04$ (see Note 22).

4. Notes

1. The use of either peripheral blood or bone marrow aspirate specimens is recommended for monitoring minimal residual disease (p210 *BCR-ABL* levels) in CML patients (1).
2. The K562 cell line is a CML-derived cell line that expresses p210 *BCR-ABL* fusion transcripts and is used as a source of RNA for p210 *BCR-ABL* and *ABL* standard curve construction (see Subheading 3.1.1, step 1) and sensitivity control production (see Subheading 3.1.2, step 4). This cell line may be obtained from the American Type Culture Collection (ATCC CCL-243).
3. The SU-DHL-16 cell line is a non-Hodgkin lymphoma-derived cell line that is used as a source of RNA that does not contain p210 *BCR-ABL* transcripts and for production of the sensitivity control (see Subheading 3.1.2, step 4). This cell line was chosen as a source of RNA because we already utilized it as a control cell line for other purposes in our molecular diagnostics laboratory. However, any cell line that does not contain the p210 *BCR-ABL* fusion transcript can be used to produce the sensitivity control. The SU-DHL-16 cell line may be obtained from the German Collection of Microorganisms and Cell Cultures (DSMZ ACC-577).
4. Target gene primers and probes should be designed according to individual needs. In this chapter, we use p210 *BCR-ABL* transcripts as an example of a target gene for RQ-PCR. The primer and probe sequences used in our assay were obtained from Gabert et al. (1). Primer working solutions are obtained by diluting stock primers in nuclease-free H_2O to achieve the desired primer concentration (e.g., 50 pmol/μl).
5. Control gene primers and probes should be designed according to individual needs. In this chapter, we use *ABL* transcripts as an example of a control gene for assay normalization in RQ-PCR. The primer and probe sequences used in our assay were obtained from Beillard et al. (2).

6. TaqMan probe and probe dilution: probes are usually received as a 100 pmol/µl stock solution. This stock solution can be diluted with nuclease-free H_2O to achieve a working solution concentration (e.g., 5 pmol/µl). Probe working solutions should be aliquoted in volumes that are sufficient for one-time use. These aliquots should be stored at –20°C and protected from light. In addition, repeated freeze-thaw cycles of probe solutions should be avoided.
7. Samples should be transported to the laboratory and processed with TRIzol reagent for RNA isolation within 8 h of collection (optimal), or up to a maximum of 48 h after collection to avoid RNA degradation. RNA integrity is critical, especially for samples used for monitoring minimal residual disease.
8. RNA extraction protocols other than the Trizol reagent method may also be used as long as RNA quality and quantity is ensured.
9. DNase treatment may not be needed if PCR primers are designed to hybridize to exon regions of the gene of interest. The amplicon resulting from the contaminating genomic DNA will be either larger than the amplicon derived from the RNA sample or no product may be produced at all due to the increased distance between the primers.
10. Our laboratory uses a NanoDrop 1000 Spectrophotometer (Thermo Fisher Scientific) to quantitate RNA samples prior to adjusting the concentration for subsequent cDNA synthesis because only one microliter of sample is required for analysis, without prior dilution.
11. cDNA or plasmid DNA containing the p210 *BCR-ABL* fusion gene may also be used to produce standard curves, but if such samples are utilized, it should be understood that this will not allow the monitoring of the efficiency of the reverse transcription step. It should also be noted that if plasmid is used for standard curve production, two different serial dilutions of plasmids (one for the *BCR-ABL* gene and one for the *ABL* gene) will be necessary to produce standard curves for both the target gene and the control gene.
12. To ensure assay accuracy, the same lot of K562 RNA should be used for patient sample analysis until it is completely exhausted. When it is anticipated that a new lot of K562 RNA will be required, it should be analyzed along with an aliquot of the old lot before being used for patient sample analysis. This process will ensure that the new lot of control RNA produces appropriate results before it is placed into service.
13. Our validation study for this assay showed that diluting K562 cell line RNA in nuclease-free H_2O containing yeast RNA helped to maintain standard curve linearity.

14. The endogenous control gene has two functions in RQ-PCR assays: (1) to monitor test sample RNA quality and quantity (low quantity [high C_T value] indicates poor RNA quality and can adversely affect result interpretation, and (2) to normalize RNA quality and quantity variance among the test samples. Housekeeping genes such as beta-actin, glyceraldehyde-3-phosphate dehydrogenase (GAPDH), or ribosomal RNA are often used as endogenous controls, but these controls should be chosen to best fit the specific needs of each assay and the sample type being analyzed. In the case of p210 *BCR-ABL* quantitation, the *BCR*, *ABL*, and beta-glucuronidase (*GUSB*) genes are the most frequently used as endogenous control genes, and any of these three genes are considered to be suitable for this purpose. However, the Europe Against Cancer program recommends using the *ABL* gene as the control of choice for RQ-PCR diagnostics and minimal residual disease detection in patients with CML (1, 2, 10).

15. If patient diagnostic samples are not available for comparison, a standardized baseline p210 *BCR-ABL* transcript level obtained from the average transcript levels of 30 untreated chronic phase CML patients may also be used (10). It is also permissible to use tumor cell line RNA (e.g., K562 cell line) as a calibrator with the assumption that this represents the equivalent of 100% tumor cells expressing p210 *BCR-ABL* transcripts.

16. It should be noted that the GeneAmp® RNA PCR Kit has limited reverse transcription capacity and can accommodate a maximum of only 1 μg of total RNA. Therefore, multiple gene RQ-PCR analysis may exhaust all the cDNA reaction volume, and additional reverse transcription reactions may be necessary if the RQ-PCR analysis needs to be repeated.

17. When the run is finished, the p210 *BCR-ABL* and *ABL* standard curves should be examined. The slope of each standard curve should ideally be close to −3.3, but in reality, a range of -3.3 ± 0.3 is acceptable. The standard curve should also be linear with a R^2 value close to 0.99.

18. Test samples should be analyzed in duplicate or triplicate according to individual laboratory statistical requirements.

19. The relative quantitative value of the test sample is a result of a comparison of the test sample p210 *BCR-ABL* to *ABL* ratio to the p210 *BCR-ABL* to *ABL* ratio of the calibrator. Assuming the calibrator is comprised of 100% tumor cells containing the p210 *BCR-ABL* transcript (and is assigned a relative quantitative value of 1), the quantitative results of the test samples are reported as a ratio that reflects the difference

in p210 *BCR-ABL* transcript levels between the test sample and the calibrator. The test sample may yield a relative quantitative value that is equal to, less than, or greater than the calibrator, depending on the actual p210 *BCR-ABL* expression levels in the test sample.

20. In an example for comparison, if the normalized p210 *BCR-ABL* quantity ÷ normalized p210 *BCR-ABL* calibrator quantity = 0.1, this means that compared to the calibrator with a relative quantitative value of 1 (or 100% tumor cells), the test sample would contain about tenfold less (or 1 log less) p210 *BCR-ABL* transcript than the calibrator (or 10% tumor cells).
21. Examples such as those described in the Applied Biosystems User Bulletin #2 Comparative C_T Method Section (6) should be used to assess PCR efficiency before using the comparative C_T method. If PCR efficiency is not satisfactory, PCR primers may need to be redesigned or the standard curve method may be used instead of relative quantification.
22. Note that the relative quantitative values for p210 BCR-ABL quantitation using the standard curve and DeltaDeltaC_T methods are very similar (1.09 for the standard curve method versus 1.04 for the DeltaDeltaC_T method).

Acknowledgments

The author would like to thank Yao Wang for her excellent technical assistance.

References

1. Gabert J, Beillard E, van der Velden VH, Bi W, Grimwade D, Pallisgaard N, Barbany G, Cazzaniga G, Cayuela JM, Cavé H, Pane F, Aerts JL, De Micheli D, Thirion X, Pradel V, González M, Viehmann S, Malec M, Saglio G, van Dongen JJ (2003) Standardization and quality control studies of 'real-time' quantitative reverse transcriptase polymerase chain reaction of fusion gene transcripts for residual disease detection in leukemia – a Europe Against Cancer program. Leukemia 17:2318–2357
2. Beillard E, Pallisgaard N, van der Velden VH, Bi W, Dee R, van der Schoot E, Delabesse E, Macintyre E, Gottardi E, Saglio G, Watzinger F, Lion T, van Dongen JJ, Hokland P, Gabert J (2003) Evaluation of candidate control genes for diagnosis and residual disease detection in leukemic patients using 'real-time' quantitative reverse-transcriptase polymerase chain reaction (RQ-PCR) – a Europe against cancer program. Leukemia 17:2474–2486
3. Heid CA, Stevens J, Livak KJ, Williams PM (1996) Real time quantitative PCR. Genome Res 6:986–994
4. Kutyavin IV, Afonina IA, Mills A, Gorn VV, Lukhtanov EA, Belousov ES, Singer MJ, Walburger DK, Lokhov SG, Gall AA, Dempcy R, Reed MW, Meyer RB, Hedgpeth J (2000) 3′-minor groove binder-DNA probes increase sequence specificity at PCR extension temperatures. Nucleic Acids Res 28:655–661

5. Arocho A, Chen B, Ladanyi M, Pan Q (2006) Validation of the 2-DeltaDeltaCt calculation as an alternate method of data analysis for quantitative PCR of BCR-ABL P210 transcripts. Diagn Mol Pathol 15:56–61

6. Applied Biosystems (1997) Relative quantitation of gene expression: ABI PRISM 7700 Sequence Detection System: User Bulletin #2: Rev B

7. Applied Biosystems (2001) ABI PRISM® 7900HT Sequence Detection System User Guide Basic Operation and Maintenance Rev A4

8. Applied Biosystems (2003) ABI PRISM® 7900HT Sequence Detection System and SDS Enterprise Database User Guide Basic Operation and Maintenance Rec C

9. Applied Biosystems (2008) Guide to performing relative quantitation of gene expression using real-time quantitative PCR Rev

10. Hughes T, Deininger M, Hochhaus A, Branford S, Radich J, Kaeda J, Baccarani M, Cortes J, Cross NC, Druker BJ, Gabert J, Grimwade D, Hehlmann R, Kamel-Reid S, Lipton JH, Longtine J, Martinelli G, Saglio G, Soverini S, Stock W, Goldman JM (2006) Monitoring CML patients responding to treatment with tyrosine kinase inhibitors: review and recommendations for harmonizing current methodology for detecting BCR-ABL transcripts and kinase domain mutations and for expressing results. Blood 108:28–37

Chapter 14

The Use of Comparative Quantitative RT-PCR to Investigate the Effect of Cysteine Incubation on GPx1 Expression in Freshly Isolated Cardiomyocytes

Nicola King

Abstract

Intracellular cysteine availability is one of the major rate limiting factors that regulate the synthesis of the major antioxidant, glutathione. Little is known, however, about the effect of cysteine upon glutathione-associated enzymes in isolated heart cells. Such knowledge is important if a full understanding and exploitation of cysteine's cardioprotective potential is to be achieved. Therefore, this study describes the use of a comparative quantitative reverse transcription polymerase chain reaction (RT-PCR) assay to investigate the effect of incubation of freshly isolated rat cardiomyocytes for 2 h at 37°C with or without 0.5 mM cysteine on the expression of cellular glutathione peroxidase (GPx1).

The main analytical method is the conventional RT-PCR in a standard thermal cycler followed by electrophoresis and scanning densitometry using the expression of the housekeeping gene, glyceraldehyde-3-phosphate dehydrogenase (GAPDH), for normalising purposes. Each step of this straight-forward and relatively inexpensive method is explained in detail to facilitate its adoption by the reader for experiments investigating the effects of any compound on any gene in any cell population. The results of the current investigation show that cysteine incubation significantly increases the expression of GPx1 in freshly isolated cardiomyocytes compared to control, suggesting the possibility of a new beneficial role for cysteine in myocardial protection.

Key words: Comparative quantitative RT-PCR, GPx1, Cysteine, Cardiomyocytes, Cardioprotection

1. Introduction

The sulphydryl-containing amino acid, cysteine, plays an important role in intermediary metabolism in the heart. Indeed, in premature and new-born infants, cysteine is considered to be an essential dietary component (1), while adults obtain cysteine from dietary sources, proteolysis and methionine catabolism (2). Thereafter, cysteine is involved in many reactions including protein synthesis

Nicola King (ed.), *RT-PCR Protocols: Second Edition*, Methods in Molecular Biology, vol. 630,
DOI 10.1007/978-1-60761-629-0_14,

(for example CoA (3), atriopeptin (4), and *S*-nitrothiols (5)); oxidative metabolism (6); sulphur provision (2) and the formation of phaeomalin and trichrome pigments (2). The intracellular concentration of cysteine in the rat heart is 0.089 mM compared to 0.097 mM in the plasma (7).

Cysteine availability is also thought to be one of the major rate limiting factors, controlling synthesis of the tripeptide, γ-glutamyl-cysteinyl-glycine (8). This antioxidant, better known as glutathione, is the main intracellular thiol where the ratio of oxidized (GSSG) to reduced (GSH) glutathione is often used as an indicator of the cell's redox status (8). GSH, together with its associated enzymes glutathione-*S*-transferase, glutathione peroxidase and glutathione reductase, forms part of the myocardium's endogenous protection against oxidative stress alongside other enzymes such as superoxide dismutase and catalase (9). Oxidative stress, involving the production of harmful reactive oxygen species (ROS) and their precursors including hydrogen peroxide, superoxide, peroxynitrite and hydroxyl radical, has been implicated in ischaemia reperfusion injury, myocardial stunning and the failing heart (9–11).

The relationship between cysteine availability and GSH synthesis has motivated strategies that test cysteine's potential as a cardioprotective agent (5, 9, 12). For example, we have shown that 0.5 mM cysteine addition improved the functional recovery of isolated and perfused adult rat hearts following ischaemia in comparison to non-cysteine perfused controls (13). However, to fully understand the relationship between cysteine, glutathione and cardiac antioxidant capacity, it is also necessary to investigate the effects of cysteine on glutathione associated enzymes.

The protocols described in this chapter outline a straightforward procedure to investigate the effect of cysteine incubation upon cellular glutathione peroxidase (GPx1) expression in freshly isolated cardiomyocytes. The approach taken is a comparative reverse transcription polymerase chain reaction (RT-PCR) end point assay, where the expression of the gene of interest is compared in cysteine-incubated and control samples using the housekeeping gene, glyceraldehyde-3-phosphate dehydrogenase (GAPDH), for normalisation. More specifically, the methods involved comprise incubation of freshly isolated samples of cardiomyocytes in the presence or absence of cysteine; isolation of RNA; reverse transcription; and a series of PCR reactions, followed by electrophoresis and analysis of band density. The quantification is made possible through the existence of a linear phase where, during the PCR reaction, amplification of the starting material is proportionate to the appearance of product (14, 15). This necessitates the addition of a small number of preliminary experiments to identify where the linear phase of amplification lies and to verify that differences in transcript number can be detected with densitometry following RT-PCR. The latter control can also provide a crude method for quantifying the level of gene expression in each sample, although in the current experiments only a simple comparison between the incubated and control samples is made.

One of the major advantages of the methods described is that they provide a quick method for assessing the effects of a compound on gene expression using relatively inexpensive and routine equipment, although the results obtained do not provide an absolute quantification of the levels of gene expression in each sample. These techniques are readily adaptable for use with any compound, any gene and any cell type which excites the researcher's interest.

2. Materials

2.1. Cysteine Incubation

1. Suspension of freshly isolated rat ventricular cardiomyocytes (see Notes 1 and 2)
2. Universal tubes (30 ml)
3. Bench-top centrifuge
4. 0.1 M L-cysteine solution (see Note 3)
5. 0.1 and 1-ml pipette and pipette tips (with last 5 mm of the 1 ml tip removed, see Note 4)
6. Shaking water-bath
7. Stop watch
8. Gauze
9. Adhesive tape.

2.2. RNA Extraction

1. Nuclease free 1.5-ml Eppendorf tubes
2. Nuclease free 1 ml and 0.2-ml pipettes and tips
3. Tri-Reagent (store at 4°C) (see Note 5)
4. Ice
5. Gloves
6. Refrigerated Eppendorf centrifuge
7. Chloroform (see Note 6)
8. Isopropanyl alcohol (see Note 7)
9. Ethanol (see Note 7)
10. –80°C freezer
11. Nuclease free water (see Note 8)

12. Heating block
13. RNA zap
14. Whirly mixer
15. Blue roll or other tissue.

2.3. Assessment of RNA Quality and Quantity

1. Spectrophotometer capable of measuring at 260 and 280 nm
2. Quartz cuvettes
3. Sterile 10 and 100-μl pipettes and pipette tips
4. Nuclease free water
5. Gloves.

2.4. Reverse Transcription

1. Ambion RETROscript® Reverse Transcription (RT) for RT-PCR kit (see Note 9)
2. RNA
3. Nuclease free Eppendorf tubes
4. Nuclease free 1, 10 and 100-μl pipette and pipette tips
5. Refrigerated Eppendorf centrifuge
6. Heating block
7. Ice
8. Whirly mixer set for low velocity
9. Gloves.

2.5. Determination of the Linear Phase of Amplification

1. PCR tubes
2. DNA zap
3. Thermal cycler
4. Nuclease free 1, 10, and 100-μl pipettes and barrier pipette tips
5. Promega PCR Master Mix
6. 10 μM of forward GPx1 primer: CCGATATAGAAGCCC TGCTG (see Notes 10–12)
7. 10 μM of reverse GPx1 primer: GAAACCGCCTTTCTTT AGGC
8. One cDNA sample from RT reaction
9. Whirly mixer
10. Bench top Eppendorf centrifuge
11. Heating block
12. Gel loading solution (supplied in Subheading 2.4, item 1)
13. DNA molecular weight markers
14. Electrophoresis equipment (for running horizontal gels)
15. Agarose

16. 50× TAE solution: dissolve 242 g of Tris(hydroxymethyl) aminomethane (TRIS), 57.1 ml of glacial acetic acid and 18.6 g of ethylenediaminetetra-acetic acid (EDTA) in 900 ml of double distilled water. Make volume up to 1 L with double distilled water.
17. 10 mg/ml ethidium bromide
18. UV transilluminator
19. Gel capturing or gel photography device
20. Densitometry software.

2.6. Determination of Sensitivity to Different Levels of Gene Expression

1. Subheading 2.5, items 1–19
2. Clean scalpel
3. QIAquick gel extraction kit (QIAgen)
4. Isopropanol
5. Ethanol
6. 1-ml pipette tips and pipette
7. 1.5-ml Eppendorf tubes
8. Spectrophotometer capable of measuring at 260 nm
9. Quartz cuvette.

2.7. Effect of Cysteine Incubation on GPx1 Expression

1. All cDNA samples from RT reactions
2. Subheading 2.5, Items 1–7, 9, 10, 12–20
3. GAPDH forward primer: GTGGACCTCATGGCCTACAT (see Note 13)
4. GAPDH reverse primer: GGATGGAATTGTGAGGGAGA
5. Appropriate statistics analysis package.

3. Methods

3.1. Cysteine Incubation

1. Warm the water-bath to 37°C.
2. Label two universal tubes (or other suitable container) with the words, “control” and “+ cysteine”.
3. Equally divide the cardiomyocytes between the two tubes. Do not cap the tubes.
4. Add sufficient of the L-cysteine solution to one of the tubes to yield a final concentration of 0.5 mM (this requires a 200-fold dilution of the stock solution).
5. Cover each tube opening with gauze and fix in place with adhesive tape (see Note 14).

6. Place the tubes into the water-bath and enable a gentle shake (see Note 15).
7. After 2-h incubation, remove the tubes and centrifuge at 900×*g* and room temperature for 2 min.
8. Discard the supernatant and continue straight onto RNA extraction.

3.2. RNA Extraction

1. All of the following procedures should be carried out on ice, wearing gloves, using nuclease free pipettes and pipette tips.
2. Cool centrifuge to 4°C and warm heating block to 55–60°C.
3. Add 1 ml of Tri-Reagent to each tube containing a cell pellet from Subheading 3.1, step 7 above.
4. Lyse the cells in each sample by repeated pipetting up and down using a 1-ml pipette and tip until a homogenous solution is obtained.
5. Transfer samples into pre-labelled sterile 1.5-ml Eppendorf tubes.
6. Stand on ice for 10 min.
7. Centrifuge at 12,000×*g* for 10 min at 4°C.
8. Transfer clear supernatant to a new tube.
9. Add 0.2 ml of chloroform and mix by inversion by hand for 20 s.
10. Stand tubes on ice for 10 min.
11. Centrifuge at 12,000×*g* for 15 min at 4°C (see Note 16).
12. Transfer the top layer in each tube into a fresh tube. Do not disturb the other two layers in the tubes (see Note 17).
13. Add 0.5 ml of isopropanyl alcohol and mix by inversion by hand for 20 s.
14. Stand tubes on ice for 5 min.
15. Centrifuge at 12,000×*g* for 10 min at 4°C.
16. Discard supernatant and add 1 ml of ethanol to each pellet.
17. Briefly whirly mix, then centrifuge at 12,000×*g* for 5 min at 4°C.
18. Discard supernatant.
19. Dry samples by uncapping the tubes and leaving them upside down on a piece of blue roll for 5 min.
20. Add 20 μl of nuclease free water to each sample.
21. Resuspend pellet by heating in the pre-warmed heating block and repeated pipetting using a nuclease free tip.
22. If samples are not to be used immediately, they can be stored at –80°C until required.

3.3. Assessment of RNA Quality and Quantity

1. Adopt appropriate precautions for handling RNA (gloves, nuclease free pipettes and tips etc).
2. If necessary thaw samples on ice.
3. Turn on and warm up the spectrophotometer.
4. Pipette 97.5 µl of nuclease free water into an appropriately sized quartz cuvette.
5. Zero the spectrophotometer at 260 nm and 280 nm using the cuvette.
6. Add 2.5 µl of sample and mix using the pipette (with tip attached).
7. Measure the absorbance at 260 nm and 280 nm, making a note of the readings.
8. Discard cuvette contents. Rinse cuvette well with nuclease free water and repeat steps 4–8 until all samples have been measured.
9. Calculate the quality of each sample using the following equation:

$$\text{Quality} = A_{260} / A_{280}.$$

10. Discard any sample which gives an answer of <1.7 to the equation above.
11. Calculate the quantity of RNA in µg/µl of each sample using the following equation:

$$\text{Quantity} = (A_{260} \times 33 \times 40)/1000 \text{ (see Note 18).}$$

3.4. Reverse Transcription

1. This is protocol 1a RT with heat denaturation of the RNA (see Note 19).
2. Adopt appropriate precautions for handling RNA.
3. Pre-label one nuclease free Eppendorf tube per sample.
4. Calculate how much of each sample is required to give an amount of RNA equal to 1 µg.
5. Calculate how much nuclease free water should be added to each sample in order to give a total volume of 12 µl containing 1 µg of RNA, 2 µl of Oligo(dT) primers (see Note 20) and the nuclease free water.
6. Add these constituents into the pre-labelled tubes.
7. Centrifuge briefly on maximum speed at 4°C.
8. Heat samples for 3 min at 80°C.
9. Remove tubes to ice, centrifuge briefly on maximum speed at 4°C, and replace on ice.

10. Prepare a cocktail (see Note 21) containing the remaining ingredients. In this case, the cocktail should consist of $n+1$ (see Note 22) of the following:

    2 µl of 10× RT buffer

    4 µl of dNTP mix

    1 µl of RNase inhibitor

    1 µl of MMLV-RT.
11. Add 8 µl of the cocktail to each tube.
12. Whirly mix on a low velocity setting and briefly centrifuge the tubes at maximum speed and 4°C.
13. Incubate samples for 60 min at 50°C.
14. Incubate samples for 10 min at 92°C.
15. Either continue directly onto PCR or store samples at −20°C.

3.5. Determination of the Linear Phase of Amplification

1. Decide on an appropriate range of cycle numbers to investigate depending on your estimate of how plentiful the gene of interest is in the samples. An example range might include 25, 30, 33, 35, 38 and 40 cycles.
2. Warm the heating block to 72°C.
3. If necessary, defrost one of the cDNA samples.
4. Label seven tubes (one for each of the different numbers of cycles and one negative control).
5. Prepare a cocktail for eight samples (i.e. $n+1$ see Note 22) consisting of:

    100 µl of Promega PCR Master Mix

    4 µl of forward GPx1 primer

    4 µl of reverse GPx1 primer

    84 µl of nuclease free water (supplied with Promega PCR Master Mix).
6. Pipette 24 µl of the cocktail into each of the seven pre-labelled tubes.
7. To six of the tubes, add 1 µl of cDNA (from RT reaction, see under Subheading 3.4 above).
8. To the other tube, add 1 µl of nuclease free water as a negative control.
9. Briefly whirly mix and centrifuge the samples on maximum power.

10. Program the thermal cycler. To run 25, 30, 33, 35, 38 and 40 cycles four programmes are required. They are:

    Programme alpha
    (a) 95°C for 90 s
    (b) 94°C for 30 s
    (c) 56°C for 40 s
    (d) 72°C for 60 s
    (e) Goto (b) 24 times

    Programme beta
    (a) 94°C for 30 s
    (b) 56°C for 40 s
    (c) 72°C for 60 s
    (d) Goto (b) four times

    Programme kappa
    (a) 94°C for 30 s
    (b) 56°C for 40 s
    (c) 72°C for 60 s
    (d) Goto (b) twice

    Programme omega
    (a) 94°C for 30 s
    (b) 56°C for 40 s
    (c) 72°C for 60 s
    (d) 94°C for 30 s
    (e) 56°C for 40 s
    (f) 72°C for 60 s
11. Place tubes in thermal cycler and run programme alpha.
12. Immediately upon conclusion, remove the 25 cycles tube and place in the pre-heated 72°C heating block for 5 min, then place at 4°C. At the same time, run the remaining tubes on programme beta.
13. Immediately upon conclusion, remove the 30 cycles tube and place in the pre-heated 72°C heating block for 5 min, then place at 4°C. At the same time run the remaining tubes on programme kappa.
14. Immediately upon conclusion, remove the 33 cycles tube and place in the 72°C heating block for 5 min, then place at 4°C. At the same time, run the remaining tubes on programme omega.
15. Immediately upon conclusion, remove the 35 cycles tube and place in the 72°C heating block for 5 min, then place at 4°C. At the same time, run the remaining tubes on programme kappa.

16. Immediately upon conclusion, remove the 38 cycles tube and place in the 72°C heating block for 5 min, then place at 4°C. At the same time run the remaining tubes on programme omega.
17. Upon completion, remove the remaining tubes (40 cycles and negative control) and place in the 72°C heating block for 5 min then place at 4°C.
18. Add 2 μl of gel loading dye to 8 μl of each sample.
19. Heat to dissolve 1 g of agarose in 100 ml of 1× TAE (1%).
20. When the gel has cooled, so that it can be comfortably placed on the back of the hand, add 4 μl of ethidium bromide from stock (see Note 23).
21. Seal up the ends of the gel tray using tape (see Note 24). Pour gel and then place a comb into the tray before placing both into the electrophoresis apparatus.
22. Once the gel has set, remove comb and tape. Pour 1× TAE solution into the electrophoresis equipment using sufficient to fill the wells at each end and to give a 0.5 cm depth over the gel.
23. Load samples and run at 90 volts for approximately 45 min.
24. Visualise gel on a UV transilluminator and capture the image.
25. Use densitometry software to investigate differences between the band's density (see Fig. 1a).

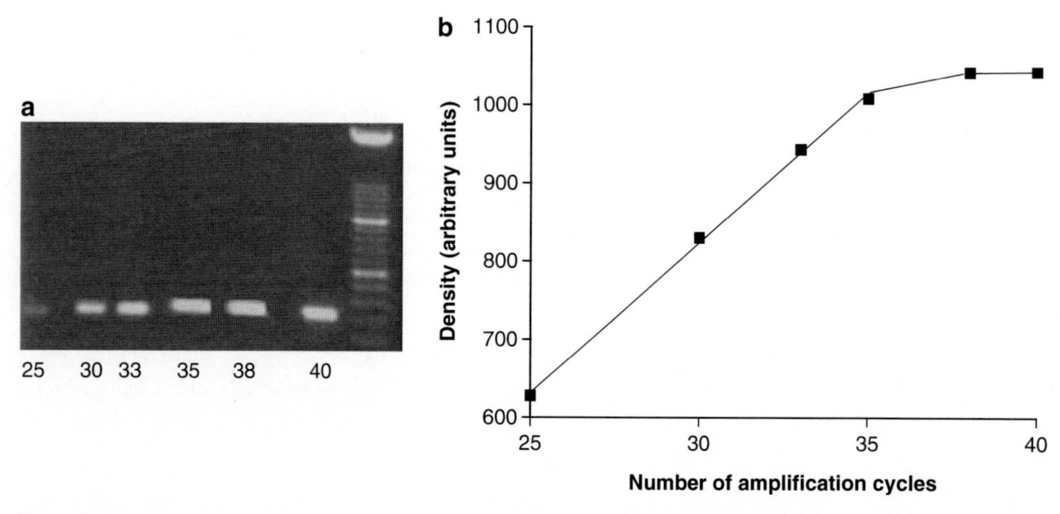

Fig. 1. Determination of the linear phase of amplification during RT-PCR of cardiomyocyte GPx1. (**a**) Photograph of a representative gel taken after RT-PCR using primers directed against GPx1 of RNA extracted from freshly isolated cardiomyocytes. Each of the samples shown underwent a different number of amplification cycles in the PCR. This number is indicated in the figures underneath the picture. (**b**) Graph of band density versus cycle number for each of the samples shown in (**a**). Least squares analysis was used to fit the best straight line to the first 4 points of the dataset as described by the equation $Y = 38.4 \times X - 326.8$ ($r^2 = 0.99$). These experiments were repeated three times with similar results.

26. If necessary, plot a graph of cycle number versus density in order to select an appropriate number of cycles within the linear phase of amplification (see Fig. 1b).

3.6. Determination of Sensitivity to Different Levels of Gene Expression

1. Prepare two 50 μl PCR reactions comprising: 25 μl of PCR Master Mix, 1 μl of forward GPx1 primer, 1 μl of reverse GPx1 primer, 2 μl or either cDNA or nuclease free water, and 21 μl of nuclease free water (see Note 25).
2. Run samples on the thermal cycler using the following programme:
   (a) 95°C for 90 s
   (b) 94°C for 30 s
   (c) 56°C for 40 s
   (d) 72°C for 60 s
   (e) Goto (b) 38 times
   (f) 72°C for 5 min
3. Add 5 μl of gel loading dye to the samples.
4. Electrophorese the entire sample following Subheading 3.5, steps 20–25 above.
5. The following procedure is the QIAquick Gel Extraction Kit Protocol using a microcentrifuge. This is described on pages 25–26 of the QIAquick Spin Handbook (see Note 26).
6. Prewarm the heating block to 50°C.
7. Use a clean scalpel to excise the band of the sample containing cDNA trying to include as little surplus gel as possible.
8. Weigh this into a clean Eppendorf tube, and add 3 volumes of solution QG (provided in the kit) to every 1 volume of gel.
9. Incubate at 50°C for 10 min (or until the gel slice has completely dissolved). To help the dissolving, whirly mix every 2–3 min.
10. Add 1 gel volume of isopropanol to the sample and whirly mix (see Note 27).
11. Place a QIAquick spin column into a collection tube (both are provided in the kit).
12. Pipette 800 μl of sample into the QIAquick column and spin at 17,900×*g* for 1 min at room temperature.
13. Discard the flow through.
14. Repeat stages 12–13 using the same QIAquick column and collection tube until the entire sample has been centrifuged.
15. Using the same QIAquick spin column and collection tube, add 500 μl of solution QG to the sample and spin at 17,900×*g* for 1 min at room temperature. Discard the flow-through.

16. Using the same QIAquick spin column and collection tube, add 750 μl of solution PE (provided in the kit) and spin at 17,900×*g* for 1 min at room temperature.
17. Discard the flow-through. Place the QIAquick spin column back into the collection tube and spin again at 17,900×*g* for 1 min at room temperature.
18. Discard the flow-through and collection tube.
19. Place the QIAquick spin column into a clean 1.5-ml Eppendorf tube. Add 50 μl of solution EB (provided in kit) and stand for 1 min at room temperature.
20. Spin sample at 17,900×*g* for 1 min at room temperature. This will transfer the DNA into the Eppendorf tube, after which the QIAquick spin column can be discarded.
21. Measure the absorption of 5 μl of the product in 100 μl of nuclease free water at 260 nm.
22. Multiply the absorption obtained by 20 and by 50 to obtain the concentration of DNA in ng/μl.
23. Work out how many copies of the PCR product are present in the sample using the average weight of a base, Avogadro's number and the expected length of the PCR product (see Note 28 for a worked example).
24. Prepare a dilution series of the sample with the highest concentration being the neat sample and the lowest sample containing 10^2 copies (see Note 29).
25. Prepare a PCR cocktail for $n+2$ (where n is the number of samples in the dilution series) samples for 25 μl reactions:

    $N+2$ times 12.5 μl of PCR Master Mix

    $N+2$ times 0.5 μl of forward GPx1 primer

    $N+2$ times 0.5 μl of reverse GPx1 primer

    $N+2$ times 10.5 μl of nuclease free water
26. Dispense the cocktail into the individual PCR tubes and add 1 μl of nuclease free water or 1 μl from the appropriate sample in the dilution series (see Note 30).
27. Run samples on the thermal cycler using the follow programme:
    (g) 95°C for 90 s
    (h) 94°C for 30 s
    (i) 56°C for 40 s
    (j) 72°C for 60 s
    (k) Goto (b) 34 times
    (l) 72°C for 5 min
28. Analyse the samples by following Subheading 3.5, steps 19–26 above.

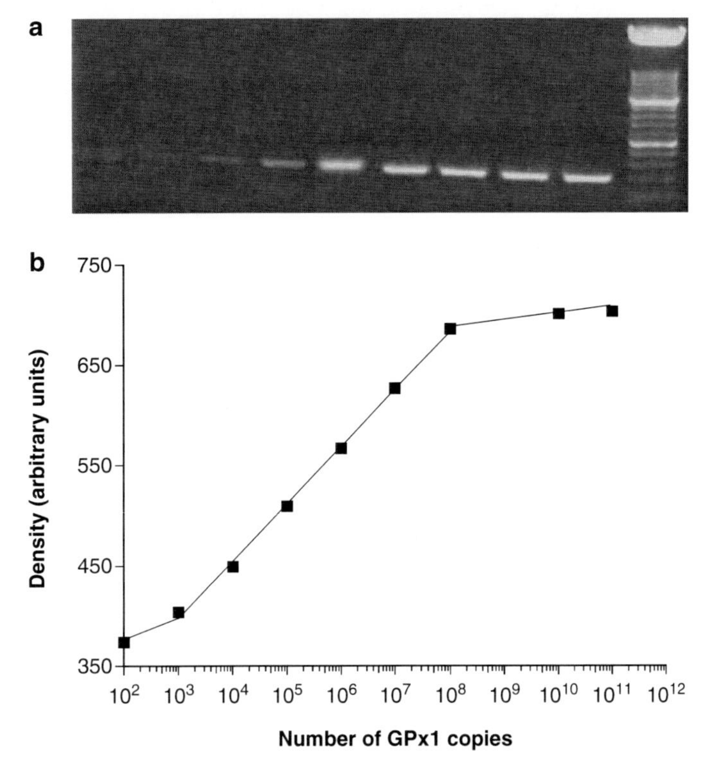

Fig. 2. Verification that differences in the level of GPx1 expression can be detected by conventional RT-PCR and scanning densitometry. (**a**) Samples containing varying amounts of cardiomyocyte GPx1 cDNA were subjected to PCR using primers directed against GPx1 followed by electrophoresis. Shown here is a representative photograph of the resultant gel. The figures underneath the picture indicate the number of GPx1 copies in the corresponding sample. (**b**) Calibration curve of band density for the samples shown in (**a**) versus the number of GPx1 copies in each sample. Least squares analysis was used to fit the data to a straight line described by the equation $Y = 24.8 \times \ln X + 225.5$ ($r^2 = 0.99$). These experiments were repeated three times with similar results.

29. Verify that there are ascending differences in the band densities for increasing concentrations in the dilution series (see Fig. 2a). If desired, plot a calibration curve of copy number versus density, which may be used as a crude means of quantifying the gene expression level in each of the cysteine-incubated and control samples (Fig. 2b).

3.7. Effect of Cysteine Incubation on GPx1 Expression

1. If necessary, defrost samples.
2. Prepare two PCR cocktails for 25 μl reactions. One cocktail will be for GPx1 and one cocktail will be for GAPDH. The cocktail should contain sufficient for $n+2$ samples (see Note 31) comprising of all the pre-prepared cDNA

from the control and cysteine-incubated samples. The cocktail should contain:

$N+2$ times 12.5 μl of PCR Master Mix

$N+2$ times 0.5 μl of forward GPx1 or GAPDH primer

$N+2$ times 0.5 μl of reverse GPx1 or GAPDH primer

$N+2$ times 10.5 μl of nuclease free water

3. Add 1 μl of the appropriate cDNA or nuclease free water (negative control) to each tube.
4. Run samples on the thermal cycler according to the following programme:

   (a) 95°C for 90 s

   (b) 94°C for 30 s

   (c) 56°C for 40 s

   (d) 72°C for 60 s

   (e) Goto (b) X number of times

   (f) 72°C for 5 min

   X should represent the number of cycles that were identified from the results from the experiments outlined under Subheading 3.5, above as falling within the linear phase of amplification.
5. Analyze the samples by following Subheading 3.5, steps 19–26 above (see Fig. 3a, b).

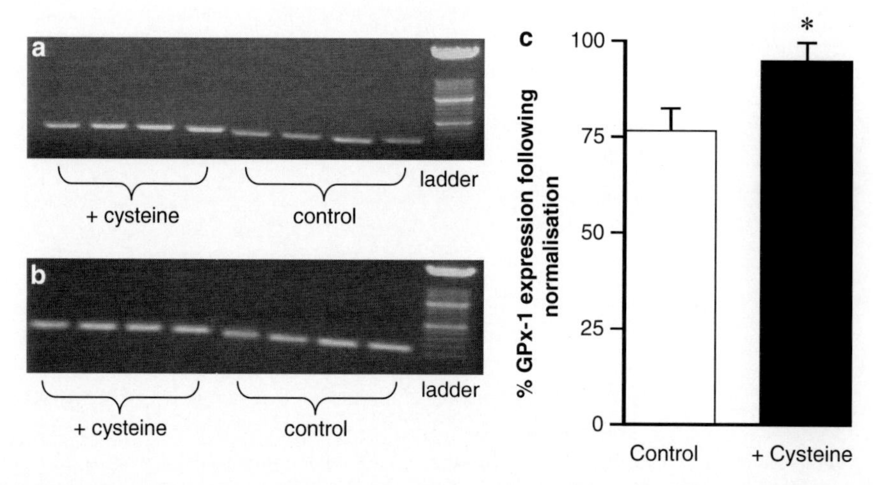

Fig. 3. Comparative quantitative RT-PCR investigation of the effect of cysteine incubation on GPx-1 expression in cardiomyocytes. (**a**) Expression of GPx1 in freshly isolated control and cysteine-incubated cardiomyocytes. This is a representative photograph of an agarose gel taken following RNA extraction, RT PCR and electrophoresis. Four separate cysteine-incubated and control samples are shown. (**b**) Expression of GAPDH in the same four cysteine-incubated and control samples as were used in (**a**). The same techniques were employed as those shown in (**a**). These experiments were repeated three times with similar results. (**c**) Comparison of the mean level of GPx-1 expression in the cysteine-incubated and control samples as determined by densitometry and with normalization for each samples' level of GAPDH expression. *$p<0.05$ versus control. Data shown in this bar graph are means ± S.E. of $n=4$.

6. Using the densitometry results, normalize the expression of GPx1 in each sample according to the same sample's GAPDH expression.
7. Sum all results and perform statistical analyses to investigate whether cysteine incubation affects GPx1 expression (Fig. 3c).

4. Notes

1. It has been assumed that the reader's main interest for selecting this chapter is to learn about comparative quantitative RT-PCR rather than the isolation of rat ventricular cardiomyocytes. Therefore, a description of the cell isolation procedure has not been included. Should the reader require these instructions, however, they will find a good description of the materials and methods involved in the following two references (16, 17).
2. This protocol is not constrained to freshly isolated rat ventricular cardiomyocytes, a suspension of cells from any species or tissue origin can be used.
3. Cysteine is prone to auto-oxidation (18). Therefore, to improve the stability of cysteine containing solutions, add 0.1 mM dithiothreitol. Alternatively, prepare a fresh cysteine stock solution for each use.
4. The final 5 mm of the tip is removed to prevent damaging the cardiomyocytes by passing them through a very small aperture.
5. Tri-Reagent contains phenol which is toxic and guanidium thiocyanate which is an irritant. Therefore, it should be handled with appropriate precautions and used only in a functional fume-cupboard.
6. Chloroform is an irritant and toxic. Therefore handle as specified under Note 5.
7. It is best to set aside stocks of these solutions which are dedicated for use in RNA work. This helps to reduce the likelihood of nuclease contamination.
8. Nowadays, commercially available Reverse Transcription and PCR kits supply an excess volume of nuclease free water. Water for molecular biology purposes can also be purchased from well-known suppliers such as Sigma. It is usually easier to use one of these kit's excess rather than preparing your own nuclease free water.
9. This is one of several reverse-transcription kits which are offered by various suppliers. Any kit which yields reliable

reproducible results can be used. If an alternative kit is selected, be sure to follow that kit's instructions since these may differ slightly from those described here.

10. The website http://www-bimas.cit.nih.gov/cgi-bin/cards provides a very good resource for locating useful general information including database accession numbers for known genes. The gene accession numbers can then be entered into the entrez pubmed website at http://www.NCBI.nlm.nih.gov/sites/entrez?db=nuccore in order to find the full mRNA sequences of the genes of interest. Both of these websites are free to use.
11. Primers were designed using the website, http://FRODO.wi.mit.edu/cgi-brn/primer3. This is a free resource.
12. *Rattus norvegicus* cellular glutathione peroxidase, alias GPx1, EC number: 1.11.1.9, NCBI accession number: NM_030826.3, expected product size after PCR: 221 base pairs.
13. *R. norvegicus* glyceraldehyde-3-phosphate, alias GAPDH, EC number: 1.2.1.12, NCBI accession number: NM_017008.3, expected product size after PCR: 156 base pairs.
14. Gauze and adhesive tape are used in preference to simply capping the universal tubes in order to reduce the effects of hypoxia. It is however important to have a covering for the tubes since this prevents contamination from undesirable atmospheric particles.
15. The force of the shake should be sufficient to gently mix the tube contents without affecting cell morphology and viability.
16. This centrifugation step should separate the sample into three phases: a red organic phase (containing protein), an interphase (containing DNA) and a colourless upper aqueous phase (containing RNA).
17. It is better to forsake a small amount of the upper colourless layer than to incorporate any of the other two layers.
18. In this equation, "40" is the dilution factor. Therefore, if it is necessary to use different amounts of nuclease free water and sample to calculate the RNA quality and quantity, this figure should be adjusted accordingly.
19. A good explanation of the rationale behind the initial heat denaturation of the RNA template step is provided on page 6 of the instruction manual that Ambion provide with their RETROscript® kit.
20. The choice to use oligo(dT) primers is personal. A good discussion of the relative merits of random sequence oligonucleotide primers versus oligo(dT) primers is provided on page 7 of the Ambion RETROscript® instruction manual.

21. Preparing and using cocktails is a good strategy to help reduce pipetting inaccuracies.
22. Using $n+1$, when there are many samples, helps counteract any problems over the volume remaining versus that required upon reaching the final sample. This is also a reflection of the possible pipetting errors that occur when handling small volumes.
23. Ethidium bromide is carcinogenic. Adopt appropriate precautions when handling it and when disposing of the used agarose gel.
24. In order to avoid the gel leaking, use two layers of tape and ensure that the tape comfortably overlaps the bottom and sides of the gel tray.
25. The following steps describe how the number of copies was calculated for a sample which has an absorbance of 0.0364:
    (a) $0.0364 \times 20 \times 50 = 36.4$ ng/μl
    (b) Average weight of a base = 221
    (c) Expected size of the PCR product (using the primers described here) = 221
    (d) ∴ molecular weight of PCR product is $660 \times 221 = 145,860$ g
    (e) ∴ 145,680 g = 6.02×10^{23} copies (Avogadro's number)
    (f) ∴ 1 g = $6.02 \times 10^{23}/145,860 = 4.13 \times 10^{18}$
    (g) ∴ 1 ng = 4.13×10^{9} copies
    (h) ∴ 36.4 ng = 1.5×10^{11} copies
    (i) The dilution series was prepared with samples from 1.5×10^{11} copies to 1.5×10^{2} copies.
26. In order to economise space whilst at the same time providing a full description of all the techniques, only the bare essentials of how to extract DNA from the gel is outlined. A more detailed description of the theory involved is provided in the QIAquick Spin Handbook.
27. The combined volume of gel, buffer QG and isopropanol may necessitate transfer to a larger volume tube. If this is the case, ensure that the new tube can be accommodated in the centrifuge.
28. Due to the sensitivity of the PCR reaction and the ever present problem of contamination, if it is possible, prepare the dilution series in a separate area of the laboratory to that used to set up PCR reactions and use separate pipettes and a separate aliquot of nuclease free water which is disposed of after this use. If these precautions are not possible, be sure to thoroughly wash the area and pipettes with DNA Zap or adopt another DNA degrading approach.

29. When setting up this PCR, be sure to work from the lowest to the highest concentration in the dilution series and to treat the bench area and pipettes afterwards with DNA Zap.
30. In theory, it is possible to extract the DNA from the PCR sample from the highest number of cycles and use this to prepare a dilution series. In practise, it is more advisable to run separate PCR experiments since it is necessary to capture a good image of the gel for each of the different cycle numbers and prolonged exposure to UV can damage the DNA.
31. $N+2$ in order to include sufficient for inadvertent pipetting mistakes and to include a negative control.

References

1. Jahoor F, Jackson A, Gazzard B, Philips G, Sharpstone D, Frazer ME, Heird W (1999) Erythrocyte glutathione deficiency in symptom-free HIV infection is associated with decreased synthesis rate. Am J Physiol 276:E205–E211
2. Cooper AJL (1983) Biochemistry of sulfur-containing amino acids. Ann Rev Biochem 52:187–222
3. Chua BHL, Giger KE, Kleinhans BJ, Robishaw JD, Morgan HE (1984) Differential effects of cysteine on protein and coenzyme A synthesis in rat heart. Am J Physiol 247:C99–C106
4. Gabrys J, Konecki J, Shani J, Durczok A, Bielacyzc G, Kosteczko A, Szewczyk H, Brus B (2003) Proteinous amino acids in muscle cytosol of rats' heart after exercise and hypoxia. Receptors Channels 9:301–307
5. Pisarenko OI (1996) Mechanisms of myocardial protection by amino acids: facts and hypotheses. Clin Exp Pharmacol Physiol 23: 627–633
6. Burns AH, Reddy WJ (1978) Amino acid stimulation of oxygen and substrate utilization by cardiac myocytes. Am J Physiol 235: E461–E466
7. Baños G, Daniel PM, Moorhouse SR, Pratt OE, Wilson PA (1978) The influx of amino acids into the heart of the rat. J Physiol 280: 471–486
8. Griffith OW (1999) Biologic and pharmacologic regulation of mammalian glutathione synthesis. Free Radic Biol Med 27:922–935
9. Dhalla NS, Elmoselhi AB, Hata T, Makino N (2000) Status of myocardial antioxidants in ischaemia-reperfusion injury. Cardiovasc Res 47:446–456
10. Ide T, Tsutsui H, Kinugawa S, Suematsu N, Hayashadini S, Ichikawa K, Utsumi H, Machida Y, Egashira K, Takeshita A (2000) Direct evidence for increased hydroxyl radicals originating from superoxide in the failing myocardium. Circ Res 86:152–157
11. Slezak J, Tribulova N, Pristacova J, Uhrik B, Thomas T, Knaper N, Kaul N, Singal PK (1995) Hydrogen peroxide changes in ischemic and reperfused heart. Am J Pathol 147:772–781
12. Tang LD, Sun JZ, Wu K, Sun CP, Tang ZM (1991) Beneficial effects of N-acetylcysteine and cysteine in stunned myocardium in perfused rat heart. Br J Pharmacol 102:601–606
13. Shackebaei D, King N, Shukla B, Suleiman MS (2005) Mechanisms underlying the cardioprotective effect of L-cysteine. Mol Cell Biochem 277:27–31
14. Wang AM, Doyle MV, Mark DF (1989) Quantitation of mRNA by the polymerase chain reaction. Proc Natl Acad Sci USA 86:9717–9721
15. Schmittgen TD, Zakrajsek BA, Mills AG, Gorn V, Singer MJ, Reed MW (2000) Quantitative reverse transcription-polymerase chain reaction to study mRNA decay: comparison of end-point and real-time methods. Anal Biochem 285:194–204
16. King N, McGivan JD, Griffiths EJ, Suleiman MS (2003) Glutamate loading protects freshly isolated and perfused adult cardiomyocytes against intracellular ROS generation. J Mol Cell Cardiol 35:975–984
17. Williams H, King N, Griffiths EJ, Suleiman MS (2001) Glutamate-loading stimulates metabolic flux and improves cell recovery following chemical hypoxia in isolated cardiomyocytes. J Mol Cell Cardiol 33:2109–2119
18. Saez G, Thornalley PJ, Hill HAO, Hems R, Bannister JV (1982) The production of free radicals during the autoxidation of cysteine and their effect on isolated rat hepatocytes. Biochim Biophys Acta 719:24–31

Chapter 15

The Renaissance of Competitive PCR as an Accurate Tool for Precise Nucleic Acid Quantification

Lorena Zentilin and Mauro Giacca

Abstract

Here, we report a detailed procedure for the exact quantification of minute amounts of nucleic acids by competitive PCR. This technique entails the co-amplification of a target DNA or cDNA in a biological sample together with a known quantity of a target-specific standard, the competitor, which is added exogenously to the sample and is almost identical to the DNA region to be amplified. Competitive PCR offers the advantage to render the PCR reaction independent of the number of amplification cycles, since any intra-assay variation has the same effect on both target and competitor. Since the final ratio between target and competitor amplification products exactly reflects the initial ratio between the two species, and being the amount of added competitor known, competitive PCR allows the determination of the exact number of molecules of target, with an accuracy that is still unsurpassed by any other alternative procedure, including real-time PCR. The protocols described in this chapter cover most of the possible applications of competitive PCR, including the quantification of an mRNA transcript and the simultaneous determination of multiple targets.

Key words: Competitive PCR, Quantification, Amplification, DNA competitor, RNA competitor, Multicompetitor

1. Introduction

The exact quantification of minute amounts of nucleic acids in biological samples is a fundamental requisite for both research and diagnosis. Although conventional PCR is a powerful technique for the amplification of low-abundance nucleic acids in complex biological samples, it has limited accuracy when used for their precise quantification. Indeed, the amount of a final PCR amplification product rarely reflects the actual concentration of the original target in the sample, since a linear relationship between input target and amplification product exists only during

Nicola King (ed.), *RT-PCR Protocols: Second Edition*, Methods in Molecular Biology, vol. 630,
DOI 10.1007/978-1-60761-629-0_15, © Springer Science+Business Media, LLC 2010

the initial, exponential phase of the PCR reaction. Moreover, the outcome of PCR is deeply influenced by a number of variables that are difficult to control, such as the relative abundance of the starting material and the existence of random tube-to-tube variations in the first phases of the amplification process.

Over the last years, several laboratories have developed different quantitative PCR methods to by-pass the limitations of conventional PCR. In particular, real-time PCR now finds widespread application (1), especially because this technique, which is based on the kinetic quantification of PCR products during amplification itself, does not require the often tedious, post-PCR analysis of the reaction fragments. Real-time PCR, however, is not immune from several of the same problems that hamper quantification by conventional PCR, such as the influence of absolute initial template concentration on final amplification yield and the occurrence of random tube-to-tube variations (2, 3). Such limitations become particularly remarkable when the sample to be amplified contains highly diluted target sequences or when dealing with RNA templates that need to be reverse-transcribed into cDNA. The latter step, indeed, introduces further variability due to the high dependence of the reverse-transcription reaction on input target RNA concentration and reaction conditions.

Competitive PCR was originally developed at the end of the 1980s as an accurate method for the absolute quantification of nucleic acids (2, 4), and still shows properties of accuracy and reproducibility unsurpassed by any other PCR-based method. The founding principle of the technique involves the amplification of the sequence under investigation in the presence of an exogenously introduced competitor molecule, which is added to the reaction in known quantity. This competitor is virtually identical to the target sequence, apart from small mutations or length variations that allow the two molecular species (target and competitor) to be easily distinguished by post-PCR analysis (2, 5–7). Since both target and competitor are affected by the same predictable and unpredictable variables that affect PCR, the ratio between the two species remains constant throughout the amplification process. As the amount of added competitor is known, the quantity of unknown target sequence in the sample can be easily calculated from the ratio between the two amplification products at the end of the reaction. An outline of the competitive PCR approach is presented in Fig. 1. In case quantification of an RNA molecule is sought, the competitor can consist of an RNA molecule, thus allowing the additional control of the reverse-transcription step (competitive RT-PCR).

The assembly of competitive PCR and RT-PCR experiments might appear cumbersome and time consuming, since it requires synthetic competitor construction, assembly of multiple reactions for a single determination and post-PCR analysis of products.

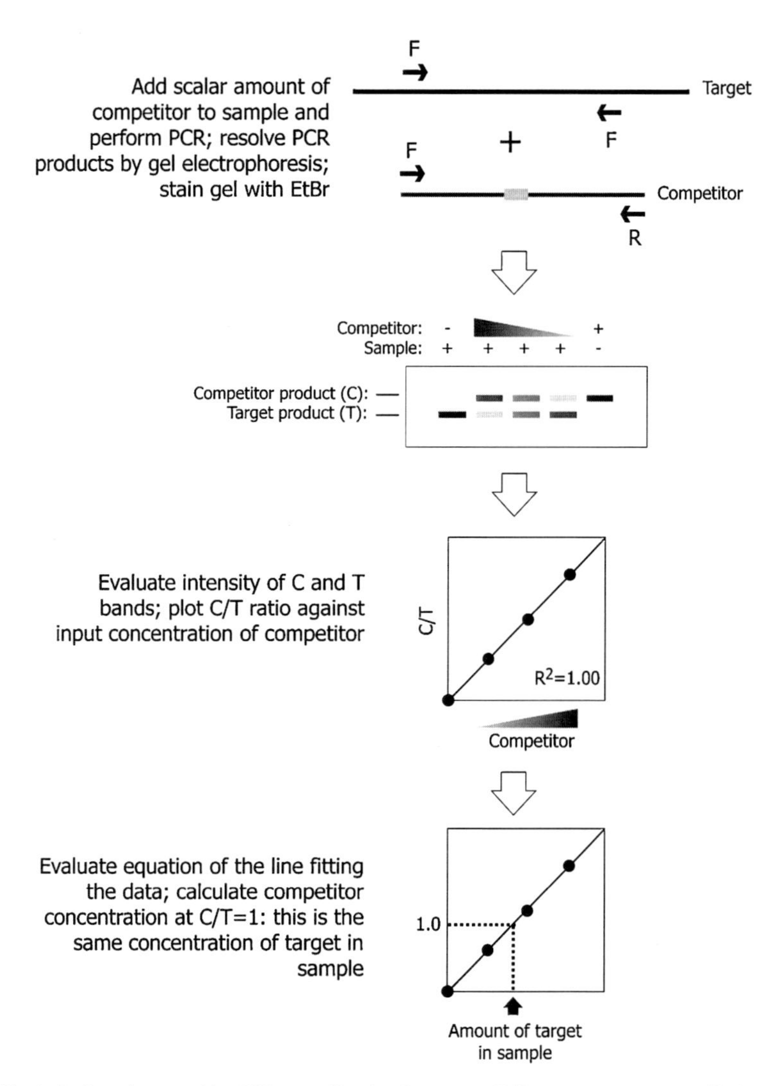

Fig. 1. Outline of competitive PCR quantification. In a target DNA sequence to be quantified, two primers are selected (*F and R*), and a competitor DNA fragment is constructed, which is also amplified by primers F and R but contains an extra nucleotide sequence in the middle (*in gray*). A fixed amount of sample (containing the target DNA) is mixed with scalar amounts of competitor, and PCR amplification is run (of any number of cycles). At the end of the amplification, PCR products are resolved by gel electrophoresis and stained with ethidium bromide. The amplification products corresponding to competitor (C) and target (T) are quantified, and the C/T ratio is plotted against the amount of input competitor. The experimental points are fitted by a line. According to the equation describing this line, the number of competitor molecules corresponding to a C/T ratio = 1 is evaluated. This number is also the number of molecules of target initially present in the reaction

However, these drawbacks are largely outweighed by the extreme sensitivity (<5 copies of input target) and accuracy (<2-fold change in target concentration easily detectable) of the procedure, which is independent of the possible presence of nonspecific PCR products and from the total yield the reaction (8–11). These features still render competitive PCR more accurate than real-time PCR in experimental settings, in which the detection of less than two to fourfold differences in nucleic acid abundance in complex biological samples is essential. Among such experimental situations are the mapping of DNA replication origins in mammalian genomes (12), the assessment of oncogene amplification in tumor samples (10, 13), the analysis of chromatin immunoprecipitation (ChIP) experiments when studying proteins loosely attached to DNA (14), the diagnosis of chromosomal aneuploidies (15), and the detection of the loss of heterozygosity in human tumor samples (16).

In this chapter, we present an exhaustive set of methods for the setting up of competitive PCR and RT-PCR reactions as well as for the construction of different types of synthetic competitor standards.

2. Materials

2.1. Equipment for RT/PCR

1. Pipettes and filter barrier pipette tips
2. Thermal cycler
3. Power supply
4. Vertical and/or horizontal electrophoresis apparatus
5. UV transilluminator and image acquisition and analysis system.

2.2. Materials for RT and PCR

1. 2′-deoxynucleoside-5′-triphosphates (dNTPs) mix
2. PCR 10× buffer, including $MgCl_2$ solution
3. Any efficient DNA polymerase, for example, AmpliTaq DNA Polymerase (Applied Biosystem)
4. Tth polymerase
5. Superscript III (Invitrogen) or eAMVTM (Sigma)
6. DuraScribe T7 transcription kit (Epicentre) or MEGA shortscript T7 kit (Ambion)
7. Oligonucleotides
8. Random hexamers or oligo-dT
9. (α-^{32}P)-dCTP; (α-^{32}P)-UTP
10. TBE buffer (10 mM Tris–HCl, 150 mM NaCl pH 7.5)
11. pGEM-T Easy Vector (Promega), GeneJET PCR cloning kit (Fermentas) or Qiagen PCR cloning kit
12. RNase-free DNase I (PCR-grade DNase I)
13. PCR product purification kit (e.g. Millipore; 100,000 MW cutoff)

14. Polyacrylamide (e.g. acrylamide 40% (w/v) stock solution) or agarose
15. 3% NuSieve agarose/1% SeaKem agarose
16. Ethidium bromide
17. TE buffer (10 mM Tris–HCl pH 8.0, 1 mM EDTA)
18. 3 M sodium acetate
19. Phenol:chloroform:isoamyl alcohol (25:24:1)
20. Ethanol
21. DNase I
22. 0.5 M EDTA
23. RNase inhibitor.

3. Methods

3.1. Competitor Construction

Five different types of competitor molecules can be designed that satisfy the requirements of competitive PCR (see Note 1) and that can be used in different major applications:

1. DNA competitor with a small insertion
2. DNA competitor with a small deletion, for the quantification of a single DNA target sequence;
3. RNA competitor with a small insertion
4. RNA competitor with a small deletion, for the quantification of a single RNA target sequence;
5. DNA (or RNA) multicompetitor, for the simultaneous quantification of multiple target sequences.

3.1.1. DNA Competitor with a Small Insertion (Fig. 2a)

1. Choose two external primers F and R in order to amplify a segment of 150–300 bp of the target DNA sequence (see Note 2). Within this amplicon, choose two other internal primers (x and y) which are contiguous sequences on opposite strands. Ensure that these internal primers bear 5′ tails of ~20 nucleotides (nt) unrelated to the target to be amplified and are complementary to each other.
2. Set up two separate standard PCR reactions in 50 μl of final volume each using one of the two external primers in combination with the internal primer of the opposite strand.
3. Resolve the amplification products (typically 10 μl out of the total amplification reaction) by electrophoresis on a polyacrylamide (usually 8%) or agarose gel (2%) appropriate for the expected PCR product size.

4. After visualization on a low-power UV transilluminator, excise the DNA bands from the gel and soak them in a small volume of TE (50 μl) in two separate tubes (see Note 3). Elute the DNA from the gel fragment at room temperature (22–24°C) for a period of time ranging from 1 h to overnight.
5. Anneal the two amplification products at the complementary 20 bp-sequence to obtain a heteroduplex molecule. For this purpose, add 1–5 μl of each elution to a standard 50 μl of amplification mixture but without the addition of oligonucleotides primers. Use the following cycling conditions for heteroduplex annealing and extension:

   Denaturation: 95°C 1 min

   Slow annealing: 95–50°C in 10 min

   Annealing: 50°C 2 min

   Elongation: 72°C 5 min
6. Amplify the annealed products by adding the appropriate amount of the two external primers (1 μM final concentration) to the reaction. A typical PCR cycling profile is as follows (see Notes 4 and 5):

   Denaturation: 95°C 1 min(1 cycle)

   Denaturation: 95°C 30 s

   Annealing: 55–60°C 30 s

   Elongation: 72°C 30 s

   (repeat 30–50 cycles)

   Elongation: 72°C 5 min

   (1 cycle)

3.1.2. DNA Competitor with 20 bp Deletion (Fig. 2b)

1. The synthesis of this competitor standard requires only three primers: one forward (F) and one reverse (R) 20 nt-long external primers, as in conventional PCR, and one 40 nt-long, internal primer in reverse orientation (primer yR). The 20 nt at the 3′ end of the internal primer corresponds to the template sequence at 20 nt downstream of the external reverse primer, while the 20 nt at the 5′ end are identical in sequence to the external reverse oligonucleotide.
2. Set up an amplification reaction as described before using primers R and yR.
3. Resolve the PCR product and purify it as described above in Subheading 3.1.1, steps 3–4, and PCR reamplify it with primers F-R to obtain the final competitor with a 20 nt deletion.

3.1.3. RNA Competitor with 20 bp Insertion or Deletion (Fig. 2c)

This competitor standard is useful for the exact quantification of RNA.

1. Reverse transcribe the target RNA using reverse transcriptase Superscript III or an RT kit and either a specific primer of opposite polarity, random primers, or oligo-dT.
2. Using a suitable primer pair (F and R in Fig. 2a), PCR-amplify the obtained cDNA, for example following the conditions described under Subheading 3.1.1 (see Note 6).
3. Following the methods described under Subheadings 3.1.1 and 3.1.2, respectively, construct a DNA competitor containing a 20 bp insertion or a 20 bp deletion.
4. Transcribe in vitro the DNA competitor fragment using an external forward primer that includes 20 nt identical to the target sequence at its 3′ end and a 5′ end tail corresponding to the recognition site for T7 RNA polymerase. Assemble the reaction starting from 50 ng of DNA template and using a T7 RNA polymerase-based in vitro transcription kit, according to the manufacturer's instructions; briefly, assemble the reaction in an RNase-free microfuge tube at room temperature by adding, in order, the T7 10× reaction buffer, the 4 NPTs (ATP, CTP, GTP and UTP), the labeled nucleotide, the template DNA, and the T7 enzyme. For convenience, the 4 NPTs can be premixed. Also add to the reaction mixture 2 μl of (α-^{32}P)-UTP. Typically, the reaction can be performed in 20 μl, but the volume can be scaled as needed.
5. Remove the DNA template from the reaction by RNase-free DNase I digestion; add 1 μl of DNase I (1 U/μl) and 2 μl of 10× DNase Reaction buffer (provided by the producer together with the enzyme) in 20 μl final volume and incubate at 37°C for 15 min. Stop the reaction by the addition of 115 μl of nuclease-free water and 15 μl of 3 M sodium acetate followed by the addition of an equal volume of Phenol/Chloroform/Isoamyl Alcohol. Transfer the aqueous phase into a new tube and precipitate the RNA by adding 2 volumes of ethanol. Wash the pellet obtained after centrifugation with 500 μl of 70% ethanol, air-dry and dissolve it in 50 μl of TE-buffer. Alternatively, elute the RNA competitor fragment after resolution by denaturing gel electrophoresis (for example, using a 8 M urea, denaturing polyacrylamide gel) (see Note 7).

3.1.4. DNA Multicompetitor for the Simultaneous Measurement of Several Targets (Fig. 2d)

This competitor is useful when multiple DNA or RNA targets will be quantified simultaneously.

1. Construct a conventional competitor molecule following one of the procedures described previously (see Subheadings 3.1.1 and 3.1.2) taking one of the target sequences to be quantified as a reference.

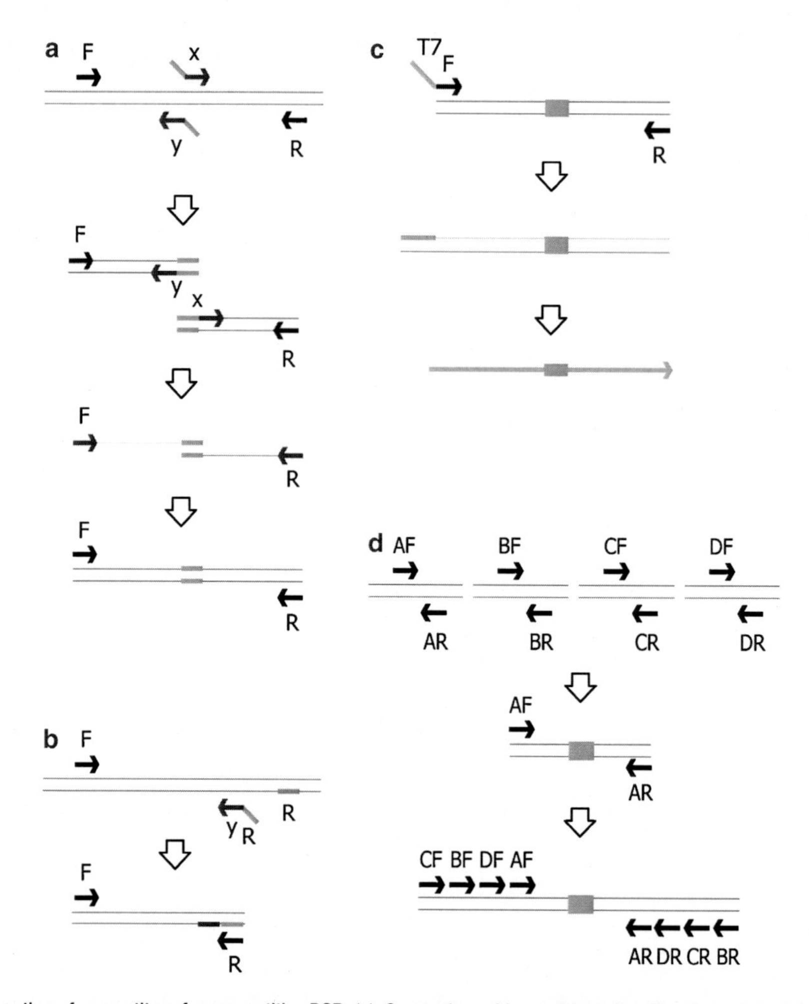

Fig. 2. Construction of competitors for competitive PCR. (**a**) *Competitor with small insertion.* The two external primers F and R amplify a target DNA segment to be quantified. Within this target amplicon, the sequences of the internal primers x and y correspond to contiguous sequences on opposite strands. In addition, the two internal primers bear 5′ tails containing two ~20 nt sequences which are unrelated to the target to be amplified and complementary to each other (*in gray*). Two amplifications are then carried out separately, using primer pairs F-y and x-R. At the end of the reaction, small aliquots of the two reaction products are mixed, DNA is denatured and left to reanneal by virtue of the two complementary tails. Finally, the annealed product is amplified using the external primers F and R. (**b**) *Competitor with small deletion.* The construction of this competitor requires three primers: the first one is a forward primer of ~20 nt corresponding to a conventional PCR oligonucleotide (*primer F*). The reverse primer (*primer yR*) is 40 nt in length, of which the 20 nt at its 3′ end (*R*) correspond to the template sequence 20 nucleotides downstream of y. The amplification product obtained with primers F-yR is re-amplified with primers F-R to obtain the final competitor with a 20 nt deletion. (**c**) *RNA competitor.* A DNA competitor obtained as shown above. A new primer is then designed, containing its 3′ end 20 nucleotides identical to the target sequence plus a 5′ end tail containing the recognition site for T7 RNA polymerase (primer T7F). The competitor is re-amplified using primers T7F-R. To obtain the competitor RNA, the resulting amplification product is transcribed in vitro using T7 RNA polymerase. (**d**) *Multicompetitor for quantification of multiple target sequences.* The example reports four different DNA target sequences, amplified by primer pairs AF-AR, BF-BR, CF-CR and DF-DR. To obtain a multi-competitor for all these four sequences, a conventional competitor for one of the regions is first obtained as outlined above. Then, two long oligonucleotides are synthesized: the forward long oligonucleotide contains all the forward primers arranged head-to-tail, while the reverse long oligonucleotide includes all the corresponding reverse primers. The two long primers are used to amplify the competitor to obtain the final multicompetitor. The relative position of each individual primer pair in the multicompetitor is chosen in order to obtain PCR products which are similar in size to the respective target amplification products, still being the two species (competitor and target amplification products) separable by gel electrophoresis. For example, if two PCR products of 200 and 220 bp are expected from the amplification of the template using two different primer pairs, the same pairs should ideally generate, when amplification is carried out from the multicompetitor, fragments of 180 or 220 bp in the first case and 200 or 240 in the second case. This requirement can be fulfilled by accurately designing the relative position of the primers on the multicompetitor

2. Subsequently, amplify this competitor using two long oligonucleotides synthesized in order to contain, arranged head-to-tail, all the forward and reverse set of primers respectively, as shown in Fig. 2d, to obtain the final multicompetitor (see Note 8).

3.2. Quantification and Cloning of the Competitor Molecule

1. Once synthesized, purify the competitor fragment from the reaction buffer, excess of primers, nucleotides, and enzymes. To this end, it is convenient to use a commercial PCR product purification kit or phenol/chloroform extraction and ethanol precipitation. When unspecific PCR products are present, it is advisable to purify the competitor molecule by directly excising the relevant band from agarose or polyacrylamide gel after electrophoresis.
2. It is essential to determine the absolute number of molecules present in the competitor preparation. The easiest way to determine the competitor concentration is to clone the obtained PCR fragment into a plasmid vector, taking advantage of the property of Taq polymerase, which leaves single 3′-dA overhangs on its reaction products. Many commercial, ready-to-use linear vectors with a 3′ terminal "T" or "U" at both ends are available from various companies. Then, purify the recombinant plasmid and determine its concentration by spectrophotometry (17, 18).
3. An alternative, however, more time consuming procedure, entails re-amplification of a diluted amount of the competitor fragment (e.g. 1 μl of a 1×10^4 fold dilution) in a standard 50 μl PCR reaction in the presence of the four dNTPs (200 nM each) plus 0.2 μl (0.66 pmol) of (α-^{32}P)-dCTP.
4. Resolve the amplification product on an 8% polyacrylamide gel, and then elute the radioactive band in 100 μl of TE. Count the radioactivity present in a small amount of the elution mixture (e.g. 5 μl); estimate the concentration of the competitor from the final specific activity of dCTP and the number of nucleotides incorporated into the competitor fragment.

3.3. Competitive PCR

3.3.1. Competitive PCR to Quantify a DNA Target

1. For the exact quantification of a DNA target, mix a fixed amount of the template with scalar amounts of the competitor DNA. Typically, use at least three scalar concentrations of competitor for quantification, even though, in principle, a single amount would be sufficient (see Note 9). After adding all the other components of the PCR mix, run a standard amplification reaction according to the cycling profile determined for competitor amplification.
2. Resolve the PCR products by gel electrophoresis and stain the gel with ethidium bromide. The two bands, target, and competitor, are easily distinguishable by their different electrophoretic mobility.

3. Determine the relative intensity of each amplification product by densitometric scanning. Alternatively, quantify the two bands on a digital photograph of the stained gel using a dedicated software (see Note 10).
4. Obtain the initial amount of the target DNA added to the amplification reaction by plotting the ratio between the amplification products (competitor/target) against input competitor concentration (Fig. 1). Since the experimental points fit a linear function, the initial amount of target DNA will correspond to the copies of competitor that give a 1:1 amplification product ratio.

3.3.2. Competitive RT-PCR

The extracted RNA to be quantified is first reverse transcribed using reverse transcriptase (RT).

1. Treat the RNA template, if needed, with DNase I to digest contaminating DNA. Treat 1–3 μg of RNA with 1 U of DNase I for 15 min at room temperature in a total 20 μl of reaction volume. Stop the reaction by adding 2 μl of 0.5 M EDTA, followed by incubation at 65°C for 15 min (see Note 11).
2. Assemble an RT Master Mix1 to include for each sample: 1–200 ng of DNase I-digested RNA (max 5 μl), scalar amounts of the RNA competitor (max 5 μl), random hexamers (50–250 ng) or oligo-dT (500 μg) or gene specific primers (2 pmol), dNTP mix (10 mM) and nuclease-free dH_2O up to 12 μl. Incubate this reaction mixture at 65°C for 5 min and then quickly chill it on ice (see Note 12).
3. In the meantime, prepare an RT Master Mix2 by combining 4 μl of 5× buffer, 2 μl of DTT 0.1 M, 1 μl of the RT enzyme, and 1 μl of an RNase inhibitor. Add 8 μl of Mix2 to Mix1 to obtain a final volume of 20 μl and incubate the reaction:

   10 min at 25°C (only for random hexamers)

   50 min at 37°C

   15 min at 70°C
4. Dilute the reverse transcribed template in order to use no more than 1/10 of the final reaction volume for the subsequent PCR amplifications. The use of larger amounts of first-strand reaction may be inhibitory and decrease the yield of the amplification (see Note 13).
5. Carry out the subsequent steps of the competitive RT-PCR reaction exactly as described above (see Note 14). Resolve the amplification products by gel electrophoresis and analyze the gel as described in Subheading 3.3.1 (see Note 15, for troubleshooting advice).

3.3.3. Example of Competitive PCR for the Quantification of Immunoprecipitated Genomic DNA in a Chromatin Immunoprecipitation (ChIP Experiment)

1. The purpose of this experiment was to verify whether the human CDC6 protein binds the Lamin B2 origin of DNA replication by ChIP, as described by Todorovic et al. (19). Briefly, human cells were treated with formaldehyde to cross-link protein to DNA, followed by chromatin sonication to obtain DNA fragments with an average length of ~500 nt with covalently attached proteins. An antibody against CDC6 was then used to immunoprecipitate CDC6-bound DNA. After immunoprecipitation, crosslinks were reversed and the amount of immunoprecipitated DNA quantified by competitive PCR (Fig. 3a).

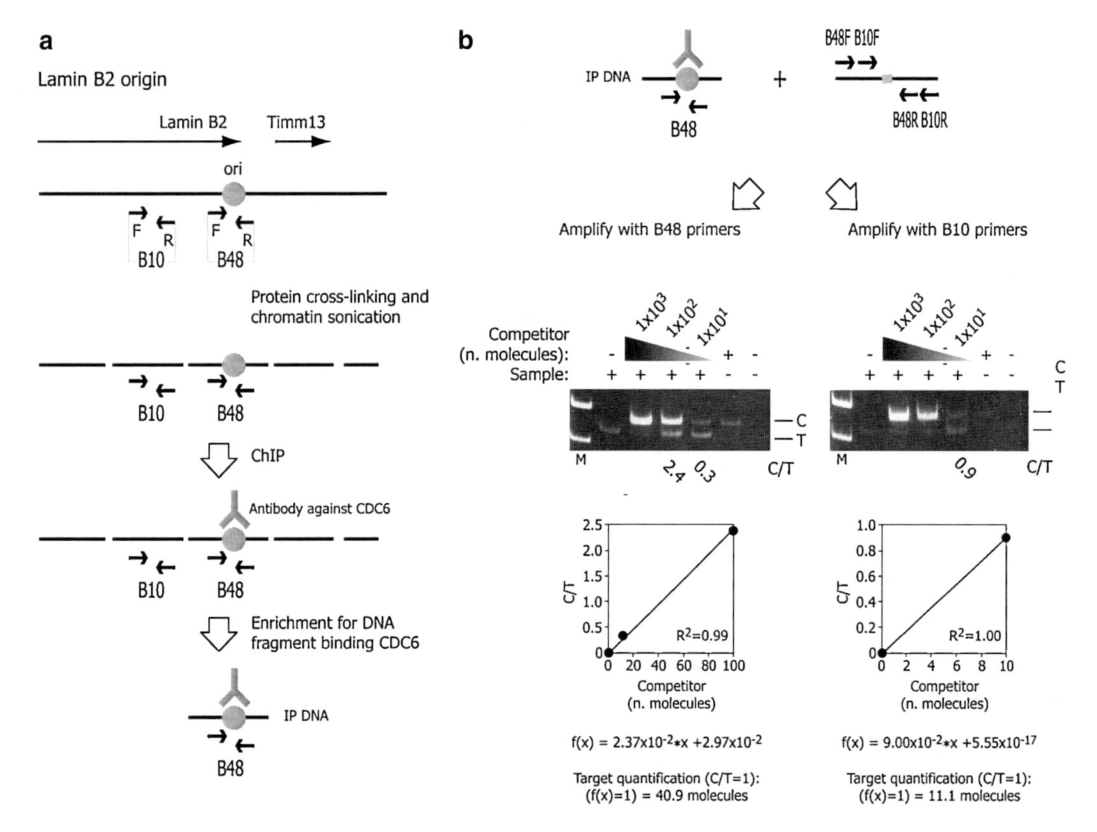

Fig. 3. Example of a competitive PCR experimnet for the quantification of genomic DNA immunoprecipitated in a Chromatin Immunoprecipitation (ChIP) experiment. (**a**) Flow chart of a ChIP experiment to detect binding of the CDC6 protein to the human lamin B2 DNA replication origin. The lamin B2 region is schematically represented along with the indication of the location of the primer pairs (B10 and B48) used for PCR quantification; the position of the origin is shown by a grey dot. Cells were treated with formaldehyde to cross-link proteins to DNA; chromatin was then sonicated down to DNA segments of <1 Kb. Immunoprecipitation using a specific antibody against CDC6 enriched a fragment containing the origin. (**b**) Schematic description of the competitive PCR experiments used for the quantification of the immunoprecipitated fragments. A multicompetitor containing the target sequences for the primer pairs B10 and B48 was used in these experiments. Three different concentrations of competitor (1×10^3, 1×10^2, 1×10^1) were mixed with the same amount of immunoprecipitated DNA and amplified with the B10 and B48 primers. The amplification products were resolved by gel electrophoresis, stained with ethidium bromide and photographed. The amount of PCR product corresponding to the target amplification product (T, corresponding to genomic DNA) and to the competitor product (C) were evaluated by densitometry; the ratios between the two values (C/T) are reported at the bottom of the gels. From the curves fitting these values (show below the gels), the numbers of DNA molecules amplified by the B48 and B13 primers were calculated

2. In the immunoprecipitated DNA, two genomic regions were investigated, one corresponding to the Lamin B2 origin (B48; primers B48F and B48R) and the other one corresponding to a DNA region spaced a few kb apart from the origin (B10; primers B10F and B10R).
3. A multicompetitor was obtained, carrying the four primers arranged in order to give rise to amplification products of comparable size to the products obtained from genomic DNA, but still distinguishable from this, according to the protocol outlined under Subheading 3.1.4. This multicompetitor was cloned into a plasmid and quantified.
4. Three aliquots, each one containing the same amount of immunoprecipitated DNA, were mixed with three different amounts of multicompetitor, corresponding to 1×10^3, 1×10^2, and 1×10^1 molecules (Fig. 3b). Half of each aliquot was then amplified with the B48 primers and the other half with the B10 primers, along with appropriate controls (sample DNA alone, competitor alone, at the lowest concentration) or water.
5. PCR was carried out for 50 cycles. The amplification products were resolved by gel electrophoresis, stained with ethidium bromide and photographed. The amount of PCR product corresponding to the target amplification product (T, corresponding to genomic DNA) and to the competitor product (C) were evaluated by densitometry. The ratio between the two values (C/T) is reported at the bottom of the gel. As shown in the graph at the bottom of Fig. 3b, the relationship between the C/T ratio and the input amount of competitor is fitted by a line. The equation describing this line is reported at the bottom of the graph. According to this equation, the number of competitor DNA molecules at C/T ratio = 1 was evaluated for the two investigated regions. This is also the number of target DNA molecules in the aliquot of sample used for PCR amplification.
6. The results obtained show that the anti-CDC6 antibody selectively enriched for the origin region compared to the non-origin genomic control, thus showing that CDC binds the lamin B2 origin of DNA replication.

4. Notes

1. In competitive PCR, the construction of the competitor molecule represents a major experimental effort. As a general rule, the ideal competitor standard must be amplified according to the same kinetic profile as the target sequence. This implies that it must

be very similar to the original endogenous amplicon in both total length and nucleotide composition. Among the different experimental strategies developed to construct the ideal competitor, the generation of molecules containing a small insertion or deletion that encompasses no more than 10% of the total length of the original template allows one to attain both accuracy of quantification and simplicity of analysis.

2. The selection of the target region and the design of the primer set for amplification is a critical step in the optimization of the PCR reaction. Computer programs (e.g. Vector NTI, PrimerQuest, Primer3, GeneFisher-interactive PCR primer design, NetPrimer) may be helpful in choosing the most favorable primer sequences and predicting the optimal annealing temperatures. It is convenient to select primers of 15–25 nt, in order to maximize the binding specificity, containing a G or a C at the 3′ position to favor stronger binding to the template and thus reduce the risk of mis-priming. Intra- and inter-complementary regions must be avoided to prevent the formation of hairpin or primer dimers. The length of the amplified segment should be 150–300 nt, with a base composition in which the GC/AT ratio is ~1. If possible, target regions that are too GC- or AT-rich have to be avoided because their annealing efficiency might be too strong or too low, respectively.
3. For the resolution of PCR products between 150 and 300 bp, an alternative to acrylamide or conventional agarose consists of a mixture of 3% NuSieve agarose/1% SeaKem agarose gel. Other, ready-to-use, reusable gels of various sizes, which guarantee excellent resolving power, are also commercially available.
4. The amplification conditions should be tested experimentally and may need optimization for different primer sets. Enough template amount should be used to visualize the PCR products after 40–50 cycles of amplification by ethidium bromide staining. Empirically, a few nanograms of genomic DNA and a few picograms of plasmid DNA are sufficient. Too much genomic DNA template may decrease PCR amplification efficiency due to the presence of contaminants in the DNA preparation, whereas too diluted templates may increase the yield of unspecific amplification products.
5. A subsequent reamplification of the competitor DNA is sometimes needed to enrich the full-length product and allow its quantification.
6. When feasible, it is recommended that the external primers for RNA transcripts quantification are designed in order to match target sequences in two contiguous but well-separated exonic regions. This renders the cDNA amplification products easily recognizable from the amplicons derived from possible

genomic DNA contaminants. In this case, the digestion step with DNase I before the RT reaction may be omitted.

7. The RNA standard obtained is identical to its original target with the exception of the deletion or insertion of 20 bp and can be directly used as a competitor. The absence of contaminating DNA from the competitor preparation can be estimated by assembling a control reaction with the omission of the RT step.

8. The design of the primers for the construction of a multicompetitor is critical and needs careful optimization in order to guarantee the precise quantification of the different targets. As a rule, each individual primer pair should amplify at the same efficiency, and its relative position should be chosen to obtain PCR products that are similar in size to the respective target amplification products. The difference in length of each competitor fragment to target sequences should be within 0.5- and 2-fold. If the size difference is above these limits, a selective advantage for the amplification of the smaller molecule is usually detected.

9. In the setting up of a competitive PCR reaction, at the beginning, tenfold scalar dilutions of competitor are used to roughly estimate the target concentration; precise quantifications are then obtained by using competitor dilutions in the target/competitor equivalence range.

10. Compared to densitometry, digital photography has a more limited dynamic range and does not allow precise quantification of bands of very different intensity. However, this might not be a problem when two bands of similar intensity are compared.

11. The DNase I digestion should not exceed 15 min of incubation at room temperature. Higher temperatures and longer time could lead to Mg^{2+}-dependent hydrolysis of the RNA. It is important to add EDTA prior to heat-inactivation to avoid this problem.

12. A DNA competitor can also be used for RNA transcript quantification. However, the accuracy of the reaction can only be considered sufficient for the relative quantification of RNA abundance. In this case, the amount of the target transcript should be calculated with respect to the abundance of an internal housekeeping gene (e.g., β-globin, β-actin, GAPDH, HPRT), paying attention to use random hexamers as primers for reverse transcription of total cellular RNA.

13. Adding a high volume of reverse transcription reaction to the PCR mix may reduce amplification efficiency. It is advisable that the volume of the reverse transcription reaction should not exceed 10% of the final PCR volume.

14. Also, set up a positive control reaction, using, for example, an RNA sample already tested for the amplification of the sequence recognized by the selected primers, and no-template and no-reverse transcription, negative control reactions.

15. A series of common problems that may be encountered when performing competitive PCR (as well as competitive RT-PCR) are listed below and, for each of them, some possible solutions are suggested.

    (a) Excessive presence of heterodimer products.

    - Reduce the concentration of the standard if it is present in great excess compared to the target
    - Reduce the number of amplification cycles

    (b) Imprecise quantification when using a multicompetitor.

    - Check the design of the standard molecule: the ratio between the length of each competitor fragment and the respective target sequence should be within 0.5–2
    - Check that each primer pair amplifies at the same efficiency and, if necessary, consider revising the design of the molecule

    (c) Lack of reproducibility when repeating quantification over time.

    - The competitor standard may be degraded; avoid repeated thawing and refreezing; store the standards in small aliquots in different dilutions and discard them if thawed too often
    - The competitor standard aliquots may be contaminated by PCR carry-over products; discard and use new or freshly prepared dilutions

    (d) Instability of RNA competitor.

    - Avoid repeated thawing and refreezing; store the standards in small aliquots in different dilutions and discard them if thawed too often

    (e) Imprecise quantification when using an RNA competitor.

    - Correct the RNA amount applied to each assay by amplifying, in competitive RT-PCR, an internal reference template such as a housekeeping gene
    - Presence of undigested contaminating DNA in the input RNA preparation, recognized by the primers
    - The reference cellular gene may be unsuitable for sample comparison due to its variable transcriptional regulation in various experimental settings or in different tissues

Acknowledgments

This work was partially supported by grants from the World Anti-Doping Agency (WADA), Montreal, Canada and by Projects Cardiocell and SeND by the Regione Friuli Venezia Giulia, Italy.

References

1. Livak KJ, Flood SJ, Marmaro J, Giusti W, Deetz K (1995) Oligonucleotides with fluorescent dyes at opposite ends provide a quenched probe system useful for detecting PCR product and nucleic acid hybridization. PCR Methods Appl 4:357–362
2. Gilliland G, Perrin S, Blanchard K, Bunn HF (1990) Analysis of cytokine mRNA and DNA: detection and quantitation by competitive polymerase chain reaction. Proc Natl Acad Sci USA 87:2725–2729
3. Siebert PD, Larrick JW (1992) Competitive PCR. Nature 359:557–558
4. Becker-Andre M, Hahlbrock K (1989) Absolute mRNA quantification using the polymerase chain reaction (PCR). A novel approach by a PCR aided transcript titration assay (PATTY). Nucleic Acids Res 17:9437–9446
5. Diviacco S, Norio P, Zentilin L, Menzo S, Clementi M, Biamonti G, Riva S, Falaschi A, Giacca M (1992) A novel procedure for quantitative polymerase chain reaction by coamplification of competitive templates. Gene 122:313–320
6. Celi FS, Zenilman ME, Shuldiner AR (1993) A rapid and versatile method to synthesize internal standards for competitive PCR. Nucleic Acids Res 21:1047
7. Henco K, Heibey M (1990) Quantitative PCR: the determination of template copy numbers by temperature gradient gel electrophoresis (TGGE). Nucleic Acids Res 18:6733–6734
8. Comar M, Simonelli C, Zanussi S, Paoli P, Vaccher E, Tirelli U, Giacca M (1997) Dynamics of HIV-1 mRNA expression in patients with long-term nonprogressive HIV-1 infection. J Clin Invest 100:893–903
9. Grassi G, Zentilin L, Tafuro S, Diviacco S, Ventura A, Falaschi A, Giacca M (1994) A rapid procedure for the quantitation of low abundance RNAs by competitive reverse transcription-polymerase chain reaction. Nucleic Acids Res 22:4547–4549
10. Sestini R, Orlando C, Zentilin L, Gelmini S, Pinzani P, Bianchi S, Selli C, Giacca M, Pazzagli M (1994) Measuring c-erbB-2 oncogene amplification in fresh and paraffin-embedded tumors by competitive polymerase chain reaction. Clin Chem 40:630–636
11. Tafuro S, Zentilin L, Falaschi A, Giacca M (1996) Rapid retrovirus titration using competitive polymerase chain reaction. Gene Ther 3:679–684
12. Giacca M, Zentilin L, Norio P, Diviacco S, Dimitrova D, Contreas G, Biamonti G, Perini G, Weighardt F, Riva S et al (1994) Fine mapping of a replication origin of human DNA. Proc Natl Acad Sci USA 91:7119–7123
13. Kato K (1997) Adaptor-tagged competitive PCR: a novel method for measuring relative gene expression. Nucleic Acids Res 25:4694–4696
14. Lusic M, Marcello A, Cereseto A, Giacca M (2003) Regulation of HIV-1 gene expression by histone acetylation and factor recruitment at the LTR promoter. EMBO J 22:6550–6561
15. Deutsch S, Choudhury U, Merla G, Howald C, Sylvan A, Antonarakis SE (2004) Detection of aneuploidies by paralogous sequence quantification. J Med Genet 41:908–915
16. Liu W, James CD, Frederick L, Alderete BE, Jenkins RB (1997) PTEN/MMAC1 mutations and EGFR amplification in glioblastomas. Cancer Res 57:5254–5257
17. Ausubel FM, Brent R, Kingston RE, Moore DD, Seidman JG, Smith JA, Struhl K (2003) Current protocols in molecular biology. Wiley, NY
18. Sambrook J, Fritsch EF, Maniatis T (1989) Molecular cloning. A laboratory manual, 2nd edn. Cold Spring Harbor Laboratory, Cold Spring Harbor, NY
19. Todorovic V, Giadrossi S, Pelizon C, Mendoza-Maldonado R, Masai H, Giacca M (2005) Human origins of DNA replication selected from a library of nascent DNA. Mol Cell 19:567–575

Part III

The RT-PCR Master Chef: Finding the Right Ingredients

Chapter 16

Skeletal Muscle RNA Extraction in Preparation for RT-PCR

Janelle P. Mollica

Abstract

Extraction of high quality RNA is paramount to successful RT-PCR, and here, a method proven optimal for skeletal muscle is described. While this method described is for use with skeletal muscle, it could be suitable for other types of mammalian tissue also. This method describes an approach to extract high quality RNA with minimal degradation and the subsequent analysis of that RNA in preparation for RT-PCR. Two separate methods of RNA quantification and analysis are described in this chapter and regardless of the method chosen for quantitation and analysis, it is imperative that the integrity of the sample be established before proceeding to RT-PCR.

Key words: RNA, RT-PCR, Gene expression, Skeletal muscle, Integrity

1. Introduction

The key to successful and accurate quantitative RT-PCR is most certainly the extraction of high quality, nondegraded RNA. Extraction of poly(A) RNA for RT-PCR analysis offers no significant advantage over extracting total RNA for RT-PCR (1). Total RNA extraction is a generally simple process, and there are many commercially available kits for such extractions. Total RNA extraction requires much less starting material than poly(A) RNA extraction and also allows scope for analysis of ribosomal RNA (rRNA) species as well as messenger RNA (mRNA). Given that 18S rRNA is a popular choice of "housekeeping" gene for RT-PCR, extraction of total RNA gives the user a choice of PCR controls. This choice of RT-PCR controls in itself requires careful consideration and information regarding this is beyond the scope of this chapter. While RNA yield from the extraction is an important consideration, increasing emphasis is now placed on the quality of RNA extracted. Several publications have been

Nicola King (ed.), *RT-PCR Protocols: Second Edition*, Methods in Molecular Biology, vol. 630,
DOI 10.1007/978-1-60761-629-0_16,

dedicated to document the necessity of high integrity RNA for such sensitive end-point analysis as RT-PCR and microarray (2–4). Yields from small starting sample sizes, while perfectly adequate for RT-PCR, were noted as the most significant limitation in being able to assess the quality of RNA. Standard denaturing RNA gels are relatively insensitive and require a substantial proportion of a given sample for the qualitative assessment of the ribosomal bands, with a much smaller proportion of the sample needed for replicate RT-PCR analyses. Consequently, the qualifying RNA integrity step has been bypassed when sample size (e.g., Human skeletal muscle biopsies) is limited. Fluorescent dyes (i.e., SYBR Green II) for RNA staining have an improved sensitivity over ethidium bromide staining; however, the introduction of automated equipment such as the Agilent 2100 Bioanalyzer and its accompanying Lab-on-a-chip technology have seen the accuracy and sensitivity of RNA quantification and integrity analysis improve markedly. Of course, with any new technology comes great expense, which is out of reach for most laboratories. Keeping this in mind, two methods of RNA quantification and analysis are described – one conventional, low-cost and the other using specialist equipment.

2. Materials

2.1. Total RNA Extraction from Skeletal Muscle

There are a number of dangerous, hazardous, and/or toxic chemicals used in the process of RNA extraction, quantification, and analysis. As is good laboratory practice, the MSDS for each chemical should be referred to before use.

1. 4-figure balance
2. Hand-held tissue homogenizer
3. Pipettes for volumes in the range 0.2 μl–1 ml.
4. Sterile filter tips 0.2 μl–1 ml
5. Autoclave
6. Sterile glassware (prepared by autoclaving)
7. Vortex
8. Refrigerated centrifuge for microfuge tubes
9. Sterile microfuge tubes (0.6 ml–1.5 ml) – these need to be compatible for use with phenol/chloroform
10. –80°C Freezer
11. Heat block (able to heat to 65°C)
12. Dry ice or Liquid nitrogen

13. Sterile 0.1% diethylpyrocarbonate (DEPC)-treated MilliQ or double-distilled H_2O: Add 2 ml of DEPC to 2 L of MilliQ or double-distilled H_2O. Incubate the treated water overnight at 37°C, then autoclave before use (see Note 1).
14. ToTALLY RNA™ Total RNA Isolation Kit (Ambion). Check the kit components and store according to manufacturer's instructions.
15. Isopropanol; molecular biology grade or better
16. Ethanol; molecular biology grade or better

2.2. RNA Quantification and Analysis: Using UV Spectrophotometry and Denaturing Gel

1. UV Spectrophotometer
2. Black-walled 0.2 ml quartz cuvettes
3. 0.2 μm syringe filters (Minsart, cat. no. 17823K)
4. 10 ml sterile luer lock syringes (Terumo, cat. no. SS-10L)
5. Sterile 10 ml polypropylene tubes
6. Microwave
7. Horizontal gel electrophoresis system and powerpack.
8. Gel documentation system, composed of a computer (with density analysis software), an ultraviolet (UV) light source, and camera.
9. Sterile DEPC-treated MilliQ or double-distilled H_2O (see Subheading 2.1).
10. DNA-grade Agarose
11. 0.5 M (EDTA) – dissolve 3.72 g of EDTA in 15 ml of sterile DEPC H_2O, adjust pH to 8.0 with 1 N NaOH, and make up to 20 ml final volume.
12. 20× MOPS buffer – dissolve 41.2 g MOPS in 400 ml of sterile DEPC H_2O, adjust pH to 7.0 with 1 N NaOH. Add 6.56 g sodium acetate and 20 ml of 0.5 M EDTA (pH 8.0). Adjust the volume to 500 ml with sterile DEPC H_2O (see Note 2).
13. 1× MOPS buffer, made fresh.
14. Filtered formaldehyde, 37% – draw 10 ml of 37% formaldehyde into a sterile 10 ml syringe. Fit a 0.2 μm syringe filter to the tip and expel the formaldehyde through the filter into a sterile 10 ml tube.
15. Ethidium Bromide, 1 mg/ml in sterile DEPC H_2O (see Note 3).
16. 1.3% Agarose gel – Add 2.6 g of DNA grade agarose to 10 ml of 20× MOPS and 182 ml of DEPC H_2O. Microwave on low for approximately 2 min (see Note 4). Check after 1 min and gently swirl. Microwave in further 30 s blocks if necessary. When the liquid gel has cooled slightly, add 8 ml of Formaldehyde (37%) and 30 μl of Ethidium Bromide (1 mg/mL) – mix by swirling gently (see Note 5).

17. Formamide.
18. Bromophenol Blue, 10 mg/ml in sterile DEPC H_2O.
19. RNA loading buffer per sample: 3.9 µl of filtered formaldehyde (37%), 12.5 µl of formamide, 1.25 µl of 20× MOPS, 1 µl of 1 mg/ml Ethidium bromide, and 1 µl of 10 mg/ml bromophenol blue (see Note 5).
20. RiboRuler™ High Range RNA Ladder (Fermentas).

2.3. RNA Quantification and Analysis: Using Agilent Bioanalyzer 2100

1. Agilent 2100 Bioanalyzer
2. Sterile DEPC-treated MilliQ or double-distilled H_2O (see subheading 2.1).
3. RNA 6000 Nano kit™ (Agilent)
4. RNA 6000 Nano ladder (Ambion)
5. RNaseZAP.

3. Methods

3.1. Total RNA Extraction from Skeletal Muscle

1. It is crucial to avoid contamination with RNAses which are inherently present everywhere. Ensure that gloves are worn at all times during RNA extraction. Make sure that all tubes and labware are sterile and RNAse-free. Autoclaving labware prior to use is an effective method of removing RNAses (ensure only compatible items are autoclaved).
2. When using phenol and chloroform, ensure appropriate protective clothing and eyewear are worn. Carry out all work in a chemical fumehood.
3. Prepare your work surface in the chemical fumehood by spraying and wiping down with a 70% ethanol solution.
4. Weigh out frozen skeletal muscle tissue; 10–50 mg of muscle (see Note 6 and Note 7)
   (a) Extract total RNA using the ToTALLY RNA™ Total RNA Isolation Kit according to the manufacturer's instructions.
   (b) Homogenize the skeletal muscle tissue using a hand held homogenizer in very short bursts (see Note 8 and Note 9). Other methods of tissue disruption have proven to be less than optimal.
5. It is important to resuspend the final RNA pellet in DEPC H_2O containing EDTA provided with the extraction kit (see Note 10). As a general guide, using 5–10 µl of DEPC H_2O/EDTA per 10 mg of starting material will yield a

reasonable RNA concentration. Completely dissolve the pellet before proceeding to the next step. It may be necessary to heat the RNA solution to 55°C for 20–30 s and vortex occasionally to dissolve the pellet (see Note 11).

6. Once RNA is dissolved store it in ≥5 µl aliquots for single use at −80°C (see Note 12).

3.2. RNA Quantification and Analysis: Using UV Spectrophotometry and Denaturing Gel

1. Turn on the spectrophotometer and set the wavelengths to 260 nm and 280 nm (see Note 13).
2. Dilute RNA sample (see subheading 3.1, step 6) by adding 1 µl to 399 µl of sterile water (1:400). Also aliquot 400 µl of sterile water into a fresh tube – this will be the blank.
3. Pipette blank sample into a cuvette and insert in the spectrophotometer.
4. At 260 nm, set the blank sample to zero. Keep the blank sample in order to re-zero the spectrophotometer between reading multiple samples.
5. Pipette diluted RNA into a cuvette and insert in the spectrophotometer. Record the absorbance at 260 nm. Using a small cuvette capable of generating a reading from <200 µl will allow duplicate analysis using the dilution at step 2.
6. Repeat steps 3–5 above, with the wavelength of the spectrophotometer set at 280 nm.
7. Record the readings at 260 nm and 280 nm for each RNA sample and calculate the RNA concentration and relative purity (see Note 14).
8. Use a horizontal agarose gel electrophoresis to analyze RNA integrity. Prepare approximately 200 ml of a 1.3% agarose gel (see subheading 2.2, step 16).
9. Wipe gel cast assembly with 70% ethanol and rinse the gel comb and gel tray with DEPC H_2O prior to assembly.
10. Slowly pour 100–150 ml of the gel solution into the prepared gel cast assembly, avoiding air bubbles. Cover with plastic film and allow the gel to set at room temperature for approximately 60 min.
11. While the gel is setting, make up enough loading buffer (see Subheading 2.2, step 19) for each sample being analyzed. Allow for the RNA ladder and an extra sample when calculating the amount of loading buffer to account for any pipetting error ($n+2$). See Table 1 for the preparation of eight RNA samples.
12. Based on the calculations for RNA concentration (see Subheading 3.2, step 7), take aliquots of RNA equal to approximately 1 µg in ≤10 µl and add it to a fresh sterile tube. Also aliquot 2 µl of the base pair ladder for sizing reference.

Table 1
Summary of the reagents and volumes required for the preparation of 8 RNA samples

Reagents	Volume for 1 sample (μl)	$n+2$	Total volume (μl)
Filtered formaldehyde (37%)	3.9	10	39
Formamide	12.5	10	125
20× MOPS buffer	1.25	10	12.5
Ethidium bromide (1 mg/ml)	1	10	10
Bromophenol blue (10 mg/ml)	1	10	10
1 μg RNA sample	≤10 μl		
	Total volume loading buffer		196.5

13. Add 19.6 μl of loading buffer to each RNA sample and ladder. The total volume of the prepared sample should equal no more than 30 μl (see Note 15).
14. Denature all the samples at 65°C for 10 mins. Place on ice immediately.
15. When the gel is set, it will solidify and become opaque. Immerse the set gel completely in 1× MOPS buffer (see subheading 2.2, step 13) in the electrophoresis chamber and load the samples and ladder in to the gel. Run the gel at 90 V for approximately 1 h or until the dye front is about two-thirds through the gel.
16. Once the gel has finished running, remove the gel from the tank and visualize using UV light. Take an image with a digital camera to record it and quantify the 28S and 18S band densities (see Note 16 and Note 17). In a skeletal muscle RNA sample, 28S and 18S ribosomal RNA bands should migrate on the gel at approximately 5,000 bp and 2,000 bp, respectively.

3.3. RNA Quantification and Analysis: Using Agilent Bioanalyzer 2100

If you have access to an automated analyzer this is the preferred alternative to the method described under subheading 3.2. It produces more accurate results due to being more sensitive.

1. The Agilent Bioanalyzer 2100 is not the only automated instrument available for RNA analysis. Keep in mind that there are alternatives to this particular piece of equipment that perform a similar function.
2. Comprehensive instructions are provided with the Agilent Bioanalyzer 2100 and the RNA 6000 Nano LabChip kit. All analyses have been optimised according to these instructions and should be completed as listed.

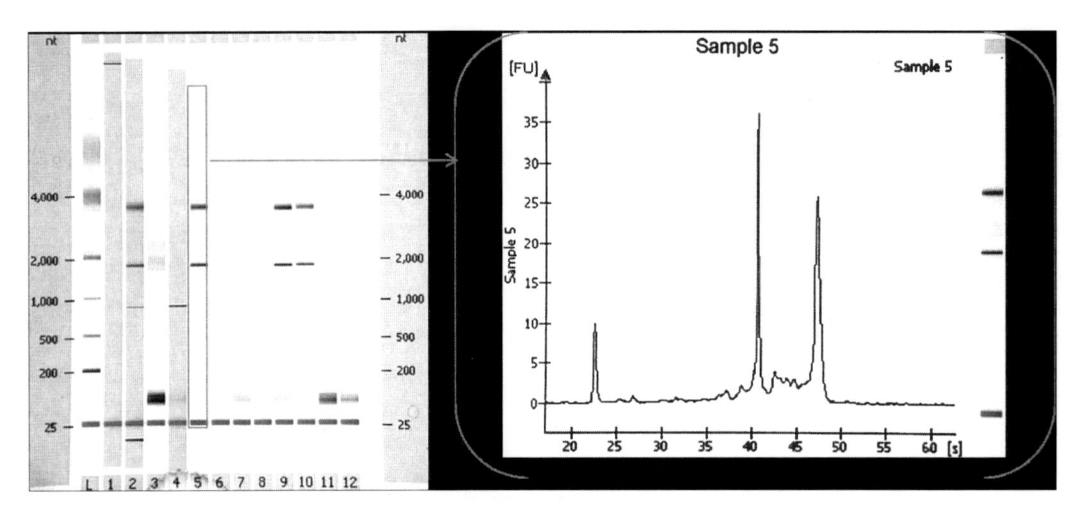

Fig. 1. Skeletal muscle RNA was extracted from 12 × 10mg samples according to the methods described (see Subheading 3.1). RNA was quantified and analyzed using the Agilent 2100 bioanalyzer and RNA 6000 Nano Labchip. On the left is a summary image of the micro-capillary gel electrophoresis of twelve samples run simultaneously for RNA quantity and quality. This shows a number of both good and poor quality RNA samples as indicated by the dominant ribosomal RNA (rRNA) 28S and 18S bands. Good quality rRNA is visible as two sharp bands half way down the gel, while poor quality RNA is inferred from fuzzy rRNA bands and the presence of low molecular weight smearing below the rRNA species. On the right is the individual electrophoreogram of sample 5, with the *y*-axis indicating the fluorescence units bound to RNA. This sample represents a good quality RNA sample, showing the large 18S and 28S peaks and no presence of small peaks above the baseline fluorescence units which would be a sign of low molecular weight RNA species indicative of sample degradation. RNA samples analyzed by conventional horizontal gel electrophoresis and stained with ethidium bromide will produce very similar results to the capillary electrophoresis shown in this figure

3. When analyzing skeletal muscle RNA, dilute 1 μl of the original RNA (see subheading 3.1, step 6) 1:4 with sterile DEPC H_2O for analysis with the RNA 6000 Nano LabChip kit (see Note 18).
4. The most recent versions of the Bioanalyzer software have an option in the results tab that calculates an RNA integrity number (RIN) for your sample as well as the concentration and 28S: 18S RNA ratio. For a detailed description of RIN see (5).
5. Figure 1 shows an example of the results which are obtained using this method.

4. Notes

1. DEPC is a chemical used to inactivate any RNAses present in the water. High temperatures are required to inactivate the DEPC so that it may be used for RNA extraction. The DEPC H_2O should be stored at room temperature, handled in an RNAse-free environment, and have an indefinite shelf life.

2. MOPS buffer is light sensitive. To protect from light, wrap the bottle in foil and store at room temperature. This solution should be kept no longer than 3 months.
3. EtBr – Make only 1 ml volumes as required and store at room temperature indefinitely protected from light.
4. Microwave power will vary depending on the model used. Heat the solution for short 30–60 s intervals and swirl to mix in between.
5. Formaldehyde and Ethidium bromide are dangerous, toxic chemicals and must be added to the solution in the fume hood. Ethidium bromide waste, that is any tips, tubes and contaminated gel, must be collected and disposed of according to laboratory regulations. After handling formaldehyde and ethidium bromide, make sure that gloves are changed and that those potentially contaminated items are disposed of with the other contaminated waste. Typically, laboratories that use ethidium bromide will have dedicated equipment and areas for use with this chemical due to its hazardous nature. Gloves used while handling ethidium bromide should be changed after contact so as not to contaminate any other nondedicated laboratory equipment.
6. 10–50 mg of tissue will provide sufficient yield of RNA for PCR analysis. More tissue may be required if the RNA is to be used for microarray analysis.
7. It is vital that the muscle does not thaw while weighing prior to extraction. Once the weight is recorded, the sample must be returned to dry ice or liquid nitrogen so that it remains frozen.
8. Homogenizing tissue for RNA extraction should be carried out in short bursts due to the generation of heat within the sample and subsequent activation of RNAses. In between short bursts, the sample should sit on ice to cool so as to protect the RNA from heat degradation.
9. Once muscle is homogenized in denaturing solution (provided in the kit), it can be processed immediately or it may be stored at –80°C for up to 3 months.
10. The addition of EDTA to the DEPC water used to re-suspend the final RNA pellet is important because EDTA is a cation chelator. Cations such as Mg^{2+} in solution cause rapid breakdown of RNA, even at –80°C.
11. When the final ethanol wash is removed from the RNA pellet, do not let the pellet dry out too much. This may make your RNA partially insoluble, and you may have to resort to heating and vigorous mixing to dissolve the pellet.
12. Avoid repeated freeze-thaw cycles of isolated RNA. If possible store aliquots of RNA so that only a single thaw cycle is required.

13. Allow the spectrophotometer UV lamp to warm up sufficiently. Generally 10–30 min prior to use is adequate depending on the model of spectrophotometer used.
14. Using the extinction coefficient for RNA; 40 μg/ml = 1.0 (OD_{260}) with a 1 cm path length, the formula used to calculate the concentration of RNA is; Concentration (μg/ml) = Absorbance 260 nm × 400 (dilution factor) × 40 (extinction coefficient). Also use the reading at both 260 nm and 280 nm to calculate relative purity. The ratio OD260: OD280 should equal 1.8–2.0. If the OD260:OD280 reading is low, this indicates there is protein contamination present in the sample.
15. The total volume of the wells in the agarose gel will depend on the size of the combs used. The indicated 30 μl volume in this method is based on the use of a 20-well comb, 1.5 mm thickness. Please check the manufacturer's specifications of your system before determining maximum final volume.
16. Band intensities of 18S and 28S are indicators of RNA integrity and these should be the only bands visualized on the gel. Any smearing in the lane and the presence of low molecular weight species are indicative of RNA degradation. Any indication of material present at the top of the gel near the wells is a sign of DNA contamination.
17. The theoretical gold standard for 28S: 18S ratio is 2:1. This value is rarely, if ever achieved, and an accepted ratio for good quality RNA is 1.6–1.8:1. Samples with values lower than this that are used for RT-PCR should be viewed with caution.
18. Diluting RNA for analysis with the Agilent Bioanalyzer gives the advantage of being able to run the sample up to four times to test reproducibility.

Acknowledgments

Thanks to Dr. Robyn M. Murphy and Heidy Latchman for their helpful feedback on the preparation and content of this document.

References

1. Liedtke W, Battistini L, Brosnan CF, Raine CS (1994) A comparison of methods for RNA extraction from lymphocytes for RT-PCR. PCR Methods Appl 4:185–7
2. Imbeaud S, Graudens E, Boulanger V, Barlet X, Zaborski P, Eveno E, Mueller O, Schroeder A, Auffray C (2005) Towards standardization of RNA quality assessment using user-independent classifiers of microcapillary electrophoresis traces. Nucleic Acids Res 33:e56
3. Fleige S, Walf V, Huch S, Prgomet C, Sehm J, Pfaffl MW (2006) Comparison of relative

mRNA quantification models and the impact of RNA integrity in quantitative real-time RT-PCR. Biotechnol Lett 28:1601–13

4. Fleige S, Pfaffl MW (2006) RNA integrity and the effect on the real-time qRT-PCR performance. Mol Aspects Med 27:126–39

5. Schroeder A, Mueller O, Stocker S, Salowsky R, Leiber M, Gassmann M, Lightfoot S, Menzel W, Granzow M, Ragg T (2006) The RIN: an RNA integrity number for assigning integrity values to RNA measurements. BMC Mol Biol 7:3

Chapter 17

Reverse Transcription of the Ribonucleic Acid: The First Step in RT-PCR Assay

Fadia Haddad and Kenneth M. Baldwin

Abstract

Reverse transcription (RT) is the synthesis of complementary deoxyribonucleic acids (DNA) from single-stranded ribonucleic acid (RNA) templates. This process is catalyzed by the reverse transcriptase enzyme, which is the replicating enzyme of retroviruses. Reverse transcriptase was discovered in 1970, and since then, it has played an instrumental role in the advancement of molecular biology and biotechnology research. In the presence of all four deoxynucleotides (dNTP: dATP, dCTP, dGTP, and dTTP) and under well-defined salt and pH conditions, the reverse transcriptase extends a primer complementary to the RNA to produce a complementary DNA (cDNA) for the RNA template. In this chapter, a simple method of reverse transcription of total cellular RNA into cDNA is described using Superscript II reverse transcriptase (Invitrogen); the resulting cDNA can be used in polymerase chain reaction (PCR).

Key words: Gene expression, Complementary cDNA, RT-PCR, Cellular RNA, mRNA, Primer extension

1. Introduction

The process of synthesis of deoxyribonucleic acid (DNA) from a ribonucleic acid (RNA) template is called "reverse transcription" because it reverses the flow of genetic information (from RNA to DNA, rather than from DNA to RNA found in normal transcription). The reverse transcription (RT) reaction is carried out by a retroviral DNA polymerase named reverse transcriptase. The reverse transcriptase is the replication enzyme within the retrovirus that converts its single-stranded genomic RNA into DNA.

In molecular biology laboratories, the reverse transcriptase is an RNA-dependent DNA polymerase enzyme used to transcribe single-stranded RNA into single-stranded complementary DNA (cDNA). During the reverse transcription process, oligonucleotide

Nicola King (ed.), *RT-PCR Protocols: Second Edition*, Methods in Molecular Biology, vol. 630, DOI 10.1007/978-1-60761-629-0_17, © Springer Science+Business Media, LLC 2010

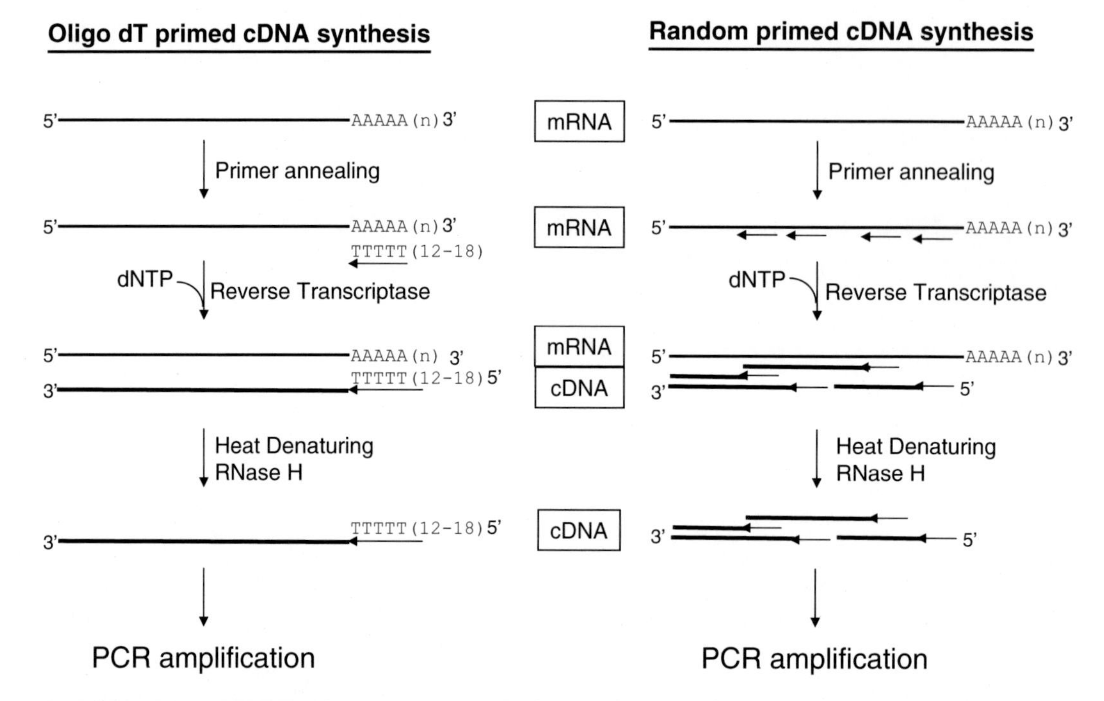

Fig. 1. Schematic representation of reverse transcription of RNA into cDNA using oligo dT, or random primers. Oligonucleotide primers (oligo dT) or random primers are annealed to the RNA template and extended by the reverse transcriptase in the presence of deoxynucleotides (dNTP) to make cDNA. RNAse H treatment can be used to degrade RNA from RNA/DNA hybrids; however, the RNase H treatment is not required before standard PCR. The denaturing step before PCR annealing and extension will make the cDNA available for the PCR primers and Taq polymerase extension

primers complementary to the RNA strand are annealed to the RNA template and extended in the presence of deoxynucleotides (dNTP) and optimal pH and salt conditions to produce cDNA (Fig. 1). The native reverse transcriptase is a multifunctional enzyme. In addition to its RNA-dependent DNA polymerase activity, it also possesses both (1) a weak DNA-dependent DNA polymerase activity that is lacking the 3′→5′ exonuclease activity and (2) an RNase H activity that degrades the RNA template from RNA/DNA hybrid strands as the cDNA synthesis proceeds.

Soon after its discovery in 1970 (1, 2), the reverse transcriptase went on to play a vital role in the advancement of molecular biology research, especially in the fields of gene discovery and biotechnology. Reverse transcriptase became an essential molecular tool used in the cloning of genes, the analysis of gene expression, and in the diagnosis of certain microbial diseases. Its influence extends from cloning to the development of microarrays to the characterization of transcriptomes and the annotation of genomes.

For molecular biology applications, the M-MLV reverse transcriptase from the Moloney murine leukemia virus, and the AMV reverse transcriptase from the avian myeloblastosis virus are most commonly used. Both enzymes have the same fundamental activities, but they differ in their optimal working condition and in a number of characteristics. For example, the M-MLV enzyme has very weak RNase H activity compared to the AMV enzyme, which makes the former more useful in the synthesis of complementary DNAs for long messenger RNAs.

Reverse transcriptase (RT) is commonly used in the first step of the assay called the reverse transcription polymerase chain reaction (RT-PCR), the most sensitive technique used in detection and quantification of RNA. The classical PCR technique can only be applied to DNA strands; however, with the use of reverse transcriptase, RNA can be transcribed into DNA, thus making PCR analysis of RNA molecules possible. Reverse transcriptase is also used to create cDNA libraries from mRNA. The commercial availability of reverse transcriptase greatly improved knowledge in the area of molecular biology as, along with other enzymes, it has allowed scientists to clone, sequence, and characterize expressed genes.

A simple method of reverse transcription of total RNA into cDNA is described in the following lines; the resulting cDNA can be used in polymerase chain reaction (PCR).

In the described method, we used the SuperScript II reverse transcriptase from Invitrogen. However, any commercial source of reverse transcriptase will work. SuperScript II is a genetically engineered form of the M-MLV reverse transcriptase. The enzyme contains a point mutation that eliminated the RNase H activity of the enzyme without reducing its DNase polymerase activity, and rendered its reverse transcriptase activity more heat stable, and fully active at 50°C (Invitrogen brochure). It is recommended that the user read the entire procedure with the related notes before starting any work.

2. Materials

1. Adjustable pipettes that are capable of accurately dispensing volumes from <1 to 20 μl.
2. Nuclease-free pipette tips that fit the pipette model, preferably filter tips to prevent cross-contamination (see Notes 1 and 2).
3. Disposable microcentrifuge tubes (0.5 ml size) that are certified to be DNase- and RNase-free.
4. Two water baths or heating blocks capable of heating to 90°C.
5. A microcentrifuge to be used for short duration spins (<10,000×*g*, 10 s) to collect the content at the bottom of the tube.

6. Nuclease free distilled water (commercially available) or autoclaved DEPC-treated distilled water (store at room temperature in a tightly closed container) (see Note 1).
7. SuperScript II (200 units/μl) reverse transcriptase enzyme (Invitrogen), which is supplied along with a vial of 5× reverse transcription buffer: 250 mM Tris–HCl (pH 8.3), 375 mM KCl, 15 mM $MgCl_2$, and a vial of 100 mM dithiotreitol (DTT) (store at –20°C) (see Notes 3 and 4).
8. dNTP mix: 10 mM dATP, 10 mM dCTP, 10 mM dGTP, and 10 mM dTTP (store at –20°C).
9. Oligodeoxynucleotide primers: Oligo dT(12–18) primers (concentration 200 ng/μl), random primers (hexamer 200 ng/μl), or gene specific primers (2 pmoles/μl) (store at –20°C) (see Note 5).
10. Total cellular RNA solution in water (concentration 0.1–2 μg/μl) (store at –80°C) (see Notes 6 and 7).
11. Optional: RNase inhibitor (40 units/μl) (store at –20°C) (Promega, RNasin)
12. Optional: RNase H (5 units/μl) (store at –20°C) (New England Biolabs)

3. Methods

The user should always wear clean disposable gloves for all procedures (see Note 1).

A 20-μl reaction volume can be used for either 1 ng–5 μg of total RNA or 1–500 ng of mRNA.

The reaction described below uses 1 μg total RNA. Reactions can be scaled up or down depending on the user's need.

1. Add the following components to the bottom of a nuclease-free microcentrifuge tube:
   1 μg of total RNA,
   1 μl of oligo(dT)12–18 (200 μg/μl),
   and/or 1 μl of random primers (200 ng/μl),
   nuclease free distilled water to a total volume of 12 μl (11 μl if using optional RNase inhibitor in Step 4).
   Gently mix with the tip of the pipette and close tightly.
2. Heat the above mixture to 70°C for 5 min (see Note 8).
3. Chill on ice for at least 1 min. Collect the contents of the tube by brief centrifugation (<10,000×*g*, 10 s).
4. Add the following components that can be prepared in a premix when processing several samples (see Note 9).

4 μl of 5× First-Strand Buffer
2 μl of DTT (100 mM)
1 μl of dNTP Mix (10 mM each)
1 μl of SuperScript™ II RT (200units/μl)
1 μl of RNase inhibitor (40 units/μl) (optional), RNase inhibitor is required if using <50 ng starting RNA.

5. Mix contents of the tube gently, and incubate at 45°C for 55 min (see Note 10).
6. Inactivate the reaction by heating at 90°C for 5 min.
7. Cool on ice, and then spin the tubes briefly (<10,000×*g*) to collect any liquid formed by condensation on the side of the tube. Store the reaction at −80°C for long term use (>5 years). (Storage at −20°C is also effective to preserve the cDNA for short term (months)).

The cDNA can now be used as a template for amplification in PCR (see Notes 11 and 12). However, amplification of some PCR targets (long target >1 kb, or low copy targets) may require the removal of RNA complementary to the cDNA to increase the sensitivity of detection.

To remove RNA complementary to the cDNA, add 1 μl of RNase H and incubate at 37°C for 20 min.

4. Notes

1. Working in an RNase-free environment and maintaining RNase-free preparations are critical for performing successful cDNA synthesis reactions. Therefore, we strongly recommend that only nuclease-free reagents, nuclease-free tubes, and pipette tips be used. Always wear clean gloves when handling kit components and samples. Change gloves frequently especially after touching any potential sources of RNase such as door knobs, pens, and pencils. The RNase A is a highly stable and active enzyme that can contaminate any laboratory environment and is present on human skin. Keep all kit components tightly sealed when not in use. Keep all tubes containing RNA tightly sealed during the incubation steps and when not being used.
2. Exercise extreme care to prevent cross contamination of stock solutions and from one sample to another. It is preferred that filter tips are used for all of these applications. The filter tips prevent contamination of the pipette and thus reduce cross-contamination due to the handling.

3. There are many commercially available reverse transcriptases, which are derived mostly from either the Moloney Murine Leukemia virus or the Avian Myeloblastosis Virus. As a result of advancement in biotechnology and genetic engineering, modifications were designed in these enzymes to inactivate their inherent RNase H activity without damaging their DNA polymerase activity. Thus they became more useful in producing high cDNA yield as well as enabling reverse transcription through a long RNA template. Other modifications also made these enzymes more thermostable. Most RT enzymes are sold in kits, which also provide the necessary buffer and additives, such as DTT for example. These are accompanied by extensive instructions and suggestions for particular needs. Most of these protocols are available online and the users must do their research and read the protocol before purchasing a kit to make sure that the enzyme fulfills their need for a specific application.
4. Avoid repeated thawing of the kit components and reagents in order to conserve their quality: place individual reagents on ice as they are needed, leaving the rest in the freezer. Return quickly to the freezer upon completion of their use. Most enzymes are stored in 50% glycerol and normally do not freeze at –20°C storage temperature. Thus, enzymes can be taken out of the freezer, used, and immediately returned to keep them at –20°C for longer shelf life. Aqueous reagents that are frozen can be thawed by incubation in water at room temperature, then placed on ice until used. Do not thaw any component in warm water to speed the process as warming will reduce the quality of the product.
5. The primers used to prime the RT reaction can be either (a) oligo dT primers which hybridize to the polyA sequences generally at the 3′end of the messenger RNA (mRNA); (b) random primers (hexamers or decamers) which anneal randomly along the entire RNA sequence; or (c) a gene specific primer complementary to a specific mRNA/RNA sequence. The latter is recommended for analyses of a specific RNA strand (sense or antisense) when analyzing overlapping transcripts (3). Oligo dT and random primers anneal at low temperature and normally the reaction is recommended to be carried out at ~42°C. However, the use of longer primers such as oligo dT (20–25) or random decamers, enables the RT reaction to occur at higher temperature (50°C). If using oligo dT primers to prime the cDNA synthesis, the PCR target should preferably be near the 3′end of the gene for higher yield of the PCR product. The RT reaction efficiency decreases for a longer template, and a much lower yield will result if the PCR were carried out at >5 kb from the priming site (Fig. 2).

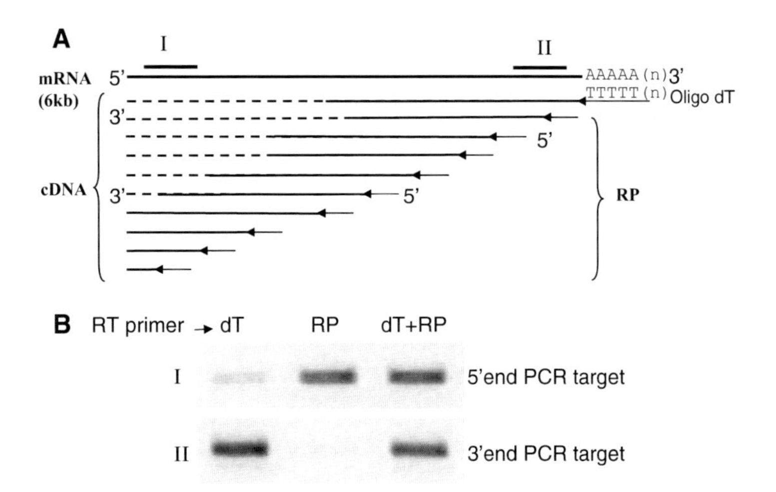

Fig. 2. The RT priming strategy can affect the efficiency of the PCR based on the PCR target location. Panel A depicts a schematic representation of cDNA synthesized from a 6kb mRNA using either oligo dT primers or random primers (RP). The oligo dT anneals to the poly A tail at the 3′ end of the mRNA and the RT extends the primer into cDNA complementary to the mRNA, which may or may not reach the 3′ end of the mRNA because of secondary structure that leads to early termination of the synthesis. Thus PCR I targeting the 5′ end of the mRNA (~5–6 kb from priming site) will have a low target and thus results in low yield as illustrated on the PCR gel image shown in panel B. In contrast, for PCR target II near the 3′end, oligo dT priming efficiently generated a cDNA corresponding for that target. When the RT reaction is primed with random primers, the primers anneal at random across the mRNA strand and extend toward its 5′ end. This results in a mix of cDNA of numerous sizes that usually span the entire mRNA sequence except the extreme 3′end near the poly A tail. PCR targeting the 3′end of the mRNA normally generates very low yield and should not be used as seen in panel B for RP primer, target II PCR product. When both oligo dT and random primers are mixed in a reaction, cDNA is synthesized to represent the entire mRNA molecule, and thus the PCR yield is acceptable regardless of the location of the PCR target. However, this is not recommended for full length cDNA synthesis needed for making cDNA libraries or for generating long PCR targets. The presence of random primers hinders the cDNA extension and results in early termination

The use of oligo dT makes any mRNA target a possibility. However, not all of the cellular RNA contains a polyA tail, for example primary transcripts (pre-mRNA) are not normally processed with polyA. Further, ribosomal RNA and some intergenic RNA or antisense RNA do not contain the polyA tail, which makes the use of oligo dT unsuitable if one was to analyze the expression of such RNAs. Similarly, the use of synthetic RNA templates that are not tailed with poly A requires the use of either random primers or gene specific primers.

In our laboratory, we routinely use a mix of oligo dT and random primers to prime RT reactions. The oligo dT primers are useful when PCR target sequences are near the poly A tail in the 3′ end of the mRNA, whereas the random primers enable us to analyze non-polyadenylated RNA such as ribosomal RNA from the same RT reaction products.

For example, in the analyses of the myosin heavy chain (MHC) gene family, the sequence specificity of the various MHC mRNA isoforms is best found in their 3′ untranslated region. As such, in order to target specific MHC mRNA isoforms, the reverse PCR primer is designed complementary to the 3′ untranslated region (4). If oligo dT primers were not added in the RT reaction, the PCR efficiency in detecting MHC mRNA may be mediocre. However, addition of random primer provides more flexibility in analyzing other transcripts (18S ribosomal RNA) and removes the 3′end bias for PCR targets.

6. The above reverse transcription procedure can be used to make cDNA to either total RNA, mRNA, or synthetic RNA. We found that the use of total RNA is sufficient to detect low copies of cellular RNA by PCR. Also using total RNA enables one to amplify other RNA products that are not poly A tailed such as nascent primary transcripts, ribosomal RNA, and other noncoding RNAs.

7. The purity of the RNA solution is an important factor that can influence the RT reaction efficiency. The presence of salt in the RNA solution may inhibit the RT reaction. In our hands, when using the Tri-reagent protocol to extract total cellular RNA from cells or tissue (Molecular Research Center), adding an extra wash step for the RNA pellet using 75% ethanol-25% nuclease free water, significantly improved the RT efficiency, because it reduces carryover of high salt from previous steps in the RNA extraction procedure. If the user is doubtful about the purity of the RNA solution, a good remedy would be to perform RNA clean up using spin-column chromatography such as using the RNeasy kit sold by Qiagen.

   A high integrity (undegraded) RNA is essential for quantitative results; however, because the PCR target is normally short (100–400 bp), the RT-PCR method can be performed on partially degraded RNA for qualitative assessment of gene expression. However, if using partially degraded RNA, the use of oligo dT may not be as efficient as the use of random primers for the broken RNA strands. The oligo dT extension products will have limited use and can work only for PCR targets near the 3′end.

8. When assembling the RT reaction, one must ensure annealing of the RT primer to the complementary RNA template. The RNA is mixed with the RT-primer and is heated to relax the RNA strands and reduce secondary structure. With heat exposure, hairpin structures are broken and the RNA molecules become more linear, and the primers can easily find their complementary match.

9. In several molecular biology methods, when processing several samples simultaneously, it is recommended to prepare a premix of several components, which not only speeds up the process by reducing the number of pipetting steps, but also increases the accuracy/reproducibility of the reaction mixture among various samples. In this case, it is suggested to prepare a premix for the number of samples+1 additional in order to provide a little extra volume. This way you will insure that there is enough volume for the individual reactions. When preparing a premix of several reagents, it is critical to mix the solution very well to insure homogeneity, and spin down briefly to collect the volume in the bottom of the tube, before allocating it into individual reactions.
10. During the RT reaction, the RT enzyme is reading the RNA and making a complementary copy of the sequence. At lower temperature, i.e., 37°C, the RNA is more subjected to secondary structure such as hairpin loops whereby the RNA folds locally on itself. These hairpin loops create a hindrance for the DNA polymerase and slow down its progress along the RNA, and lead to early termination of the cDNA synthesis. Thus, for long cDNA synthesis, one must keep the RNA in a more relaxed configuration to allow cDNA synthesis to occur without hindrance, and this is achieved by carrying out the reaction at higher temperature. We found the SuperScript II retains its full activity at 50°C. For specific applications, Invitrogen designed another form of M-MLV-RT named Superscript III. SuperScript III Reverse Transcriptase has been engineered to reduce its RNase H activity and provide increased thermal stability. The enzyme is used to synthesize cDNA at a temperature up to 55°C, providing increased specificity, higher yields of cDNA, and more full-length product than other reverse transcriptases (Invitrogen).

    Furthermore, in 1998, it was discovered that RT becomes significantly thermostable and active in the presence of a disaccharide, trehalose (5). With increased thermostability, the RT reaction could be carried out at a higher temperature (60°C), at which the formation of fewer RNA secondary structures can increase the number of full-length cDNAs. However, at higher temperature, one needs to make sure that the primers can efficiently anneal. The use of long oligo dT(20-25) primers or gene specific primers are most suitable for this purpose.
11. The preparation of cDNA is a procedure for many molecular biology applications. It is an essential tool for PCR amplification of RNA, cDNA cloning, in the preparation of cDNA libraries, and in microarray technology. Each application requires its own specific condition. The cDNA prepared as described in this chapter can be used directly in end point PCR or real time PCR for the analyses of specific gene expression.

12. It is noted throughout this procedure, in performing the RT-reaction to prepare the cDNA prior to PCR, there are a number of options and variations that can be optimized depending on specific needs. Please note that for quantitative comparisons of gene expression across many samples, the RT conditions must be optimized and kept unchanged for all the samples to be compared. These conditions include the type and the amount of RNA being reverse transcribed, the priming strategy, inclusion or exclusion of RNase inhibitor, the reaction incubation temperature, and duration. It is known that small variations in RT condition can be amplified many fold in subsequent PCR analyses (exponential factor). Thus, for accurate RT-PCR data interpretation it is critical to keep the RT conditions the same in order to minimize experimental variability.

Acknowledgments

This work was supported by grants HL-73473 and AR-30346 for KMB from the National Institute of Health, of which F Haddad is a researcher and a co-investigator.

The authors would like to thank Anqi Qin, Weihua Jiang, and Li Ying Zhang for skillful technical assistance.

References

1. Baltimore D (1970) RNA-dependent DNA polymerase in virions of RNA tumour viruses. Nature 226:1209–1211
2. Temin HM, Mizutani S (1970) RNA-dependent DNA polymerase in virions of Rous sarcoma virus. Nature 226:1211–1213
3. Haddad F, Qin A, Giger J, Guo H, Baldwin K (2007) Potential pitfalls in the accuracy of analysis of natural sense-antisense RNA pairs by reverse transcription-PCR. BMC Biotechnology 7:21
4. Wright C, Haddad F, Qin AX, Baldwin KM (1997) Analysis of myosin heavy chain mRNA expression by RT-PCR. J Appl Physiol 83: 1389–1396
5. Carninci P, Nishiyama Y, Westover A, Itoh M, Nagaoka S, Sasaki N, Okazaki Y, Muramatsu M, Hayashizaki Y (1998) Thermostabilization and thermoactivation of thermolabile enzymes by trehalose and its application for the synthesis of full length cDNA. Proc Natl Acad Sci USA 95:520–524

Chapter 18

Primer Design for RT-PCR

Kelvin Li and Anushka Brownley

Abstract

Primer design is a crucial initial step in any experiment utilizing PCR to target and amplify a known nucleotide sequence of interest. Properly designed primers will increase PCR amplification efficiency as well as isolate the targeted sequence of interest with higher specificity. Many factors that may limit the success of a primer pair can be detected a priori with computational methods. For example, primer dimer detection, amplification of alternative products, stem loop interference, extreme melting temperatures, and genotype-specific variations in the target sequence can all be considered computationally to minimize subsequent PCR failures. The use of computational sequence analysis tools to select the best primer pair from the available candidates will not only reduce experimental rates of failure but also avoid the generation of misleading results arising from the amplification of alternative products.

Key words: Primer design, Targeted amplification, Primer dimer, Alternative products

1. Introduction

The selection of primer pairs for reverse transcription polymerase chain reaction (RT-PCR) is critical in the specific amplification of a target sequence. The sequence flanked by the annealed primer pairs is amplified during the RT-PCR reaction. If the chosen primer pair sequences do not anneal to the sample RNA efficiently or anneal to a sequence that is inhibitory to any of the steps of the PCR reaction, then the efficiency of the RT-PCR reaction is decreased and will result in little or no product yield. If the selected primer pairs also anneal to and flank sequence that was not targeted, then the RT-PCR reaction will also amplify unwanted regions of RNA, leading to a heterogeneous population of products. It is possible to computationally predict when these situations will occur, a priori, so they may be avoided.

A variety of tools exist for performing primer design computationally. Please see Notes 7–9 for their descriptions in greater depth.

Nicola King (ed.), *RT-PCR Protocols: Second Edition*, Methods in Molecular Biology, vol. 630,
DOI 10.1007/978-1-60761-629-0_18, © Springer Science+Business Media, LLC 2010

In this chapter, a theoretical explanation of the sequence analyses that should be considered and the practical application of those theories with computational methods will be presented. RT-PCR primer design builds upon the same theoretical and computational foundations that apply to general PCR primer design, so the tools and analyses employed are identical.

2. Materials

1. Computer with Internet access: Computational sequence analysis is most conveniently performed on a UNIX operating system using the existing bioinformatics tools that are available for download freely through the Internet. Many of the bioinformatics software packages that are used to design PCR primers can be downloaded from SourceForge.net (1).
2. Reference Sequence: In order to conduct PCR primer design, a reference sequence is needed to identify the exact sequence being targeted and from where to select primer pair candidates. Ideally, the reference sequence consists of not only the regions being targeted for amplification but also all the nucleotide sequences that will be present in the sample set used during the amplification process. This is crucial to ensure that the designed primers do not amplify alternative products.
3. Primer3: a widely used computer program for generating a list of primer pair candidates to target and amplify a specified nucleotide sequence (2). Primer3 assigns a score to each primer pair candidate by evaluating various characteristics based on user-specified design goals. Primer3 is available for download at: http://primer3.sourceforge.net/.
4. EMBOSS: an Open Source software package containing numerous bioinformatics tools which are extremely useful for sequence analysis (3). The following are the EMBOSS tools that will be used in primer design.
   (a) `palindrome`, a tool for detecting inverted repeats in a nucleotide sequence, which can be used to search for stem loop structures.
   (b) `dreg`, a tool for searching through a nucleotide sequence with a regular expression, which can be used to validate the specificity of degenerate primers.
   (c) `consambig`, a tool for creating an ambiguous consensus sequence from a multiple sequence alignment (MSA), which can be used to generate a reference sequence representing multiple genotypes.
   The EMBOSS suite of tools can be downloaded from: http://emboss.sourceforge.net/.

5. BLAST (Basic Local Alignment Search Tool): a program used to search for similarities between two sequences (4). In primer design, BLAST may be configured to search for primer binding sites on a set of sequences to predict alternative product amplification. BLAST is available at the NCBI website: http://www.ncbi.nlm.nih.gov/BLAST/download.shtml.
6. CLUSTAL: a software tool for performing MSA (5). Clustalw is the command line version of the program, while clustalx provides a graphical user interface. Both versions are available from the following website: http://www.clustal.org/.

3. Methods

Primers can be designed as regular or degenerate. When regular primers are synthesized, only two populations of primers are present in the PCR reaction: the forward primer and the reverse primer populations. When degenerate primers are synthesized, there may be multiple forward and reverse primer variants corresponding to an enumeration of the ambiguity codes (see Note 1) represented in the degenerate primer. For example, if a forward primer is designed to contain an N at a single position (see Note 2), then all four forward primer variants are synthesized and will be present in the PCR reaction, in addition to the reverse primer. Degenerate primers are used to amplify targeted sequence across multiple genotypes by allowing variants to occur in the primer binding site.

The steps for designing regular primers are different from those used to design degenerate primers. To clarify when these differences occur, the two methods will be separated by a header for 'Regular' and 'Degenerate' primer design.

3.1. Reference Sequence Generation

The reference sequence is obtained in two different ways depending on the type of primer design being conducted: regular or degenerate.

3.1.1. Regular Primer Design Reference Sequence

Regular primer design requires a set of reference sequences that contains only the nonambiguous nucleotides (T, G, A, and C) or N. The reference sequences should include all nucleotide sequences that will be present in the sample during the PCR amplification. These may include mRNA, genomic DNA (gDNA), or any other expected sequences that will be present. The sequences can be retrieved from prior sequencing experiments or from online sequence repositories. Some useful online repositories include the National Center for Biotechnology Information (NCBI), http://www.ncbi.nlm.nih.gov/, and the European Bioinformatics Institute, http://www.ebi.ac.uk/.

3.1.2. Degenerate Primer Design Reference Sequence

Degenerate primer design expects a set of references sequences containing degenerate nucleotides that represent a sample with multiple genotypes. These sequences can be retrieved from prior sequencing experiments, from online sequence repositories or by generating a consensus with ambiguity codes based on an alignment of sequences from a representative population.

The following outlines how to generate a consensus with ambiguity codes based on a set of representative sequences from a population.

1. Format the set of representative sequences into a FASTA file to be used as input into `clustalw`. The following is an example of an input FASTA file named, `sequences.fasta`.

```
>sequence A
ctgctataggtgcagctcttggtgactgcaaattacttac
taatctctgt
>sequence B
attggtgaagctcttggtaactggaaattacttac
taatctctgt
>sequence C
gatggaaatactgctattggtgcggctcttggtaactg
caaattacttactaa
>sequence D
gatggaaatactgctattggtgcggctcttggtaactg
caaatta
>sequence E
gatggaaatactgctattggtgcggctcttggtaactg
>sequence F
Gatggaaatactgctattggtgcggctcttggtaa
ctgcaaattacttactaatctctgt
```

2. Execute `clustalw` on the command line as follows (see Table 1, for a line by line explanation):

```
1. clustalw\
2. sequences.fasta
```

It is important that the sequences within the input FASTA file are highly similar. If the contents of the FASTA file are a mixed

Table 1
Explanation of command line invocation of clustalw

Line number	Value	Description
1	`clustalw`	Program name for computing a multiple sequence alignment given a set of sequences
2	`sequences.fasta`	Sequence input file name. This is the set of similar sequences from which you are trying to generate a consensus.

population of unrelated sequences, `clustalw` will include all sequences in the MSA, even if the global alignment is poor and unusable.

The `clustalw` output is automatically saved into a file called `sequences.aln` in the same directory as the input file. The following is a sample output file that corresponds to running `clustalw` on the `sequences.fasta` sample input file.

```
CLUSTAL W (1.83) multiple sequence alignment

sequence_D      GATGGAAATACTGCTATTGGTGCGGCTCTTGGTAACTGCAAATTA---------------
sequence_E      GATGGAAATACTGCTATTGGTGCGGCTCTTGGTAACTG----------------------
sequence_C      GATGGAAATACTGCTATTGGTGCGGCTCTTGGTAACTGCAAATTACTTACTAA-------
sequence_F      GATGGAAATACTGCTATTGGTGCGGCTCTTGGTAACTGCAAATTACTTACTAATCTCTGT
sequence_A      ----------CTGCTATAGGTGCAGCTCTTGGTGACTGCAAATTACTTACTAATCTCTGT
sequence_B      ---------------ATTGGTGAAGCTCTTGGTAACTGGAAATTACTTACTAATCTCTGT
                                ** ****  ********* ****
```

3. Compute the consensus sequence by collapsing all the nucleotide sequences in the MSA into a single representative sequence with the help of IUPAC ambiguity codes (see Note 1). Execute `consambig` from the command line using the sample input file, as follows (see Table 2, for a line by line explanation):

```
1 consambig \
2 -sequence sequences.aln \
3 -outseq sequences.cons
```

Table 2
Explanation of command line invocation of consambig

Line number	Value	Description
1	`consambig`	Program name for computing a consensus sequences using the ambiguity codes to represent variations across the input MSA.
2	`-sequence sequences.aln`	Input file name containing the MSA. This is the same file that was outputted from `clustalw`.
3	`-outseq sequences.cons`	Output consensus sequence file name. If you omit this option, `consambig` may overwrite your original `sequences.fasta` file.

The output of `consambig` will be saved in `sequences.cons` and will contain the consensus sequence based on the input MSA. Below is the output consensus sequence derived from the sample input file, `sequences.aln`:

```
>EMBOSS_001
gatggaaatactgctATWGGTGMRGCTCTTGGTRACTG
saaattacttactaatctctgt
```

This sequence can be used as the reference for degenerate primer design. It is important to note that where the depth of coverage is sparse (i.e. where only a few sequences cover a particular region) the consensus sequence may have lower integrity. It will be at the experimenter's discretion to decide if the consensus is reliable enough to be a template for primer design.

3.2. Primer Pair Candidate Selection

For a specified target region, multiple primer pairs can be chosen as candidates to amplify the sequence. Characteristics of both the primers and their resultant amplicon can disqualify primer pairs as successful candidates for amplifying the target region. In order to maximize the number of primer pair candidates to choose from, it is important to ensure that there is a sufficient amount of flanking sequence around the target region from which to search for primers. The target region and the primer search space flanking it are considered part of the "template" sequence. Primers should not overlap the target region since allelic dropout could occur at the primer binding site when variations are present. Assuming there is sufficient flanking search space, the greater the amplicon size, with respect to the targeted region, the greater the number of possible candidate primer pairs. The recommended flanking search space on each side of the target is (Max Amplicon Length – Max Primer Length – Target Length), where the maximum amplicon/primer length is the longest desired amplicon/primer length. See Fig. 1.

3.2.1. Primer3

Primer3 is a widely used program for generating PCR primer pair candidates.

1. Generate an input file called `input.bio`. This input file uses the "BoulderIO" format and requires that each line contains exactly one '`field name=field value`' pair. This file defines

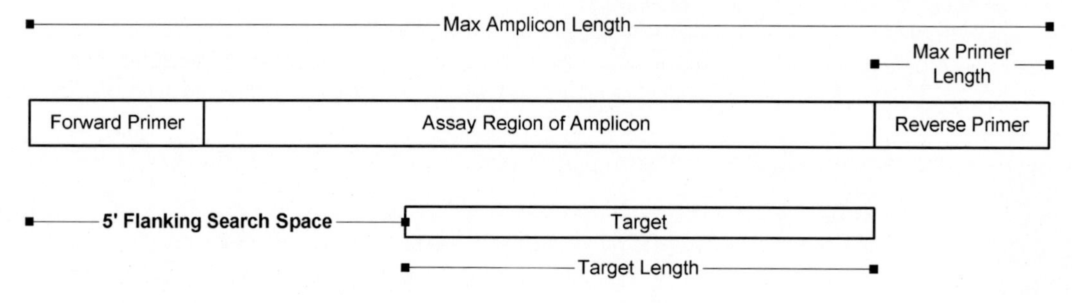

Fig. 1. Description of relationship between primer, amplicon, and target region

the template sequence, the targeted region on the template sequence, and the primer design constraints, based on laboratory protocols (see Note 6), that will be used by Primer3. If the template sequence that is to be inputted into Primer3 has degenerate bases, then those bases MUST be masked with N's (see Note 4). Also for degenerate primer design, the input parameter PRIMER_NUM_NS_ACCEPTED should be set to the maximum number of acceptable degenerate bases in each primer. The last line of the input file should be a single equal sign ('=') on its own. The following is an example of an input.bio file. Only the most frequently used parameters have been included in the example below (see Table 3, for a line by line explanation):

Table 3
Explanation of the BolderIO format required for input into primer3

Line number	Value	Description
1	SEQUENCE	Template sequence, including flanking regions, on a single line
2	TARGET	Start position and length of targeted region within the input template sequence
3 4 6 7 8	PRIMER_PRODUCT_MIN_TM PRIMER_PRODUCT_MAX_TM PRIMER_MIN_TM PRIMER_OPT_TM PRIMER_MAX_TM	Melting temperature constraints for amplicon products and primers
5	PRIMER_PRODUCT_SIZE_RANGE	Size constraints for the amplicon product
9 10 11	PRIMER_MIN_SIZE PRIMER_OPT_SIZE PRIMER_MAX_SIZE	Length, in base pairs, of the minimum, optimal, and maximum primer size
12 13	PRIMER_SALT_CONC PRIMER_DNA_CONC	Salt and DNA concentrations used during PCR amplification. Primer3 uses this information to calculate primer and amplicon melting temperatures
14	PRIMER_EXPLAIN_FLAG	Flag that sets whether to report the number of eliminated candidates and an explanation of what caused their elimination.
15	PRIMER_NUM_RETURN	The number of candidates to be returned. Since additional criteria downstream of Primer3 may eliminate candidates, it is important to try to receive as many candidates as possible from Primer3
16	PRIMER_NUM_NS_ACCEPTED	The number of N's to allow in the design of the primer. This may be set to a nonzero value when designing degenerate primers

```
1  SEQUENCE=TAGAAAGAT...entire sequence not included...
   GAACCTTGAAAACATTA
2  TARGET=950,329
3  PRIMER_PRODUCT_MIN_TM=65
4  PRIMER_PRODUCT_MAX_TM=85
5  PRIMER_PRODUCT_SIZE_RANGE=550-800
6  PRIMER_MIN_TM=55
7  PRIMER_OPT_TM=65
8  PRIMER_MAX_TM=75
9  PRIMER_MIN_SIZE=18
10 PRIMER_OPT_SIZE=20
11 PRIMER_MAX_SIZE=25
12 PRIMER_SALT_CONC=50.0
13 PRIMER_DNA_CONC=120.0
14 PRIMER_EXPLAIN_FLAG=1
15 PRIMER_NUM_RETURN=10000
16 PRIMER_NUM_NS_ACCEPTED=0
17 =
```

2. Run primer3 on the command line as follows:

```
1 primer3 < input.bio > output.bio
```

Primer3 outputs a list of candidate primer pairs in the same BolderIO format. The following is an example of the file, output.bio, which contains the resultant candidate primer pairs (see Table 4, for a line by line explanation):

...Input is repeated...

```
1 PRIMER_LEFT_EXPLAIN=considered 7324, GC content failed 1114, low tm 3132, high tm 4, high any compl 78, high end compl 369, long poly-x seq 22, ok 2605
2 PRIMER_RIGHT_EXPLAIN=considered 7441, GC content failed 604, low tm 2388, high tm 62, high end compl 125, long poly-x seq 81, ok 4181
3 PRIMER_PAIR_EXPLAIN=considered 492922, unacceptable product size 465261, high end compl 1080, ok 26581
4 PRIMER_PAIR_PENALTY=0.3059
5 PRIMER_LEFT_PENALTY=0.103719
6 PRIMER_RIGHT_PENALTY=0.202185
7 PRIMER_LEFT_SEQUENCE=AGCACCACCACTACCGATGC
8 PRIMER_RIGHT_SEQUENCE=TTTCCACAATGG
  CATCACCA
```

Table 4
Explanation of selected output in BolderIO format produced by primer3

Line number	Value	Description
1 2 3	PRIMER_LEFT_EXPLAIN PRIMER_RIGHT_EXPLAIN PRIMER_PAIR_EXPLAIN	Useful fields for determining why candidate primers and primer pairs were eliminated
9 10	PRIMER_LEFT PRIMER_RIGHT	Candidate primer pair position information needed for primer design. The actual primer sequences are extracted from the reference sequence, since the input template may be N-masked
11 12	PRIMER_LEFT_TM PRIMER_RIGHT_TM	Forward and reverse primer melting temperatures

```
 9 PRIMER_LEFT=813,20
10 PRIMER_RIGHT=1573,20
11 PRIMER_LEFT_TM=64.896
12 PRIMER_RIGHT_TM=64.798
13 PRIMER_LEFT_GC_PERCENT=60.000
14 PRIMER_RIGHT_GC_PERCENT=45.000
15 PRIMER_LEFT_SELF_ANY=3.00
16 PRIMER_RIGHT_SELF_ANY=5.00
17 PRIMER_LEFT_SELF_END=2.00
18 PRIMER_RIGHT_SELF_END=2.00
19 PRIMER_LEFT_END_STABILITY=8.1000
20 PRIMER_RIGHT_END_STABILITY=8.2000
21 PRIMER_PAIR_COMPL_ANY=5.00
22 PRIMER_PAIR_COMPL_END=3.00
23 PRIMER_PRODUCT_SIZE=761
24 PRIMER_PRODUCT_TM=71.8832
25 PRIMER_PRODUCT_TM_OLIGO_TM_DIFF=7.0854
26 PRIMER_PAIR_T_OPT_A=54.8576
   ... similar information reported for each candidate primer pair
   suggested...
27 =
```

The other information in this file is not considered because subsequent steps will use the values in Table 4 to filter out potentially unsuccessful primer pairs by applying more specialized methods to evaluate additional primer pair characteristics.

3. For degenerate primer design, extract the original primer sequences from the reference to ensure that the original ambiguity codes in the reference sequence are used in the primer sequence. If running a regular primer design, this is not an issue.

3.3. Melting Temperatures

There are two types of melting temperatures (T_m) considered during primer design: the primer T_m, and the amplicon T_m. An extreme primer or amplicon T_m is a common explanation for PCR failure. Fortunately, existing equations can be used to easily estimate T_m's for the primer and amplicon sequences. One problem with these calculations is that both the DNA and salt concentrations are necessary inputs into the equation. However, many other reagents are present in a PCR reaction that are not accounted for in the equation and may shift the actual melting temperatures away from the predicted ones. To resolve this issue, the GC content of the primers and amplicons should be tracked along with their success rates to determine the optimal GC content range for the specific amplification protocol (see Note 6). The following two sections provide information about how to use the melting temperature estimations to predict potential PCR failures.

3.3.1. Primer Melting Temperatures

1. For each primer, retrieve the primer melting temperatures calculated by `primer3`. Eliminate a primer if its T_m is out of the desired melting temperature range. For the "Standard" protocol described in (10), we found that a melting temperature range between 60 and 68°C results in a reliable set of primers. The "High GC" protocol, also described in (10), is able to support primer melting temperatures up to 85°C.

   `Primer3` uses a technique proposed by Rychlik et al. (6) that is based on empirical values from the nearest neighbor thermodynamic calculations determined by Breslauer et al. (7). Melting temperatures for sequences that are less than or equal to 36 bp are more accurately calculated with this technique compared to the technique used for amplicon-length sequences, described in the next section.

2. Eliminate primer pairs that have a T_m difference of greater than 7°. By minimizing the difference between the forward and reverse primer melting temperatures, imbalances in strand synthesis, which may decrease PCR efficiency, are avoided.

3.3.2. Amplicon Melting Temperatures

1. Calculate the amplicon melting temperature using the formula by Sambrook et al. (8) that is based on the measurements performed by Bolton et al. (9). The formula is as follows:

$$T_m = 81.5 + 16.6\left(\log_{10}\left(\left[Na^+\right]\right)\right) + 0.41 \times (\%GC) - 600 / \text{length}$$

   Although this is a convenient formula for estimating the melting temperature of a nucleotide sequence on average in an isolated environment, it has a few important shortcomings. Using the simplified average %GC computation hides important information about localized high and low GC regions along the amplicon. These variations could result in poor PCR efficiency due to incomplete melting of the copy strand from the template strand.

2. In order to detect these local irregularities, use a sliding window of 20 bps to calculate the GC content along the length of the amplicon. Use the %GC measurement in each of these windows to calculate the standard deviation of GC content across the whole amplicon. Use the equation above to calculate the upper and lower bound of T_ms using the average %GC ± the standard deviation. The amplicon is considered acceptable if both of these T_ms fall within the desired amplicon melting temperature range. For the "Standard" and "High GC" protocols described in (10), the amplicon temperature ranges supported were between 70–81.5°C and 70–95°C, respectively.

3.4. Detection of Alternative PCR Products

The detection of potential alternative PCR products is essential to ensure that the amplification of nontarget sequence during PCR is avoided. Alternative PCR products can be computationally calculated by searching for primer and amplicon sequences in the full set of nucleotide sequences present in the sample that is used during PCR. Of course, this means that the nucleotide sequences in the sample must be known. Detection of alternative PCR products for RT-PCR samples may be more complicated than for whole gDNA PCR, depending on the composition of the initial mRNA sample. An understanding of what other sequences may exist in the sample, in addition to the sequence that is being targeted, is crucial for ensuring the specificity of the designed primers. In the simplest case involving the RT-PCR of bacterial mRNA, a set of the bacterial mRNA sequences would be sufficient to test for primer specificity. In the case of eukaryotic mRNA, a library of alternate splice variants would be necessary. In an even more complicated scenario, where genomic contamination of the mRNA sample is expected, the gDNA of the organism from which the mRNA's were collected could be used to increase primer pair specificity.

3.4.1. Primer Binding Search

The determination of primer specificity can be parameterized by two criteria:

(a) Minimum 3′ critical cutoff

This value represents the minimum length of the alignment between the 3′ end of the primer and the reference sequence necessary for the location to be considered a potential primer binding site. This is because only the 3′ end of the primer needs to anneal for the polymerase to bind to the primer/template complex and initiate the copy strand's synthesis. The 5′ end of the primer improves the thermodynamic stability of the primer/template complex and increases the primer's efficiency, but is not sufficient to guarantee specific binding. Decreasing the minimum 3′ critical cutoff will increase the number of false positive primer binding sites detected. This increase in sensitivity for determining potential primer binding sites will also increase the number of false positive alternative products that are detected. Over-calling alignments as potential primer binding sites may needlessly disqualify the potentially successful primer pair candidates.

(b) Maximum amplifiable amplicon length

This value represents the longest expected PCR product that could be amplified with the PCR protocol being used (see Note 6).

3.4.1.1. Regular Primer Binding Search

1. For each primer, use BLAST to search for alignments in the reference sequences. The following is an example of how BLAST should be executed on the command line (see Table 5, for a line by line explanation):

```
1 blastall \
2 -p blastn \
3 -W 7 \
4 -G 3 \
5 -e 7000 \
6 -z 3100000000 \
7 -b 1000000 \
8 -i primer.fasta \
9 -d transcripts.fasta \
10 > blast_results
```

2. Filter BLAST results that do not meet the following criteria: 100% identity along 11 bps on the 3′ end of the alignment

3.4.1.2. Degenerate Primer Binding Search

1. Use `dreg` to search for primer binding locations in the reference with ambiguity codes. `Dreg` searches through a sequence

Table 5
Explanation of command line invocation of blastall

Line number	Value	Description
1	`blastall`	BLAST program with both protein and nucleotide search capabilities combined in one executable.
2	`-p blastn`	Program Name: Specifies that a nucleotide sequence (query) will be searched against another nucleotide sequence (database).
3	`-W 7`	Word Size: Specifies the length of the exact match seed sequence that must be found in order to extend an alignment computation around it. Smaller word sizes slow down the BLAST search, but are required for the sensitivity we need to search for short primer sequences.
4	`-G 3`	Gap Penalty: A positive gap penalty encourages gaps to be opened.
5	`-e 7000`	Expect Value: This is a score that BLAST assigns to the alignment that predicts the likelihood that the sequence is similar to the database sequence by chance. A lower likelihood is a better alignment. Alignments with E-values above the specified cutoff are filtered out. Since we perform additional downstream checks, the cutoff value should be set high enough to ensure that no alignments that may meet our criteria are discarded prematurely.
6	`-z 3100000000`	Database Size: The Expect Value is based on the database (transcript set) size. In order to have consistent Expect Value scores across alternative runs, this value should remain constant. If you chose an alternative database size, you will also need to determine the appropriate Expect Value cutoff to achieve the same result set.
7	`-b 1000000`	Maximum Number of Alignments to Show: Since you may be expecting the primer to align more frequently than the default setting, you should set this to a sufficiently large number so no potential primer binding sites are discarded.
8	`-i primer.fasta`	Name of the query (primer) sequence FASTA file.
9	`-d transcripts.fasta`	Name of the database (transcripts) sequence FASTA file.
10	`>blast_results`	Name of the output file to which the BLAST results should be saved.

for a user-specified pattern, called a "regular expression." A regular expression is capable of unambiguously describing complex string patterns (see Note 3). When matching degenerate primers to a reference, only the first 11 nucleotides from the 3′ end of the primer need to be aligned. For example, to search for the primer, AGTTGGCGACCCTTTCATMCGH, you will need to construct a regular expression only for the underlined portion of the primer to search for its binding sites.

Table 6
Explanation of command line invocation of dreg

Line number	Value	Description
1	dreg	Regular expression search tool program name.
2	-sequence transcripts.fasta	Sequence input file name: This should be the reference sequences of your transcripts.
3	-pattern	Regular expression qualifier: The regular expression should follow this flag.
4	"[BCHMNSVY][BCHMNSVY] [BDHKNTWY][BDHKNTWY] [BDHKNTWY][BCHMNSVY] [ADHMNRVW][BDHKNTWY] [ABCDHMNRSVWY][BCHMNSVY] [BDGKNRSV][ABCDHKMNRSTVWY]"	Regular expression representation of the 3′ end of the primer sequence: CTTTCATMCGH.
5	-rdesshow3	Indicates that the description following the ID in the input sequence's defline should be reported. See Note 5 for a description of the FASTA file format and an explanation of its components.
6	-stdout	Send output to standard output. In this case, the standard output will be saved to the file following the > sign.
7	-auto	Turns off the prompting for each default parameter.
8	> dreg.results	Saves results to the specified file, dreg.results.

The following is an example of how to run Dreg on the command line (see Table 6, for a line by line explanation):

```
1 dreg \
2 -sequence transcripts.fasta \
3 -pattern \
4 "[BCHMNSVY][BCHMNSVY][BDHKNTWY][BDHKNTWY]
  [BDHKNTWY][BCHMNSVY][ADHMNRVW]
  [BDHKNTWY][ABCDHMNRSVWY][BCHMNSVY]
  [BDGKNRSV][ABCDHKMNRSTVWY]" \
5 -rdesshow3 \
6 -stdout \
7 -auto \
8 > dreg.results
```

(a) Note that `dreg` only searches on the forward strand. In order to search for primer binding sites on the reverse strand, the 3′ end of the primer sequence must be reverse complemented and another `dreg` search must be performed with a new regular expression.

(b) The `dreg` results from both the forward and reverse search must be stored for downstream filtering. The orientation of the reverse results will need to be reversed when combined with the forward results in order to have primers properly mated for alternative product detection. See Subheading 3.4.2 for completing the remaining primer specificity calculations.

3.4.2. Mating Primer Pairs

Candidate primer pairs are eliminated if more than one amplification product is computationally predicted. If the following three conditions are met by any pair of primers, then that pair is likely to amplify a product. Note that these conditions should also be applied to forward–forward and reverse–reverse primer pair combinations.

(a) Primers must bind proximally to each other within the maximum PCR product size.

(b) Primers must bind on opposite strands

(c) The 5′ end of each primer should flank the PCR product.

1. For each contiguous reference sequence, sort the list of forward and reverse primers by the start position of each primer's alignment to the reference according to the primer binding search results.
2. For every primer on the forward strand in the sorted list, count the number of potential amplification products that may exist. A potential amplification product may exist if a primer on the forward strand is proximal (*see* Subheading 3.4.1) to any downstream primer that aligns on the reverse strand.
3. If more than one potential amplification product is identified, then eliminate the candidate primer pair, because it is nonspecific.

3.5. Primer Dimers

Primer dimers occur when primers anneal to each other and, as a result, are unable to bind to the template strand to prime the PCR reaction. Primer dimers lower PCR efficiency, thus reducing the concentration of the amplified sequence. There are two types of primer dimer that need to be detected: end and internal. End primer dimer occurs between the 3′ ends of the PCR primer. Internal primer dimers occur when a significant enough number of internal bases are complementary, but this is less problematic than end primer dimers.

1. If necessary, tail the forward and reverse PCR primers with any sequencing primers or adaptor sequences that will be used in downstream processing. Primer3 will eliminate PCR primer candidates that result in primer dimers, but will not take into account additional sequences that may be appended to the PCR primers. Primer3 also does not take into account degenerate bases when computing primer dimers.
2. If degenerate primers are being designed, then each disambiguated primer (see Note 2) should be enumerated and used to detect primer dimers.
3. Compute a 2-D matrix that captures the complementary relationship between bases for each of the forward–forward, reverse–reverse, and forward–reverse primer combinations.
   (a) Arrange one primer at the top of the matrix from left to right and the other primer on the left of the matrix from bottom to top, where both sequences occur in the 5′–3′ direction. In Fig. 2, the forward primer sequence is AAAGGCA, and the reverse primer sequence is CAATGCC.
   (b) Mark positions in the matrix where the nucleotides are complementary according to Watson-Crick pairing rules. Watson-Crick pairing rules insist that As and Ts, and Gs and Cs, will always pair together.
   (c) After the entire matrix has been marked, look for the longest consecutive stretch of complementary base pairing, proceeding diagonally from top left to bottom right. In Fig. 2, four bases at the end of the primers are complementary and, therefore, may result in end primer dimer.

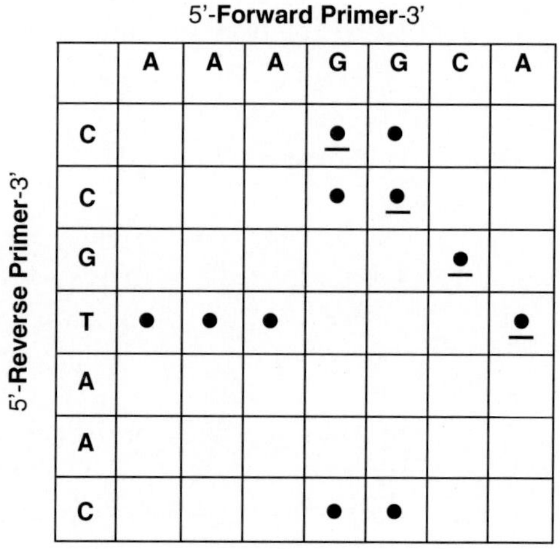

5'-Forward Primer-3'

5'-Reverse Primer-3'

	A	A	A	G	G	C	A
C				●	●		
C				●	●		
G						●	
T	●	●	●				●
A							
A							
C				●	●		

Fig. 2. Primer dimer calculation matrix

4. For internal primer dimers, eliminate the primer pair candidate if a complementary stretch of eight or more base pairs is detected anywhere away from the 3′ end.
5. For end primer dimer, eliminate the primer pair candidate if a complementary stretch of three or more base pairs is detected at the 3′ ends of the primers, as in Fig. 2.

3.6. Stem Loop Interference

Stem loop interference occurs when the DNA or mRNA sequence anneals to itself. This can take place anywhere along the length of the amplicon, but when it occurs at the primer binding site, PCR amplification is inhibited. Stem loop interference can be detected by searching for pairs of palindromic sequences since these can anneal to each other and form a stable secondary structure.

Once palindromic elements have been detected in the amplicon sequence, if the following criteria are met, then stem loop interference will occur:

1. The stem of the structure contains greater than 14 hydrogen bonds. The number of hydrogen bonds is calculated by multiplying 3 or 2 times the number of G/C or A/T residues in the stem, respectively. In Fig. 3, there are 3×3 H-bonds (1G+2C's)+2×5 H-bonds (3T's+2A's), or a total of 19 hydrogen bonds involved in the stem formation.
2. The loop structure contains less than 12 residues.

3.6.1. Stem Loop Detection

1. Detect palindromes with the help of an application from the EMBOSS suite of bioinformatics tools named, palindrome. The amplicon sequence should be used as input into palindrome. The following is an example of how to run palindrome on the command line (see Table 7, for a line by line explanation):

```
1 palindrome \
2 -sequence amplicon.fasta \
3 -minpallen 7 \
4 -gaplimit 11 \
5 -outfile results.pal \
6 -filter
```

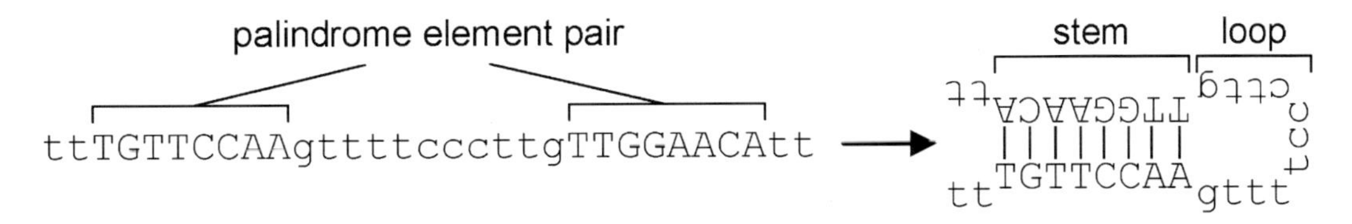

Fig. 3. Palindromic elements lead to stem loop formation

Table 7
Explanation of command line invocation of palindrome

Line number	Value	Description
1	`palindrome`	Program name from the EMBOSS suite of bioinformatics tools which will search for palindromes.
2	`-sequence amplicon.fasta`	Sequence input file name: This should be the input sequence of the amplicon you are analyzing. You may pass in a multi-FASTA of sequences.
3	`-minpallen 4`	Minimum nucleotide length of palindrome to report: You will need to set this to your (minimum stem H-bond cutoff (?))/3. This will ensure that the program will properly report the case of only G/C nucleotides in the stem.
4	`-gaplimit 11`	Maximum distance between palindrome pairs to report. This will represent the maximum loop size to accept.
5	`-outfile results.pal`	Palindrome report output file name: This is the name of the output file you wish to save the results into.
6	`-filter`	This forces `palindrome` to take its own default values, instead of prompting the user to replace them.

Table 8
Explanation of output from palindrome

Line number	Description
`1 - 8`	Summary of input data and run parameters
`11,15,21`	Start position, sequence, and stop position of first element found in a palindromic pair
`13,17,23`	Stop position, sequence, and start position of second element found in a palindromic pair

The following is an example of the palindrome output (see Table 8, for a line by line explanation):

```
1 Palindromes of: amplicon
2 Sequence length is: 770
```

```
3  Start at position: 1
4  End at position: 770
5  Minimum length of Palindromes is: 4
6  Maximum length of Palindromes is: 100
7  Maximum gap between elements is: 11
8  Number of mismatches allowed in
   Palindrome: 0
9
10 Palindromes:
11 26    caaa    29
12       ||||
13 37    gttt    34
14
15 36    tgta    39
16       ||||
17 51    acat    48
18
19 ... similar information reported for each palindrome pair
   detected...
20
21 710   tagc   713
22       ||||
23 718   atcg   715
```

2. Eliminate the primer pair if, for any pair of alignments in the palindrome result set, all of the following conditions are met:
   (a) The coordinates of the forward or reverse palindrome mates overlap with the forward and reverse primer binding sites.
   (b) The number of hydrogen bonds involved in the stem structure exceeds the pre-defined cutoff.
   (c) The distance between the two palindrome sequences is less than the loop length cutoff.

3.7. Low Complexity

Low-complexity regions are defined as sequences with a nonrandom distribution of a restricted number of bases. Most low complexity primers will be eliminated by the check for extreme melting temperatures or nonspecificity. However, since these primers may still bind to other low-complexity regions of the genome and decrease PCR efficiency, regardless of their primer mate, it is still important to eliminate them. Low complexity must be detected for both forward and reverse primers.

Table 9
Explanation of command line invocation of dust

Line number	Value	Description
1	`dust`	Low complexity sequence analyzer program name.
2	`primer.fasta`	Sequence input file name: This should be the input sequence of your primer. You have to test both forward and reverse primers for low complexity.
3	`> primer.fasta.lc_masked`	By default, the low complexity masked sequence will be sent to standard output. Pipe it into your output file for further analysis.

1. Generate a FASTA file containing the primer sequence as input into `dust`. The following is a sample input file named *primer.fasta*:

```
>primer
agcaaaacaaaaacaaacaaaacca
```

2. Execute `dust`, a subprogram of BLASTN, to search for low complexity regions. The following is an example of how `dust` is executed on the command line (see Table 9, for a line by line explanation):

```
1 dust \
2 primer.fasta \
3 > primer.fasta.lc_masked
```

The dust output file will have the primer sequence with the low complexity region masked with N's, as follows

```
>primer
agNNNNNNNNNNNNNNNNNNNNcca
```

3. Eliminate the primer if the number of N's is greater than or equal to 80% of the primer sequence.

3.8. Donor Specific Variations

Differences between the genotype of the reference transcript that the primers were designed on and the genotype of the samples being amplified may lead to primer failures, especially if variations are present toward the 3′ end of the primer binding site. Allelic dropout can also occur in metagenomic or nonhaploid samples if the primers are biased toward one population of genotypes. One way to avoid these problems is to use a priori information about ultravariable sequences, or conversely, highly conserved sequences, in order to limit the regions where primer pair candidates can fall. The validity of this application does, however, depend on whether or not the genotypes studied in previous samples sufficiently represent the current target sample.

For regular primer design, if any known variation is present in a primer binding site, then that primer should be eliminated. For degenerate primer design, no additional steps need to be performed if the variations are strictly substitutions. However, primers must be eliminated if they overlap with variations that are insertions or deletions (indels), since these indels cannot be represented with ambiguity codes.

3.9. Favoring Splice Sites

Genomic contamination is a frequent problem in RT-PCR. Fortunately, the differences between mRNA and gDNA can be exploited to minimize the effects. Two techniques in primer selection can be used to favor the amplification of transcripts over gDNA:

The first technique is to design primers that bind at the splice site between two exons. In the example illustrated below, the Forward Primer 1 binds at the splice site between exon 1 and exon 2. The mating primer may bind to both mRNA and gDNA, but since both primers must bind to ensure that an amplicon will form, the mRNA sequence will be uniquely amplified. The shortcoming of solely employing this technique is that the complete PCR of large transcripts with multiple tiling amplicons will be limited if exons are sufficiently larger than the achievable amplicon size. Also, since only the 3′ end of a primer is involved in specificity, the degree of latitude that a designed primer may have to span a splice site is very low.

The second technique to improve the amplification of transcripts over gDNA is to select primers that mate across an intron sufficiently larger than the PCR protocol can amplify. In Fig. 4, the distance between Forward Primer 2 and Reverse Primer 2 is

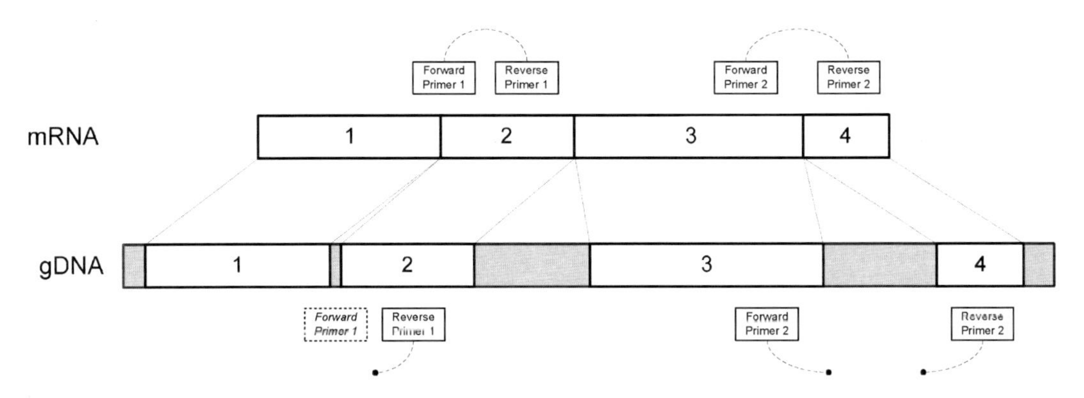

Fig. 4. Splice sites can be used to avoid genomic contamination

much greater in the genomic sequence, due to an intron, than it is in the mRNA. If the PCR range is shorter than the intron distance, then the gDNA will not be amplified. Since shorter amplicons tend to outcompete larger amplicons, this technique is likely to be effective even when the PCR protocol has a range long enough to amplify the gDNA.

To implement both techniques, the search for primer specificity needs to include not only the set of transcript sequences that are expected to be in the sample, but also the genomic sequence of the organism from which the sample was taken. By including both transcripts and gDNA as references in the primer binding site search, candidate primer pairs that amplify both genomic and transcript sequences will be detected by the same mate distance criteria and 3′ exact alignment characteristics employed to determine alternative amplifications. Also, by including alternatively spliced transcripts into the reference sequence, primer pairs can be selected to unambiguously amplify specific splice form junctions.

4. Notes

1. Table 10 lists the IUPAC ambiguity codes and the nucleotides that are represented.
2. The process of disambiguating a degenerate sequence involves enumerating all of the possible combinations represented by the degenerate sequence using only the standard nucleotides {C, A, T, G}. The IUPAC ambiguity codes {M, R, W, S, Y, K}, {V, H, D, B}, and {N}, each represent 2, 3, and 4 underlying nucleotides, respectively. For, example, to enumerate all

Table 10
IUPAC ambiguity code representations

Ambiguity code	Nucleotides represented	Ambiguity code	Nucleotides represented
M	A/C	V	A/C/G
R	A/G	H	A/C/T
W	A/T	D	A/G/T
S	C/G	B	C/G/T
Y	C/T	N	A/C/G/T
K	G/T		

possibilities for a sequence containing an H, requires three sequences, one for each case where the H is replaced by an A, C, and T. Each ambiguity code in a sequence will multiplicatively increase the number of enumerated sequences, depending on the number of standard nucleotides represented by the ambiguity code. For example, if a degenerate sequence is CATMCGH, the following sequences would be enumerated in the disambiguation process: CATACGA, CATACGC, CATACGT, CATCCGA, CATCCGC, and CATCCGT. Since M = {A, C} and H = {A, C, T}, there are $2 \times 3 = 6$ disambiguated sequences based on the degenerate sequence.

3. To write a regular expression that describes a sequence *without* ambiguity codes to search through a reference *without* ambiguity codes, the regular expression is the search sequence, verbatim. This represents both the order of and the specific nucleotides that need to be matched. To search for degenerate primers on a consensus reference sequence with ambiguity codes, the left bracket ('[') and right bracket (']') construct needs to be employed. The brackets delimit the nucleotides that represent a single nucleotide match in the reference sequence. For example, to allow for the matching of the D ambiguity code, the following letters in the reference sequence must be counted as a match: A, B, D, G, H, K, M, N, R, S, T, V, W, or Y. Since the D ambiguity code matches with A, G, or T, any other ambiguity code that represents an underlying A, G, or T, should be considered a match. Table 11 lists the reference letters that should be matched for each primer letter:

Table 11
IUPAC code match possibilities for every IUPAC code

Code	Reference match	Code	Reference match	Code	Reference match
A	ADHMNRVW	M	ABCDHMNRSVWY	V	ABCDGHKMNRSVWY
C	BCHMNSVY	R	ABDGHKMNRSVW	H	ABCDHKMNRSTVWY
G	BDGKNRSV	W	ABDHKMNRTVWY	D	ABDGHKMNRSTVWY
T	BDHKNTWY	S	BCDGHKMNRSVY	B	BCDGHKMNRSTVWY
		Y	BCDHKMNSTVWY	N	ABCDGHKMNRSTVWY
		K	BDGHKNRSTVWY		

Table 12
Explanation of command line invocation of PERL code that converts ambiguity codes to N

Line number	Value	Description
1	`perl`	Program name for the popular scripting processor
2	`-e`	Tells perl to execute the command following this flag
3	`'while(<>){if(!(/^>/)){$_=uc; ~s/[MRWSYKVHDB]/N/g;} print;}'`	For each line in the input, if the line is not a defline, uppercase the sequence and then convert all ambiguity codes to N. Output the converted sequence
4	`< inseq.fasta`	Input FASTA file name
5	`> outseq.fasta`	Output FASTA file name

For example, to search for the 3′ end of the example primer sequence, CTTTCATMCGH, the regular expression would be: [BCHMNSVY][BDHKNTWY][BDHKNTWY][BDHKNTWY][BCHMNSVY][ADHMNRVW][BDHKNTWY][ABCDHMNRSVWY][BCHMNSVY][BDGKNRSV][ABCDHKMNRSTVWY]. Note: The number of bracketed sets of letters is identical to the length of the primer sequence.

4. A quick way of converting ambiguity codes into N's is provided below in Perl (see Table 12, for a line by line explanation). Perl is a widely used scripting language that allows users to quickly manipulate text files and perform the string manipulations commonly used in bioinformatics sequence analysis. Having a basic fluency in Perl will save countless hours in the analysis and manipulation of data.

```
1  perl \
2  -e \
3  'while(<>){if(!(/^>/)){$_=uc;~s/[MRWSYKVHDB]/
   N/g;} print;}' \
4  < inseq.fasta \
5  > outseq.fasta
```

5. FASTA is a text-based file format commonly used in sequence analysis. A FASTA file may consist of one or more (multi-FASTA) nucleotide or peptide sequences. Each sequence in a FASTA file is represented by a sequence descriptor on a single line, followed by the sequence itself, on one or more lines.

The sequence descriptor is called the "defline," and it must start with a greater-than-sign (>) immediately followed by a string which should be a unique identifier for that sequence within the FASTA file. In the defline, the unique identifier may be followed by additional descriptive text. The following is a sample FASTA file consisting of two sequence records.

```
>sequence1

AAAGAACTTGTTATTAATTAACTTATTTGAA
AAGAACTTGTTATTAATTA

ACTTATTTGAAAAGAACTTGTTATTAATTAACTT
ATTTGA

>sequence2

GCTAATGTTTCTCATAAAGCTTAAGTTCGTCAAT
GGAATCTTGAATATTA

AGCTAATGTTTCTCATAAAGCTTAAGTTCGTCAATG
GAATCTTGAATATT

AA
```

6. There are numerous parameters involved in specifying the desired criteria for designing primers. The desired values of these parameters are primarily dependent on the PCR protocols being used. Table 13 is a list of some of the primer design parameters that need to be calibrated to the laboratory protocols.

Table 13
Description of primer design parameters that may need to be calibrated according to specific laboratory protocols

Type	Parameter(s)
Melting temperature	Primer min, max, and optimal T_m Amplicon min, max, and optimal T_m Primer T_m difference
Amplicon size	Amplicon min, max, and optimal size
Primer size	Primer min, max, and optimal size
Primer dimer	Minimum length of complimentary sequence between primer ends and internally
Stem loop interference	Minimum length of loop size and number of hydrogen bonds in stem structure
Primer binding	Minimum length of 3′ perfect match needed for an alignment to be considered a primer binding site

7. The JCVI High Throughput Primer Design Pipeline (10) was designed and implemented at the J. Craig Venter Institute with the goal of maximizing the success rate of designed primers for high throughput directed sequencing. Since most of the important criteria for designing successful RT-PCR primers are similar to those for designing general PCR primers, the pipeline is suitable for solving both problems.

   The primer design pipeline is composed of two key software components: the Coverage Manager (CM) and the Primer Critiquor (PC).

   The Coverage Manager (CM) is responsible for generating a dynamic amplicon tiling across the supplied target regions. Its input parameters include the target sequence, the optimal, minimum, and maximum amplicon size, the minimum number of overlapping base pairs required between amplicons, and the depth of redundant amplicon coverage required. The CM employs Primer3 to generate a list of primer pair candidates. After the first amplicon has been accepted, the length and position of every proceeding amplicon will depend on the length and position of the prior upstream amplicon and the required minimum required amplicon overlap. Target regions are left uncovered by amplicons only if all primer pair candidates have been rejected in that region.

   The Primer Critiquor (PC) is responsible for determining whether a primer pair passes or fails the selection criteria. The PC was designed to have each criterion specified as a separate module. In the event that a new failure pattern is discovered, a new criterion can easily be incorporated as an additional module. The PC may be invoked iteratively by the CM during a primer design run, or it may be called upon independently to evaluate an arbitrary set of primer pairs generated by methods that are external to the primer design pipeline. The criteria that the PC can test for include all of the suggested checks described in the Methods section.

8. The JCVI High Throughput Primer Design Pipeline can be downloaded from: http://sourceforge.net/projects/primerdesigner

   (a) Click on the *Download* link, which can be located at the top of the page.

   (b) When the new page loads, you will find instructions on how to uncompress and install the software package. Click on the *Download primer_design.tar.gz* link to begin the download. Upon completion, follow the installation instructions contained on the same page.

Table 14
Comparison of software features in publicly available primer design tools

Software feature		Primer design assistant (PDA)[11]	SNP box[12]	Mut screener[13]	EasyExon primer[14]	PrimerZ[15]	JCVI primer design tool
Year		2003	2005	2006	2006	2007	2008
Reference Sequence Identifiers	Ensembl ID	-	-	-	-	✓	✓
	dbSNP/AFFY Probe ID	-	-	-	-	dbSNP/ AFFY Probe ID	dbSNP ID
	Genomic coordinates	-	-	UCSC annotation	-	-	Ensembl coordinates
	RefSeq ID / Genbank ID	-	Genbank ID	-	Refseq mRNA ID	-	-
	gene name / keyword	-	-	-	gene name and key-word	gene name	-
	Sequence	✓	✓	✓	-	-	✓
	other	-	-	Dnannotator annotation	-	-	-
Primer pair candidate selection method		complete enuma-ration	primer3	primer3	primer3	primer3	primer3
Checks for primer dimer occurrence		-	✓	-	-	-	✓
Checks for alternate products		-	-	-	✓†	-	✓
Checks for stem-loop structures		✓	-	-	-	-	✓
Able to use sequence annotation		-	✓	✓	✓	✓	✓
Primer pairs validated in the laboratory		-	-	-	✓	-	✓
Provides graphical web interface		✓	✓	✓	✓	✓	-

✓Feature supported
-Feature not supported
†Searches for the full length of the primer, instead of just the critical region on the 3' end

(c) After the installation has been completed, review the `PCR_Primer_Design-README` file, which contains detailed instructions on how to run an example primer design. This can be found in the `primer_design` directory.

(d) Once you have become familiar with the software's general functionality, you may find more information in the `primer_design/PrimerDesigner/Manuals` directory.

9. Additional computational primer design tools which are also available include Primer Design Assistant (11), SNPbox (12), MutScreener (13), EasyExon Primer (14), and PrimerZ (15). Each tool provides a graphical web interface for low throughput experiments and has the ability to locate your reference sequence if a public sequence identifier has been assigned to it. Supported identifiers include Genbank/Refseq IDs (SNPbox/EasyExon Primer), UCSC Annotation (MutScreener), dbSNP/AFFY Probe IDs (PrimerZ), and Ensembl IDs (PrimerZ). With the exception of EasyExon Primer, most software tools have not validated their designed primers in a laboratory nor do they perform the important alternative amplification check. If the primers you have designed with these alternative tools are not successful, you may need to manually perform the additional checks described in this chapter to determine if their mode of failure could have been avoided and to perform a more stringent redesign. A more detailed comparison of the described tools can be found in Table 14.

Acknowledgements

The authors would like to thank colleagues at the J. Craig Venter Institute for their contributions that led to the success of the software implementation of the primer design methods described in this chapter. Sean Murphy and Samuel Levy helped to conceive the project and participated in its design. Tim Stockwell aided in the analysis of data, evaluation of the results, and the management of the authors' time. Karen Beeson, Tina C. McIntosh, Dana Busam, and Steve Ferriera designed the laboratory protocols, ran the experiments, and generated the data. These concerted efforts were necessary to bring the empirical results closer with the theoretical calculations.

References

1. SourceForge.net: Open Source Software, http://sourceforge.net/
2. Rozen S, Skaletsky H (2000) Primer3 on the WWW for general users and for biologist programmer. Met Mol Biol 132:365–386
3. Rice P, Longden I, Bleasby A (2000) EMBOSS: The European molecular biology open software suite. Trends Gen 16:276–277
4. Altschul SF, Gish W, Miller W, Myers EW, Lipman DJ (1990) Basic local alignment search tool. J Mol Biol 215:403–410
5. Chenna R, Sugawara H, Koike T, Lopez R, Gibson TJ, Higgins DG, Thompson JD (2003) Multiple sequence alignment with the Clustal series of programs. Nucleic Acids Res 31:3497–3500
6. Rychlik W, Spencer WJ, Roads RE (1990) Optimization of the annealing temperature for DNA amplification in vitro. Nucleic Acids Res 18:6409–6412
7. Breslauer KJ, Frank R, Blocker H, Marky LA (1986) Prediction DNA duplex stability from the base sequence. Proc Natl Acad Sci U S A 83:3746–3750
8. Sambrook J, Fritsch EF, Maniatis T (1989) Molecular cloning: a laboratory manual. Cold Spring Harbor Laboratory Press. Cold Spring Harbor, NY
9. Bolton ET, McCarthy BJ (1962) A general method for the isolation of RNA complementary to DNA. Proc Natl Acad Sci U S A 48:1390–1397
10. Li K, Brownley A, Stockwell T, Beeson K, McIntosh TC, Busam D, Ferriera S, Murphy S, Levy S (2007) Novel computational methods for increasing PCR primer design effectiveness in directed sequencing. BMC Bioinformatics 9:191
11. Chen SH, Lin CY, Cho CS, Lo CZ, Hsiung CA (2003) Primer design assistant (PDA): a web-based primer design tool. Nucleic Acids Res 31:3751–3754
12. Weckx S, Rijk PD, Van Broeckhoven C, Del-Favero J (2005) SNPbox: a modular software package for large-scale primer design. Bioinformatics 21:385–387
13. Yao F, Zhang R, Zhu Z, Xia K, Liu C (2006) MutScreener: primer design tool for PCR-direct sequencing. Nucleic Acids Res 34:W660–W664
14. Wu X, Munroe D (2006) EasyExonPrimer: automated primer design for exon sequences. Appl Bioinformatics 5(2):119–120
15. Tsai MF, Lin YJ, Cheng YC, Lee KH, Huang CC, Chen YT, Yao A (2007) PrimerZ: streamlined primer design for promoters, exons and human SNPs. Nucleic Acids Res 35:W63–W65

Chapter 19

Hot Start PCR

Natasha Paul, Jonathan Shum, and Tony Le

Abstract

Hot Start activation approaches are increasingly being used to improve the performance of PCR. Since the inception of Hot Start as a means of blocking DNA polymerase extension at lower temperatures, a number of approaches have been developed that target the essential reaction components such as magnesium ion, DNA polymerase, oligonucleotide primers, and dNTPs. Herein, five different Hot Start activation protocols are presented. The first method presents the use of barriers as a means of segregating key reaction components until a Hot Start activation step. The second and third protocols demonstrate Hot Start approaches to block DNA polymerase activity through the use of anti-DNA polymerase antibodies and accessory proteins, respectively. The fourth and fifth protocols utilize thermolabile chemical modifications to the oligonucleotide primers and dNTPs. The results presented demonstrate that all protocols significantly improve the specificity of traditional thermal cycling protocols.

Key words: PCR, Hot Start PCR, Primer dimer, Mispriming, Thermal cycler, DNA polymerase, dNTPs

1. Introduction

The polymerase chain reaction (PCR) is a thermal cycling process used to amplify a nucleic acid target of interest (1, 2). Typical PCR setups employ a fine-tuned mixture of two target-specific oligonucleotide primers, a thermostable DNA polymerase, deoxynucleoside 5′-triphosphates (dNTPs), and a magnesium-containing reaction buffer. The copied amplicon can arise from a number of template sources, including DNA (e.g. genomic DNA, plasmid DNA, synthetic DNA, and mitochondrial DNA) and RNA (e.g. total RNA, messenger RNA, and micro RNA). As is described in greater detail throughout this volume, RNA samples are copied by an RT-PCR (reverse transcription PCR) protocol that uses reverse transcriptase to replicate the RNA into a cDNA

Nicola King (ed.), *RT-PCR Protocols: Second Edition*, Methods in Molecular Biology vol. 630,
DOI 10.1007/978-1-60761-629-0_19,

(complementary DNA) copy, which is then used as the template for PCR amplification.

The versatile application of PCR for sequence detection has found utility in assays that encompass the fields of forensics, biodefense, and molecular diagnostics (3–8). The need for improved PCR performance in these high stringency applications has prompted the development of methods that improve the reproducibility, selectivity, and sensitivity of PCR. As has been described by Chou et al., one common cause of unreliable PCR performance is off-target amplifications, such as primer dimer and mispriming products, that result from nonspecific primer binding and extension along regions of lesser complementarity (9). These off-target amplicons are thought to originate during the lower temperature, less stringent conditions of sample preparation, and thermal cycler ramping to the initial denaturation temperature of PCR (~94°C). At these cooler temperatures, the primers which are in considerable excess over the desired target can form primer dimer and mispriming extension products, which will ultimately compete with amplification of the desired target during the thermal cycling process. A number of methods, collectively termed "Hot Start," have been devised to reduce off-target amplifications. The goal of these Hot Start approaches is to block DNA polymerase extension at lower temperatures, until the Hot Start conditions of the initial denaturation temperature in PCR are reached (9). At the elevated temperatures of PCR thermal cycling, the DNA polymerase extension resumes to allow for primer extension during conditions where the stringency of primer hybridization is optimal to favor predominant formation of the desired primer/target complex.

As first practiced, Hot Start activation in PCR was achieved by withholding a key component during reaction setup (e.g., magnesium cofactor or DNA polymerase) and introducing it once the reaction mixture reaches the elevated, thermal cycling temperatures (10, 11). Since the introduction of the manual Hot Start technique, a number of approaches have emerged that sought to further streamline the Hot Start process to minimize the number of handling steps (Table 1). These have included the implementation of barriers, which first introduced the creation of a separation between essential reagents by melting wax beads to create a barrier (9, 12). This concept of a barrier was further simplified by the introduction of preprepared beads which encapsulated reagents (13) and by the use of reagents that precipitate the essential magnesium ion (14, 15). More recent developments have included the implementation of microfluidic devices to precisely control the addition of reagents into the reaction (16, 17). The majority of the developments in improved Hot Start technologies have been made to alter the activity of the DNA polymerase by the use of antibodies (18, 19), aptamers (oligonucleotide molecules that bind to the DNA polymerase) (20),

Table 1
Summary of different approaches to hot start activation in PCR

Hot start approach	Examples (with references)	Commercially available as an individual component
Manual hot start	Pre-amplification heating (10)	Can be easily performed using available reagents
Barriers/encapsulated reagents	Wax barrier – separates key components until elevated temperatures are reached (9, 12) Encapsulated DNA polymerase – released at elevated temperatures (13) Microfluidic devices to create a barrier within a reaction (17)	AmpliWax® PCRGems (Applied Biosystems) *Taq*Bead™ Hot Start DNA polymerase (Promega)
Controlled addition of magnesium into the reaction	Magnesium precipitate; magnesium dissolves at elevated temperatures (14, 15) Required metal ion generated by a thermally-induced redox reaction (43) Microfluidic devices – introduce magnesium at elevated temperatures (16)	Rockstart PCR buffer (DNA Polymerase Technology)
DNA polymerase-mediated	Antibodies – block DNA polymerase activity at lower temperatures; antibodies dissociate at elevated temperatures (18, 19) Aptamers – block DNA polymerase activity at lower temperatures; aptamers dissociate at elevated temperatures (20) Chemical-modifications – block DNA polymerase activity at lower temperatures; modifications dissociate at elevated temperatures (21, 22) Amino acid mutations – provide reduced activity at lower temperatures (24) Fusion protein – includes hyperstable DNA binding domains and topoisomerase (23) Temperature dependent inhibitor (ligand)	Platinum® Taq (Invitrogen Corporation) (see Subheading 4, Note 10) AmpliTaq® Gold (Applied Biosystems) (see Subheading 4, Note 16) Cesium Taq (DNA Polymerase Technology) TopoTaq DNA polymerase (Fidelity Systems) HotMaster *Taq* DNA polymerase (Eppendorf and 5 Prime)
Accessory proteins	Single-stranded binding protein – sequesters primers at low temperature (27) Thermostable pyrophosphatase – hydrolyzes inorganic pyrophosphate at elevated temperatures (26) Thermolabile blocker, such as a blocking polymerase protein (25)	Hot Start-IT® (USB) Paq5000™ Hotstart (Agilent Technologies)
Modified primers	Competitor sequences (30, 31) Primers with secondary structure (28, 29) Primer modifications that improve hybridization selectivity (33, 34) 3′ modifications that block primer extension until 3′-5′ exonuclease removal (35, 36) Nucleobase blocking modifications that are removable by UV irradiation (37)	EnzyStart™ (GeneCopoeia, Inc.)

(continued)

Table 1 (continued)

Hot start approach	Examples (with references)	Commercially available as an individual component
	Thermal deprotection of glyoxal-modified dG (38) Thermal deprotection of phosphotriester modification groups (39) Hairpin sequences (32)	CleanAmp™ Primers (TriLink BioTechnologies, Inc.)
Modified dNTPs	Thermal deprotection of glyoxal-modified dG (38) Thermolabile 3′-protected dNTPs (40)	CleanAmp™ dNTPs (TriLink BioTechnologies, Inc.)

chemical modifications (21, 22), protein fusions (23), temperature-sensitive amino acid point mutants (24), and temperature dependent ligands that block polymerase activity until a Hot Start activation step. These polymerase-based approaches have further been developed to use accessory proteins that provide additional Hot Start benefit by their differential interaction with the PCR reagents at varying temperatures (25–27). Other developments have included the use of modified primer constructs (28, 29) or competitor sequences (30–32). Notable primer modifications have included modifications that improve hybridization selectivity (33, 34), and modifications that are removable by 3′-5′ exonuclease action (35, 36), by UV irradiation (37), or by thermal deprotection (38, 39). More recent developments have included the modification of the dNTPs with thermolabile modification groups (38, 40).

The methods described herein will describe protocols for PCR experiments using a number of commercially available Hot Start approaches. In these studies, we will employ a model 365 base pair (bp) primer/template system from HIV-1 genomic DNA that is prone to primer dimer formation (9). The examples presented will include a procedure that employs a Hot Start barrier (9, 12), a procedure that employs a modified DNA polymerase (18, 19), a procedure that employs an accessory protein (27), a procedure that employs modified primers (39), and a procedure that employs modified dNTPs (40). Each of the described Hot Start approaches was investigated in standard thermal cycling protocols, and similar improved performance was found for each. Although traditional PCR setups were investigated herein, further guidance on the utility of each of these Hot Start approaches to more advanced applications can be found in Notes 1–8.

2. Materials

2.1. Conventional Hot Start PCR Using a Hot Start Barrier Approach

1. Pipettes capable of dispensing volumes <1 μL
2. Sterile filter pipette tips
3. Benchtop centrifuge
4. Thermal cycler
5. Electrophoresis equipment
6. UV transilluminator
7. Sterile 1.5- and 0.5-mL microcentrifuge tubes
8. Thin-walled 200 μL PCR Tubes
9. 1 μg/μL Molecular weight 50 bp ladder
10. dNTP mix: 10 mM each dNTP
11. Primer, cartridge purified, Forward: 5′-GAATTGGGTGTC-AACATAGCAGAAT-3′, Reverse: 5′-AATACTATGGTCCA-CACAACTATTGCT-3′) (see Note 9)
12. HIV-1 genomic DNA (Applied Biosystems – component of GeneAmplimer HIV-1 control reagents kit)
13. 100 μg of Human genomic DNA
14. HPLC water
15. 5U/μL Taq DNA polymerase (Invitrogen)
16. AmpliWax® PCR Gem 50 (Applied Biosystems)

2.2. Conventional Hot Start PCR Using a Modified DNA Polymerase Approach

1. Items 1–15 from Subheading 2.1
2. 5U/μL Platinum® *Taq* DNA polymerase (Invitrogen) (see Note 10)

2.3. Conventional Hot Start PCR Using an Accessory Protein Approach

1. Items 1–15 from Subheading 2.1
2. 1.25U/μL HotStart-IT® Taq DNA polymerase (USB)

2.4. Conventional Hot Start PCR Using a Modified Primer Approach

1. Items 1–15 from Subheading 2.1
2. CleanAmp™ Precision Primers (TriLink BioTechnologies, Inc.) (see Note 11)

2.5. Conventional Hot Start PCR Using a Modified dNTP Approach

1. Items 1–15 from Subheading 2.1
2. CleanAmp™ dNTP Mix (TriLink BioTechnologies, Inc.) (see Note 12)

3. Methods

3.1. Conventional Hot Start PCR Using a Hot Start Barrier Approach

In this experiment, the performance of a Hot Start barrier protocol that employed AmpliWax® PCR Gem 50 was compared to a normal PCR protocol (Fig. 1a). For each procedure, two reactions were set up – one that included template and one without template (NTC: no template control). For further details on the recommended use of this approach (see Note 1).

1. Prepare a 5× reaction buffer for *Taq* DNA polymerase by adding 500 µL of 10× PCR Buffer (supplied with *Taq* DNA polymerase, Subheading 2.1, item 15), 250 µL of 50 mM $MgCl_2$ (supplied with *Taq* DNA polymerase, Subheading 2.1, item 15), and 250 µL of HPLC water.
2. Make three master mixes, one for the lower reagent mix (below the wax), one for the upper reagent mix with template, and one for the upper reagent mix without template, by adding each component as listed in Table 2 (sufficient lower reagent master mix was made for 4.5 reactions, while sufficient upper reagent master mix was made for 2.5 reactions) (see Note 13).
3. Mix the master mixes gently by pipetting up and down (do not vortex). Pulse centrifuge if necessary.
4. Pipette 25 µL of the Lower Reagent Mix into four thin-walled PCR tubes, and place one AmpliWax® PCR Gem 50 (Subheading 2.1, item 16) in each of two PCR tubes.
5. Place the four tubes in a thermal cycler at 80°C for 5 min followed by 2 min at 25°C.
6. Pipette 25 µL of Upper Reagent Mix (w/template) into two of the PCR tubes (one with AmpliWax and one without). Then, pipette 25 µL of the Upper Reagent Mix (w/out template) into the remaining two thin-walled PCR tubes. All PCR tubes should have a final reaction volume of 50 µL.
7. Place tubes in a thermal cycler, preset with the following program: 94°C for 1 min; 35 cycles of 94°C for 45 s, 56°C for 30 s, and 72°C for 1 min; 72°C for 7 min.
8. At the end of the PCR reaction, load 20 µL of each sample into a 2% agarose gel for analysis (see Subheading 2.1, item 5; see Note 14). On the same gel, load 20 µL of a 50 bp ladder (diluted 1:50) as a reference. Run the gel for 30 min and visualize with a UV transilluminator.

3.2. Conventional Hot Start PCR Using a Modified DNA Polymerase Approach

In this experiment, an antibody-based Platinum® *Taq* DNA polymerase protocol was compared to a normal PCR protocol with *Taq* DNA polymerase (Fig. 1b). For each enzyme, two reactions were set up – one that included template and

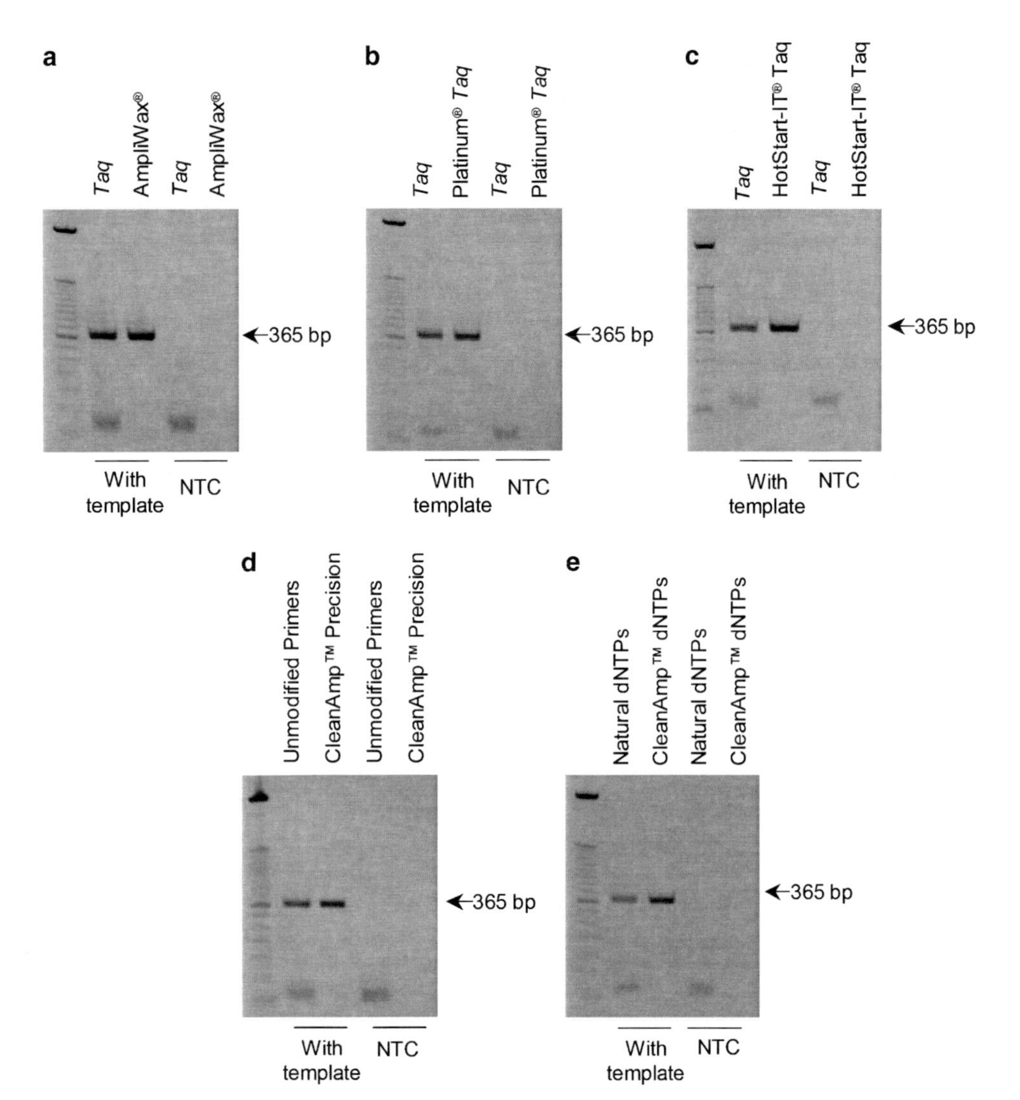

Fig. 1. Comparison of traditional PCR protocols to those employing different Hot Start activation protocols. Agarose gel analysis from PCR protocols using primers targeting a 365 bp fragment of HIV-1 DNA. For each Hot Start approach, Lane 1 is a 50 bp ladder, lanes 2 and 3 have five copies of HIV-1 recombinant DNA template, and lanes 4 and 5 are nontemplate controls (NTC). (**a**) Evaluation of the use of a Hot Start barrier – comparison of protocols employing AmpliWax® PCR Gems (lanes 3 and 5) to standard PCR setups with *Taq* DNA polymerase (lanes 2 and 4). (**b**) Evaluation of the use of a modified DNA polymerase – comparison of protocols employing Platinum® *Taq* (lanes 3 and 5) to standard PCR setups with *Taq* DNA polymerase (lanes 2 and 4). (**c**) Evaluation of the use of an accessory protein approach – comparison of protocols employing HotStart-IT® Taq (lanes 3 and 5) to standard PCR setups with *Taq* DNA polymerase (lanes 2 and 4). (**d**) Evaluation of the use of a modified primer approach – comparison of protocols employing CleanAmp™ Precision Primers (lanes 3 and 5) to standard PCR setups with *Taq* DNA polymerase and unmodified primers (lanes 2 and 4). (**e**) Evaluation of the use of a modified dNTP approach – comparison of protocols employing CleanAmp™ dNTPs (lanes 3 and 5) to standard PCR setups with *Taq* DNA polymerase and natural (unmodified) dNTPs (lanes 2 and 4)

one without template (NTC: no template control). For further details on the recommended use of this approach, see Notes 2, 3, and 4.

Table 2
Reaction setup for hot start barrier experiments

Component	Lower reagent mix		Upper reagent mix (w/template)		Upper reagent mix (no template)	
	Final concentration (50 µL reaction)	Volume (4.5 reaction)	Final concentration (50 µL reaction)	Volume (2.5 reaction)	Final concentration (50 µL reaction)	Volume (2.5 reaction)
Sterile de-ionized water		60.50 µL		48.13 µL		49.38 µL
5× PCR buffer (*Taq*)	1×	45 µL	1×	12.5 µL	1×	12.5 µL
dNTP mixture (10 mM each)	0.2 mM	4.5 µL				
Forward Primer (50 µM)	500 nM	1.25 µL				
Reverse Primer (50 µM)	500 nM	1.25 µL				
Taq DNA polymerase (5 units/µL)			0.025 units/µL	0.625 µL	0.025 units/µL	0.625 µL
HIV-1 template DNA (1 copy/µL in 10 ng/µL Human genomic DNA)			5 copies/50 µL	1.25 µL		
Total volume		112.5 µL		62.5 µL		62.5 µL

1. Prepare two different 5× reaction buffers. Prepare a 5× buffer (Platinum® *Taq*) by adding 500 μL of 10× PCR Buffer (supplied with Platinum® *Taq* DNA polymerase, Subheading 2.2, item 2), 150 μL of 50 mM $MgCl_2$ (supplied with Platinum® *Taq* DNA polymerase, Subheading 2.2, item 2), and 350 μL of HPLC water. Prepare a second 5× buffer (*Taq*) by adding 500 μL of 10× PCR Buffer (supplied with *Taq* DNA polymerase, Subheading 2.1, item 15), 250 μL of 50 mM $MgCl_2$ (supplied with *Taq* DNA polymerase, Subheading 2.1, item 15), and 250 μL of HPLC water.
2. Make two master mixes, one for ordinary *Taq* DNA polymerase and another for Platinum® *Taq* DNA polymerase, by adding the components as listed in Table 3 (sufficient master mix was made for 2.5 reactions) (see Note 13).
3. Mix the master mixes gently by pipetting up and down (do not vortex). Pulse centrifuge if necessary.
4. Pipette 45 μL of the *Taq* DNA polymerase master mix into two thin-walled PCR tubes and pipette 45μL of the Platinum® *Taq* DNA polymerase master mix into two thin-walled PCR tubes.
5. Then, pipette 5 μL of HPLC water into the NTC controls for each enzyme and 5 μL of HIV-1 template DNA (1 copy/μL

Table 3
Reaction setup for modified DNA polymerase experiments

Component	*Taq* DNA polymerase		Platinum® *Taq* DNA polymerase	
	Final concentration (50 μL reaction)	Volume (2.5 reaction)	Final concentration (50 μL reaction)	Volume (2.5 reaction)
Sterile de-ionized water		83.38 μL		83.50 μL
5× PCR buffer (*Taq*)	1×	25 μL		
5X PCR buffer (Platinum® *Taq*)			1×	25 μL
dNTP mixture (10 mM each)	0.2 mM	2.5 μL	0.2 mM	2.5 μL
Forward Primer (50 μM)	200 nM	0.5 μL	200 nM	0.5 μL
Reverse Primer (50 μM)	200 nM	0.5 μL	200 nM	0.5 μL
Taq DNA polymerase (5 units/μL)	0.025 units/μL	0.625 μL		
Platinum® *Taq* DNA polymerase (1.25 units/μL)			0.020 units/μL	0.5 μL
Total volume		112.5 μL		112.5 μL

in 10 ng/μL Human genomic DNA) into the remaining PCR tubes. All PCR tubes should have a final reaction volume of 50 μL.

6. Pulse centrifuge the samples to settle their contents.
7. Place these tubes in a thermal cycler, preset with the following program: 94°C for 2 min; 35 cycles of 94°C for 30 s, 55°C for 30 s, and 72°C for 1 min.
8. At the end of the PCR reaction, load 20 μL of each sample in a 2% agarose gel for analysis (see Subheading 2.1, item 5). On the same gel, load 20 μL of a 50 bp ladder (diluted 1:50) as a reference. Run the gel for 30 min and visualize with a UV transilluminator.

3.3. Conventional Hot Start PCR Using an Accessory Protein Approach

In this experiment, an accessory protein-based HotStart-IT® Taq DNA polymerase protocol was compared to a normal PCR protocol with *Taq* DNA polymerase (Fig. 1c). For each enzyme, two reactions were set up – one that included template and one without template (NTC: no template control). For further details on the recommended use of this approach see Notes 3, 4, and 5.

1. Prepare a 5× reaction buffer for *Taq* DNA polymerase by adding 500 μL of 10X PCR Buffer (supplied with *Taq* DNA polymerase, Subheading 2.1, item 15), 250 μL of 50 mM $MgCl_2$ (supplied with *Taq* DNA polymerase, Subheading 2.1, item 15), and 250 μL of HPLC water. Use the 10× PCR buffer supplied by the manufacturer for reactions employing HotStart-IT® Taq DNA polymerase (see Subheading 2.3, item 2).
2. Make two master mixes, one for ordinary *Taq* DNA polymerase and another for HotStart-IT® Taq DNA polymerase, by adding the components as listed in Table 4 (sufficient master mix was made for 2.5 reactions) (see Note 13).
3. Mix the master mixes gently by pipetting up and down (do not vortex). Pulse centrifuge if necessary.
4. Pipette 45 μL of the *Taq* DNA polymerase master mix into two thin-walled PCR tubes and 45 μL of the HotStart-IT® Taq DNA polymerase master mix into two thin-walled PCR tubes.
5. Then, pipette 5 μL of HPLC water into the NTC controls for each enzyme and 5 μL of HIV-1 template DNA (1 copy/μL in 10 ng/μL Human genomic DNA) into the remaining PCR tubes. All PCR tubes should have a final reaction volume of 50 μL.
6. Pulse centrifuge the samples to settle their contents.
7. Place these tubes in a thermal cycler, preset with the following program: 94°C for 2 min; 35 cycles of 94°C for 30 s, 56°C for 30 s, and 72°C for 1 min; and a final extension at 72°C for 5 min.

Table 4
Reaction setup for accessory protein experiments

	Taq DNA polymerase		HotStart-IT® *Taq* DNA polymerase	
Component	**Final concentration (50 μL reaction)**	**Volume (2.5 reaction)**	**Final concentration (50 μL reaction)**	**Volume (2.5 reaction)**
Sterile de-ionized water		83.375 μL		94.0 μL
5× PCR buffer (*Taq*)	1×	25 μL		
10× PCR buffer (HotStart-IT®)			1×	12.5 μL
dNTP mixture (10 mM each)	0.2 mM	2.5 μL	0.2 mM	2.5 μL
Forward Primer (50 μM)	200 nM	0.5 μL	200 nM	0.5 μL
Reverse Primer (50 μM)	200 nM	0.5 μL	200 nM	0.5 μL
Taq DNA polymerase (5 units/μL)	0.025 units/μL	0.625 μL		
HotStart-IT® *Taq* DNA polymerase (1.25 units/μL)			0.025 units/μL	2.5 μL
Total volume		112.5 μL		112.5 μL

8. At the end of the PCR reaction, load 20 μL of each sample in a 2% agarose gel for analysis (see Subheading 2.1, item 5). On the same gel, load 20 μL of a 50 bp ladder (diluted 1:50) as a reference. Run the gel for 30 min and visualize with a UV transilluminator.

3.4. Conventional Hot Start PCR Using a Modified Primer Approach

In this experiment, a procedure employing chemically modified CleanAmp™ Precision Primers was compared to a normal PCR procedure with unmodified primers (Fig. 1d). For each approach, two reactions were set up – one that included template and one without template (NTC: no template control). For further details on the recommended use of this approach see Notes 6 and 7.

1. Prepare a 5× reaction buffer for *Taq* DNA polymerase by adding 500 μL of 10× PCR Buffer (supplied with *Taq* DNA polymerase, Subheading 2.1, item 15), 250 μL of 50 mM $MgCl_2$ (supplied with *Taq* DNA polymerase, Subheading 2.1, item 15), and 250 μL of HPLC water.
2. Make two master mixes, one for ordinary *Taq* DNA polymerase and one for CleanAmp™ Precision Primers

Table 5
Reaction setup for modified primer experiments

Component	*Taq* DNA polymerase		CleanAmp™ precision primers	
	Final concentration (50 µL reaction)	Volume (2.5 reaction)	Final concentration (50 µL reaction)	Volume (2.5 reaction)
5× PCR buffer (*Taq*)	1×	25 µL	1×	25 µL
dNTP mixture (10 mM each)	0.2 mM	2.5 µL	0.2 mM	2.5 µL
Forward primer (50 µM)	200 nM	0.5 µL		
Reverse primer (50 µM)	200 nM	0.5 µL		
CleanAmp™ precision forward primer (10.4 µM)			200 nM	
CleanAmp™ Precision reverse primer (18 µM)			200 nM	
Taq DNA polymerase (5 units/µL)	0.025 units/µL	0.625 µL	0.025 units/µL	0.625 µL
Sterile de-ionized water		83.375 µL		82.135 µL
Total volume		112.5 µL		112.5 µL

(see Subheading 2.4, item 2), by adding the components as listed in Table 5 (sufficient master mix was made for 2.5 reactions) (see Notes 11, 13, and 15).

3. Mix the master mixes gently by pipetting up and down (do not vortex). Pulse centrifuge if necessary.
4. Pipette 45 µL of the *Taq* DNA polymerase master mix into two thin-walled PCR tubes and 45 µL of the CleanAmp™ Precision Primers master mix into two thin-walled PCR tubes.
5. Pipette 5 µL of HPLC water into the NTC controls for each reaction and 5 µL of HIV-1 template DNA (1 copy/µL in 10 ng/µL Human genomic DNA) into the remaining PCR tubes. All PCR tubes should have a final reaction volume of 50 µL.
6. Pulse centrifuge the samples to settle their contents.
7. Place these tubes in a thermal cycler, preset with the following program: 94°C for 10 min; 35 cycles of 94°C for 40 s, 56°C for 30 s, and 72°C for 1 min; and a final extension at 72°C for 7 min.

8. At the end of the PCR reaction, load 20 μL of each sample in a 2% agarose gel for analysis (see Subheading 2.1, item 5). On the same gel, load 20 μL of a 50 bp ladder (diluted 1:50) as a reference. Run the gel for 30 min and visualize with a UV transilluminator.

3.5. Conventional Hot Start PCR Using a Modified dNTP Approach

In this experiment, protocols employing chemically modified CleanAmp™ dNTPs were compared to a normal PCR procedure with unmodified dNTPs (Fig. 1e). For each type of dNTP, two reactions were set up – one that included template and one without template (NTC: no template control). For further details on the recommended use of this approach (see Notes 7 and 8).

1. Prepare a 5× reaction buffer for *Taq* DNA polymerase by adding 500 μL of 10× PCR Buffer (supplied with *Taq* DNA polymerase, Subheading 2.1, item 15), 250 μL of 50 mM $MgCl_2$ (supplied with *Taq* DNA polymerase, Subheading 2.1, item 15), and 250 μL of HPLC water.
2. Make two master mixes, one for *Taq* DNA polymerase (see previous comments) and another for CleanAmp™ dNTPs (see Subheading 2.5, item 2), by adding the components as listed in Table 6 (sufficient master mix was made for 2.5 reactions) (see Notes 12 and 13).
3. Mix the master mixes gently by pipetting up and down (do not vortex). Pulse centrifuge if necessary.
4. Pipette 45 μL of the *Taq* DNA polymerase master mix into two thin-walled PCR tubes and 45 μL of CleanAmp™ dNTPs master mix into two thin-walled PCR tubes.
5. Then, pipette 5 μL of HPLC water into the NTC controls for each type of dNTP and 5 μL of HIV-1 template DNA (1 copy/μL in 10 ng/μL Human genomic DNA) into the remaining PCR tubes. All PCR tubes should have a final reaction volume of 50 μL.
6. Pulse centrifuge the samples to settle their contents.
7. Place these tubes in a thermal cycler, preset with the following program: 94°C for 10 min; 35 cycles of 94°C for 40 s, 56°C for 30 s, and 72°C for 1 min; and a final extension at 72°C for 7 min.
8. At the end of the PCR reaction, load 20 μL of each sample in a 2% agarose gel for analysis (see Subheading 2.1, item 5). On the same gel, load 20 μL of a 50 bp ladder (diluted 1:50) as a reference. Run the gel for 30 min and visualize with a UV transilluminator.

Table 6
Reaction setup for modified dNTP experiments

Component	*Taq* DNA polymerase		CleanAmp™ dNTPs	
	Final concentration (50 μL reaction)	Volume (2.5 reaction)	Final concentration (50 μL reaction)	Volume (2.5 reaction)
Sterile de-ionized water		83.375 μL		81.375 μL
5× PCR buffer (*Taq*)	1×	25 μL	1×	25 μL
dNTP mixture (10 mM each)	0.2 mM	2.5 μL		
CleanAmp™ dNTPs mixture (10 mM each)			0.4 mM	4.5 μL
Forward primer (50 μM)	200 nM	0.5 μL	200 nM	0.5 μL
Reverse primer (50 μM)	200 nM	0.5 μL	200 nM	0.5 μL
Taq DNA polymerase (5 units/μL)	0.025 units/μL	0.625 μL	0.025 units/μL	0.625 μL
Total volume		112.5 μL		112.5 μL

4. Notes

1. Although it provides a similar benefit to each of the protocols examined, the AmpliWax® PCR Gem 50 protocol requires the greatest number of manipulation steps. Therefore, this Hot Start approach is not the best choice for individuals who routinely set up multiple reactions. However, it has recently shown applicability in more advanced protocols, such as one-step RT-PCR. In this approach, the use of the wax barrier allows the RT reaction to occur independent of the DNA polymerase and PCR primers (41, 42).
2. Platinum® *Taq* DNA polymerase provides improved specificity and yields of PCR compared to *Taq* DNA polymerase without Hot Start.
3. Both Platinum® *Taq* DNA polymerase and HotStart-IT® Taq DNA polymerase provide a fast (2 min) Hot Start activation, allowing for faster completion of thermal cycling protocols and reduced heat-damage to DNA samples.
4. Platinum® *Taq* DNA polymerase and HotStart-IT® Taq DNA polymerase each can be purchased as Master Mixes (Platinum® PCR SuperMix and HotStart-IT® Taq Master Mix (2×),

respectively). These ready to use PCR cocktails allow convenience and speed in PCR setup, as the user only needs to add the primers and template.

5. The HotStart-IT® single-stranded binding protein is available separately. It can be added to PCR setups to introduce Hot Start activation, with particular utility in multiplex situations, where the probability of primer dimer formation is higher.
6. CleanAmp™ Primers come in two varieties that show benefit in more advanced applications. CleanAmp™ Turbo Primers are the best choice for mulitplexing and fast thermal cycling. CleanAmp™ Precision Primers provide the greatest specificity and show benefit in one-step RT-PCR protocols.
7. CleanAmp™ Primers and dNTPs can be employed with a variety of different thermostable DNA polymerases, allowing for application with a user's DNA polymerase of choice. This option is important when a DNA polymerase which has not been prepared as a Hot Start version is desirable for performance reasons (i.e., ease of processing, error rate, etc.)
8. CleanAmp™ dNTPs provide great flexibility, as they can easily be introduced into existing protocols by substitution of CleanAmp™ dNTPs for the natural dNTPs without the need to change DNA polymerases or primer sequences. CleanAmp™ dNTPs also work well in combination with other Hot Start approaches. A Hot Start DNA polymerase or primer coupled with CleanAmp™ dNTPs enhances the selectivity of PCR even more than when used alone (unpublished data).
9. Primers can be ordered from a number of commercial sources.
10. A number of companies also offer antibody-based Hot Start versions of *Taq* DNA polymerase. These include BioCat, Bioron, Clontech, Genaxxon Bioscience, Genesis Biotech Inc., Invitrogen, Kapa Biosystems, Inc., Metabion international AG, Minerva Biolabs, OmniBioLabs, PEQLAB Biotechnologie GMBH, Promega, Sigma, and TaKaRa Bio USA.
11. CleanAmp™ Turbo and Precision Primers are provided as a ~200 mM stock in DMSO. Thaw the CleanAmp™ Primer DMSO stock solution at room temperature, vortex and pulse centrifuge two to three times to thoroughly mix. Do NOT heat the stock solution to thaw, as heating will result in removal of the CleanAmp™ Primer modification group. If the setup requires a lower primer concentration, remove a small aliquot of CleanAmp™ Primers (1 mL), diluting with either water or buffer (pH 7–9) to the desired working concentration and discard after use.
12. Thaw the CleanAmp™ dNTP stock solution at room temperature, vortexing to thoroughly mix. Do NOT heat the

stock solution to thaw, as heating will result in removal of the CleanAmp™ dNTP modification group. Do not expose the stock solution to room temperature for more than 10 total hours.

13. Keep DNA polymerases on ice and place frozen reagents on the bench at room temperature to thaw.
14. Use a pipette tip to puncture a small hole in the wax layer. Then, use a fresh tip to pipette out the desired amount of sample.
15. For setups including CleanAmp™ primers, add the components of each master mix in the order as shown in the table.
16. A number of companies also offer chemically-modified Hot Start versions of Taq DNA polymerase. These include Biomatik Corp, Biotium, Fermentas Life Sciences, MIDSCI, Molecular Cloning Laboratories, MP Biomedicals (Qbiogene), Novagen, Roche Applied Science, Thermo Scientific, and Qiagen.

Acknowledgments

The authors would like to thank R. Hogrefe, G. Zon, E. Ashrafi, S. Shore, and A. Lebedev for helpful discussions and for critical reading of the manuscript.

References

1. Mullis KB (July 28, 1987) Process for amplifying nucleic acid sequences. US4683202
2. Saiki RK, Gelfand DH, Stoffel S, Scharf SJ, Higuchi R, Horn GT, Mullis KB, Erlich HA (1988) Primer-directed enzymatic amplification of DNA with a thermostable DNA polymerase. Science 239:487–91
3. Budowle B, Schutzer SE, Einseln A, Kelley LC, Walsh AC, Smith JA, Marrone BL, Robertson J, Campos J (2003) Public health. Building microbial forensics as a response to bioterrorism. Science 301:1852–3
4. Dahiya R, Deng G, Selph C, Carroll P, Presti J Jr (1998) A novel p53 mutation hotspot at codon 132 (AAG->AGG) in human renal cancer. Biochem Mol Biol Int 44:407–15
5. Elnifro EM, Ashshi AM, Cooper RJ, Klapper PE (2000) Multiplex PCR: optimization and application in diagnostic virology. Clin Microbiol Rev 13:559–70
6. Kolmodin LA, Williams JF (1997) PCR. Basic principles and routine practice. Met Mol Biol 67:3–15
7. Saldanha J, Minor P (1994) A sensitive PCR method for detecting HCV RNA in plasma pools, blood products, and single donations. J Med Virol 43:72–6
8. Sato Y, Hayakawa M, Nakajima T, Motani H, Kiuchi M (2003) HLA typing of aortic tissues from unidentified bodies using hot start polymerase chain reaction-sequence specific primers. Leg Med (Tokyo) 5(Suppl 1):S191–3
9. Chou Q, Russell M, Birch DE, Raymond J, Bloch W (1992) Prevention of pre-PCR mis-priming and primer dimerization improves low-copy-number amplifications. Nucl Acids Res 20:1717–23
10. D'Aquila RT, Bechtel LJ, Videler JA, Eron JJ, Gorczyca P, Kaplan JC (1991) Maximizing sensitivity and specificity of PCR by pre-amplification heating. Nucl Acids Res 19:3749
11. Newton CR, Graham A, Heptinstall LE, Powell SJ, Summers C, Kalsheker N, Smith JC, Markham AF (1989) Analysis of any point mutation in DNA. The amplification refractory mutation system (ARMS). Nucl Acids Res 17:2503–16

12. Tanzer LR, Hu Y, Cripe L, Moore RE (1999) A hot-start reverse transcription-polymerase chain reaction protocol that initiates multiple analyses simultaneously. Anal Biochem 273:307–10
13. Setterquist RA, Smith KG (January 4, 1996) Encapsulated PCR reagents. WO9600301 (A1)
14. Barnes WM, Rowlyk KR (2002) Magnesium precipitate hot start method for PCR. Mol Cell Probes 16:167–71
15. Barnes WM, Rowlyk KR (February 13, 2003) Magnesium precipitate hot start method for molecular manipulation of nucleic acids. WO03012066
16. Andersen MR (February 5, 2004) Mg-mediated hot start biochemical reactions. WO2004011666 (A2)
17. Cottingham HV (September 9, 1999) Device and method for DNA amplification and assay. US5948673 (A)
18. Eastlund E, Mueller E (2001) Hot Start RT-PCR results in improved performance of the enhanced avian RT-PCR system. LifeSci Quarterly 2:2–5
19. Mizuguchi H, Nakatsuji M, Fujiwara S, Takagi M, Imanaka T (1999) Characterization and application to hot start PCR of neutralizing monoclonal antibodies against KOD DNA polymerase. J Biochem (Tokyo) 126:762–8
20. Dang C, Jayasena SD (1996) Oligonucleotide inhibitors of Taq DNA polymerase facilitate detection of low copy number targets by PCR. J Mol Biol 264:268–78
21. Birch DE, Laird WJ, Zoccoli A (June 30, 1998) Nucleic acid amplification using a reversibly inactivated thermostable enzyme. US5773258
22. Moretti T, Koons B, Budowle B (1998) Enhancement of PCR amplification yield and specificity using AmpliTaq Gold DNA polymerase. Biotechniques 25:716–22
23. Pavlov AR, Belova GI, Kozyavkin SA, Slesarev AI (2002) Helix-hairpin-helix motifs confer salt resistance and processivity on chimeric DNA polymerases. Proc Natl Acad Sci USA 99:13510–5
24. Kermekchiev MB, Tzekov A, Barnes WM (2003) Cold-sensitive mutants of Taq DNA polymerase provide a hot start for PCR. Nucl Acids Res 31:6139–47
25. Borns M (January 11, 2007) Hot start polymerase reaction using a thermolabile blocker. US2007009922 (A1)
26. Clark DR, Vincent SP (March 16, 2006) Amplification process. US2006057617
27. Kubu CJ, Muller-Greven JC, Moffett RB (June 12, 2008) Novel hot start nucleic acid amplification. US2008138878 (A1)
28. Ailenberg M, Silverman M (2000) Controlled hot start and improved specificity in carrying out PCR utilizing touch-up and loop incorporated primers (TULIPS). Biotechniques 29 (1018–20):22–4
29. Kaboev OK, Luchkina LA, Tret'iakov AN, Bahrmand AR (2000) PCR hot start using primers with the structure of molecular beacons (hairpin-like structure). Nucl Acids Res 28:E94
30. Puskas LG, Bottka S (1995) Reduction of mispriming in amplification reactions with restricted PCR. Genome Res 5:309–11
31. Vestheim H, Jarman SN (2008) Blocking primers to enhance PCR amplification of rare sequences in mixed samples – a case study on prey DNA in Antarctic krill stomachs. Front Zool 5:12
32. Wangh LJ, Rice J, Sanchez JA, Pierce K, Salk J, Reis A, Hartshorn, C (August 10, 2006) Reagents and methods for improving reproducibility and reducing mispriming in PCR amplification. US2006177842 (A1)
33. Laird WJ, Niemiec JT (September 21, 2004) Amplification using modified primers. US6794142 (B2)
34. Will SG (December 14, 1999) Modified nucleic acid amplification primers. US6001611
35. Ankenbauer W, Heindl D, Laue F, Huber N (June 6, 2003) Composition and method for hot start nucleic acid amplification. US2003119150
36. Ullman EF, Lishanski A, Kurn N (November 19, 2002) Method for polynucleotide amplification. US6482590
37. Young DD, Edwards WF, Lusic H, Lively MO, Deiters A (2008) Light-triggered polymerase chain reaction. Chem Commun (Camb) 28:462–4
38. Bonner AG (August 28, 2003) Reversible chemical modification of nucleic acids and improved method for nucleic acid hybridization. US2003162199 (A1)
39. Lebedev AV, Paul N, Yee J, Timoshchuk VA, Shum J, Miyagi K, Kellum J, Hogrefe RI, Zon G (2008) Hot start PCR with heat-activatable primers: a novel approach for improved PCR performance. Nucl Acids Res 36:e131
40. Koukhareva I, Haoqiang H, Yee J, Shum J, Paul N, Hogrefe RI, Lebedev AV (2008) Heat activatable 3′-modified dNTPs: synthesis and application for hot start PCR. Nucl Acids Symp Ser (Oxf); 259–60.
41. Fan XY, Lu GZ, Wu LN, Chen JH, Xu WQ, Zhao CN, Guo SQ (2006) A modified single-tube one-step product-enhanced reverse transcriptase (mSTOS-PERT) assay with heparin as DNA polymerase inhibitor for specific detection of RTase activity. J Clin Virol 37:305–12

42. Sears JF, Khan AS (2003) Single-tube fluorescent product-enhanced reverse transcriptase assay with Ampliwax (STF-PERT) for retrovirus quantitation. J Virol Met 108:139–42

43. Ignatov K, Kramarov V (November 11, 2007) Method for performing the hot start of enzymatic reactions. US2007254327 (A1)

Chapter 20

Real-time RT-PCR for Automated Detection of HIV-1 RNA During Blood Donor Screening

Jens Müller

Abstract

Real-time RT-PCR has become the method of choice for automated detection of viral RNA target sequences in the clinical laboratory. Besides commercially available certified test systems, a variety of so-called in-house methods have been described in the literature. Generally, appropriate validation and continuous quality control are mandatory if these in-house-developed assays are used in clinical diagnostics. In this chapter, an in-house HIV-1 real-time RT-PCR assay for blood donor screening is described. The procedure includes the pooling of plasma samples, viral RNA isolation, and subsequent detection of amplification in real-time one-step RT-PCR. The validation considers the specificity, the sensitivity on HIV-1 genomic variants, and the robustness of the assay.

Key words: NAT, Real-time PCR, Real-time RT-PCR, HIV-1, Internal control, Blood donor screening, Validation

1. Introduction

The screening of blood donors for Hepatitis C Virus (HCV) genomes by nucleic-acid amplification techniques (NAT) has been mandatory in Germany since 1999. This measure was justified to reduce the diagnostic window period between infection and diagnosis of infection by HCV antibody detection. Under the same point of view, screening for human immunodeficiency virus-1 (HIV-1) group M NAT has been obligatory since May 2004 (1).

Due to the observed high HIV-1 viral load during the pre-seroconversion phase, the minimum sensitivity (95% cut-off) was set to 10,000 IU/ml (~5,000 copies/ml) for single donation testing (2). The high sensitivity of NAT usually allows the blood donor samples to be pooled to increase throughput and reduce costs.

Nicola King (ed.), *RT-PCR Protocols: Second Edition,* Methods in Molecular Biology vol. 630,
DOI 10.1007/978-1-60761-629-0_20,

With obligatory NAT testing, the residual risk of unrevealed HIV-1 window period donations is now estimated to be less than 1 in 4,000,000 (1, 3).

HIV-1 is characterized by a high genetic diversity and classified into groups M (main), O (outlier), and N (non-M, non-O). While HIV-1 group M infections are the major cause of the HIV pandemic, group O and N infections are mainly restricted to Central Africa. HIV-1 group M subtype B is the predominant subtype in Western Europe, North America, and Australia (4). However, with respect to global traveling and migration, NAT assays used for HIV-1 screening in these areas should also be sensitive to all relevant HIV-1 variants. This view is supported by several reports that describe HIV-1 group O infections outside endemic regions (5–7). Automated detection of HIV-1 RT-PCR products is a prerequisite for increasing the throughput of routinely performed clinical testing. In recent years, automated heterogeneous assay formats have been challenged by the advent of homogenous real-time RT-PCR applications (8, 9). In this chapter, an internally controlled real-time RT-PCR assay for the detection of HIV-1 group M, O, and N genomes and its validation and use for the screening of blood donors is described. The presented principles, however, are adaptable to similar applications for the detection of viral infectious diseases in clinical diagnostics.

2. Materials

2.1. Blood Drawing and Pooling of Plasma Samples

1. EDTA blood-drawing tubes.
2. Secondary tubes for pooling of plasma samples.
3. Barcode label printer/labels.
4. Pipetting robot with barcode reader, disposable tip functionality, and liquid level detection.
5. Software for sample pooling and protocol recording.
6. Conductive pipette tips with aerosol barriers.

2.2. Viral RNA Purification

1. QIAamp® Viral RNA Mini Kit (Qiagen). Store at room temperature.
2. Ethanol (96–100%). Store at room temperature.
3. Nuclease- and pyrogen-free water. Aliquot and store at –20°C until use.
4. 1.5-ml RNase-free microcentrifuge tubes.
5. Sterile, RNase-free pipette tips with aerosol barriers.
6. Microcentrifuge (with rotor for 1.5-ml and 2-ml tubes).

2.3. One Step Multiplex HIV-1/GAPDH Real-Time RT PCR

1. HotStarTaq® DNA Polymerase, 5 U/μl (Qiagen). Store at –20°C.
2. SuperScript™ III Reverse Transcriptase (RT), 200 U/μl (Invitrogen, see Note 1). Store at –20°C.
3. Storage buffer for pre-dilution of the SuperScript™ III RT enzyme: 20 mM Tris–HCl (pH 7.6), 100 mM NaCl, 0.1 mM EDTA, 1 mM DTT, 0.01% (v/v) Igepal CA-630 (Nonidet P-40 substitute), 50% (v/v) Glycerol. Store at –20°C (see Note 2).
4. ROX Reference Dye, 50× (Invitrogen, see Note 3). Store at –20°C
5. RNase Out, 40 U/μl (Invitrogen). Store at –20°C.
6. Primer (desalted) and dual-labelled hydrolysis probes (HPLC-purified) for detection of HIV-1-RNA and GAPDH-mRNA sequences. Store working and stock solutions at –20°C (see Note 4). HIV-1: Forward primer 1 (group M; 5′-GGTTTATTACAGGGACAGCAGAGA-3′; nucleotides (nt) 4901–4924 of HXB2 reference sequence [GenBank accession no. K03455)), Forward primer 2 (group O; 5′-GGTCTATTACAGAGACAGCAGAGA-3′ nt 4901–4924), Reverse primer (group M/O; 5′-ACCTGCCATCTGTTTTCCATA-3′; nt 5060–5040), Probe 1 (group M; 5′-FAM-ACTACTGCCCCTTCACCTTTCCAGAG-BHQ1-3′; nt 4978–4953), and Probe 2 (group O; 5′-FAM-ACTACTGCTCCCTCACCTTTCCAC-AG-BHQ-1-3′; nt 4978–4953) (see Note 5).

   GAPDH: Forward primer (5′-GAAGGTGAAGGTCGGAGTC-3′; nt 81–99, according to GenBank accession no. NM_002046), Reverse Primer (5′-GAAGATGGTGATGGGATTTC-3′; nt 306–287), Probe (5′-CAAGCTTCCCGTTCTCAGCC-3′; nt 277–258), labeled with the fluorescence dye Yakima Yellow (VIC® substitute) at the 5′ end and the eclipse dark quencher at the 3′ end (see Note 6).
7. dNTP-mix (25 mM each). Store at –20°C.
8. Nuclease- and pyrogen-free water. Aliquot and store at less than or equal to –15°C until use.
9. Optical PCR tubes (strips)/caps (e.g. Applied Biosystems [ABI])
10. Sterile, RNase-free pipette tips with aerosol barriers
11. Real-time thermocycler (e.g. ABI, SDS 7700 instrument).
12. Microsoft Excel® software.

2.4. Assay Validation

1. HIV-1 RNA, 2nd International WHO Standard (NIBSC). Store at less than –70°C until use.
2. HIV-1 RNA Genotype panel for NAT Assays (NIBSC). Store at less than –70°C until use.

3. HIV-1 RNA-negative EDTA plasma (in-house preparation). Store at less than −30°C until use.
4. Software for Probit Analysis.

3. Methods

Applying an in-house NAT method for the screening of blood donors for viral infectious diseases requires an extensive validation process to verify the performance of the assay according to regulatory guidelines (10, 11). The most important characteristics to be examined for validation include the specificity of the assay and its sensitivity to relevant genomic variants. To ensure steady sensitivity of the test system, all assay components must be carefully chosen with respect to both reliability and reproducibility. Preventive maintenance of instruments and laboratory equipment should be performed. Before use, all new batches of reagents must be checked for reliability to ensure uniform assay performance (11). To prevent handling errors and increase throughput, an automated system should be used if pooling of plasma samples is intended. Enrichment of viral particles by e.g. ultracentrifugation may be established to improve the sensitivity of the assay. Isolation of nucleic acids is performed directly from (pooled) plasma samples or pelleted virus particles using either manual or automated applications (12).

Homogenous NAT assay formats for the detection of virus-specific genome sequences in nucleic acid isolates are preferable to conventional heterogeneous methods. Owing to the closed tube format, the risk of reaction mixtures being contaminated by previously handled amplification products is minimized. Furthermore, no postamplification steps are needed, thereby eliminating the risk of sample mix-up if manual handling is performed (13, 14).

To avoid false-negative test results due to failure during nucleic acid extraction or inhibition of the (RT)-PCR reaction, the use of an internal control reaction is highly recommended (15–17). The use of differently labeled fluorogenic primers or probes for the generation of amplification-dependent fluorescence signals allows the combination of two or more parallel NAT reactions in a so-called multiplex NAT. Furthermore, fluorogenic probes provide an additional sequence-specific hybridisation step that improves the specificity of the assay (18).

The described one-step real-time RT-PCR assay for the detection of HIV-1 RNA utilizes the so-called hydrolysis probes, dually labeled with a reporter dye and a quenching molecule that suppresses the fluorescence of the reporter (17). These probes are generally designed to hybridize to a region within the amplicons. During the extension step of amplification, the 5′–3′ exonuclease activity of the Taq DNA polymerase degrades the probe, thereby leading to fluorescence generation that is proportional to the amount of template present (18). Target-specific detection of amplification in the multiplex-assay is discriminated by the use of different reporter dyes. As an internal control, the detection of endogenous glyceraldehyde-3-phosphate de-hydrogenase (GAPDH) mRNA is used (17).

The entire methodology was validated according to German legal guidelines and was demonstrated to be suitable for blood donor screening. Nevertheless, implementation of the methodology in a third-party laboratory requires appropriate validation, primarily with respect to assay sensitivity.

3.1. Blood Drawing and Pooling of Plasma Samples

1. Draw blood either directly from the punctured vein or the predonation pouch (19) into EDTA-collection tubes. The tubes should have been labeled with a donor-specific barcode for automated sample pooling.
2. Centrifuge tubes at 2,500 × *g* for 10–15 min at room temperature.
3. Transfer the centrifuged tubes to appropriate primary racks for automated sample pooling.
4. Label secondary tubes with barcodes for definite sample assignment and place them in secondary sample racks.
5. Read in barcode labels and refer samples to pools via the pooling software application. Choose maximum pool size according to the determined detection limit of the applied HIV-1 NAT assay (see Note 7).
6. Start pipetting of plasma pools. Transfer equal volumes (e.g. 100 μl) from each sample to the corresponding pool.
7. Export data on sample assignment and generate an appropriate pooling protocol.

3.2. Isolation of Viral RNA from Pooled Plasma Samples

1. Introduce at least one negative control and one positive run-control to the extraction process (see Notes 8 and 14).
2. Create an extraction protocol to assign samples and controls to ascending numbers starting from 1 (see Note 9).
3. Isolate viral RNA from 140 μl of samples and controls using the QIAamp® Viral RNA Mini Kit with the spin procedure according to the manufacturer's instructions (see Note 10). Elute purified nucleic acids into 1.5-ml RNase-free microcentrifuge tubes using 50 μl of nuclease-free water (see Note 11).

3.3. One-Step Multiplex HIV-1/ GAPDH Real-Time RT-PCR and Interpretation of Collected Data

1. Include no template controls (NTC) in each run to reveal potential contaminations.
2. Create a pipetting protocol to assign extracts and NTCs to specific positions of correspondingly arranged optical PCR tubes.
3. Arrange samples, pipette tips, and optical PCR tubes in an identical configuration to minimize the risk of mix-up during pipetting.
4. Prepare a master mix according to Table 1 (see Notes 2 and 12).
5. Add 30 μl of the prepared master mix to each of the optical PCR tubes.
6. Add 20 μl of each sample and NTC to the dedicated tubes, cap, and centrifuge briefly at room temperature to collect reaction mixtures at the bottom of the tubes.
7. Place tubes in the SDS 7700 real-time cycler. Open a preset real-time (single reporter) masterfile with the following thermal cycling program (see Note 13):

Stage 1:	60°C for 20 min
Stage 2:	95°C for 15 min
Stage 3:	10 cycles of 94°C for 20 s (step 1) and 64°C for 30 s (step 2)
Stage 4:	40 cycles of 94°C for 20 s (step 1) and 57°C for 30 s (step 2).
Select stage 4, step 2 for collection of fluorescence emission data.	

8. Define the sample setup on the FAM and VIC® dye layers and start the program.
9. Save data after the run is completed.
10. Analyze and view normalized amplification plots (see Fig. 1).
11. Set baseline areas for calculation of threshold values between amplification cycles 3 and 20 for both the FAM and the VIC® layers and accept suggested values.
12. Analyze the plotted data. Fluorescence signals that exceed the calculated threshold value on the FAM-layer (Ct < 40) indicate an HIV-1-RNA-reactive result (see Fig. 1a). The included positive run-control(s) (see Note 8) must yield a reactive result in order to prove the validity of the test run (see Note 14). Samples are assessed as nonreactive if adequate amplification of GAPDH-mRNA sequences is detected on the VIC®-layer (see Note 15 and Fig. 1b) and the FAM fluorescence signal does not exceed the threshold value. No amplification signal

Table 1
Pipetting scheme for preparation of the HIV-1/GAPDH Real-time RT-PCR Mastermix

Reagents [µl]	Number of reactions													
	1	10	12	14	16	18	20	22	24	26	28	30	32	34	200
H_2O	10.95	110	131	153	175	197	219	241	263	285	307	329	350	372	2,190
10× PCR buffer[a]	5.0	50.0	60.0	70.0	80.0	90.0	100	110	120	130	140	150	160	170	1,000
$MgCl_2$ (25 mM)[a]	5.0	50.0	60.0	70.0	80.0	90.0	100	110	120	130	140	150	160	170	1,000
dNTPs (25 mM each)	0.4	4.0	4.8	5.6	6.4	7.2	8.0	8.8	9.6	10.4	11.2	12.0	12.8	13.6	80.0
HIV-1 FW Primer 1/2-Mix (10 µM)[b]	2.0	20.0	24.0	28.0	32.0	36.0	40.0	44.0	48.0	52.0	56.0	60.0	64.0	68.0	400
HIV-1 Rev Primer (10 µM)	2.0	20.0	24.0	28.0	32.0	36.0	40.0	44.0	48.0	52.0	56.0	60.0	64.0	68.0	400
HIV-1 Probe 1 (FAM, 5 µM)	1.0	10.0	12.0	14.0	16.0	18.0	20.0	22.0	24.0	26.0	28.0	30.0	32.0	34.0	200
HIV-1 Probe 2 (FAM, 5 µM)	1.0	10.0	12.0	14.0	16.0	18.0	20.0	22.0	24.0	26.0	28.0	30.0	32.0	34.0	200
GAPDH Primer-Mix (10 µM)[b]	0.5	5.0	6.0	7.0	8.0	9.0	10.0	11.0	12.0	13.0	14.0	15.0	16.0	17.0	100
GAPDH Probe (VIC, 5 µM)	0.25	2.5	3.0	3.5	4.0	4.5	5.0	5.5	6.0	6.5	7.0	7.5	8.0	8.5	50.0
ROX (50×)	1.0	10.0	12.0	14.0	16.0	18.0	20.0	22.0	24.0	26.0	28.0	30.0	32.0	34.0	200

(continued)

Table 1 (continued)

Reagents [µl]	Number of reactions														
	1	10	12	14	16	18	20	22	24	26	28	30	32	34	200
Sub-volume	29.1	291	349	407	466	524	592	640	698	757	815	873	931	989	5,820
RNase Out (40 U/µl)	0.25	2.5	3.0	3.5	4.0	4.5	5.0	5.5	6.0	6.5	7.0	7.5	8.0	8.5	
HotStarTaq® DNA Polymerase (5 U/µl)	0.25	2.5	3.0	3.5	4.0	4.5	5.0	5.5	6.0	6.5	7.0	7.5	8.0	8.5	
SuperScript™ III RT (20 U/µl)[c]	0.40	4.0	4.8	5.6	6.4	7.2	8.0	8.8	9.6	10.4	11.2	12.0	12.8	13.6	
Final volume	30.0	300	360	420	480	540	600	660	720	780	840	900	960	1,020	

[a]PCR-buffer and $MgCl_2$-solution are delivered with the HotStarTaq® DNA polymerase. The 10× PCR-buffer contains 15 mM $MgCl_2$
[b]Prepare equimolar mixtures (10 µM) of HIV-1 FW-Primers and GAPDH FW and Rev-Primers
[c]Dilute SuperScript™ III RT (200 U/µl) with the described storage buffer to reach a final concentration of 20 U/µl (see Note 2)

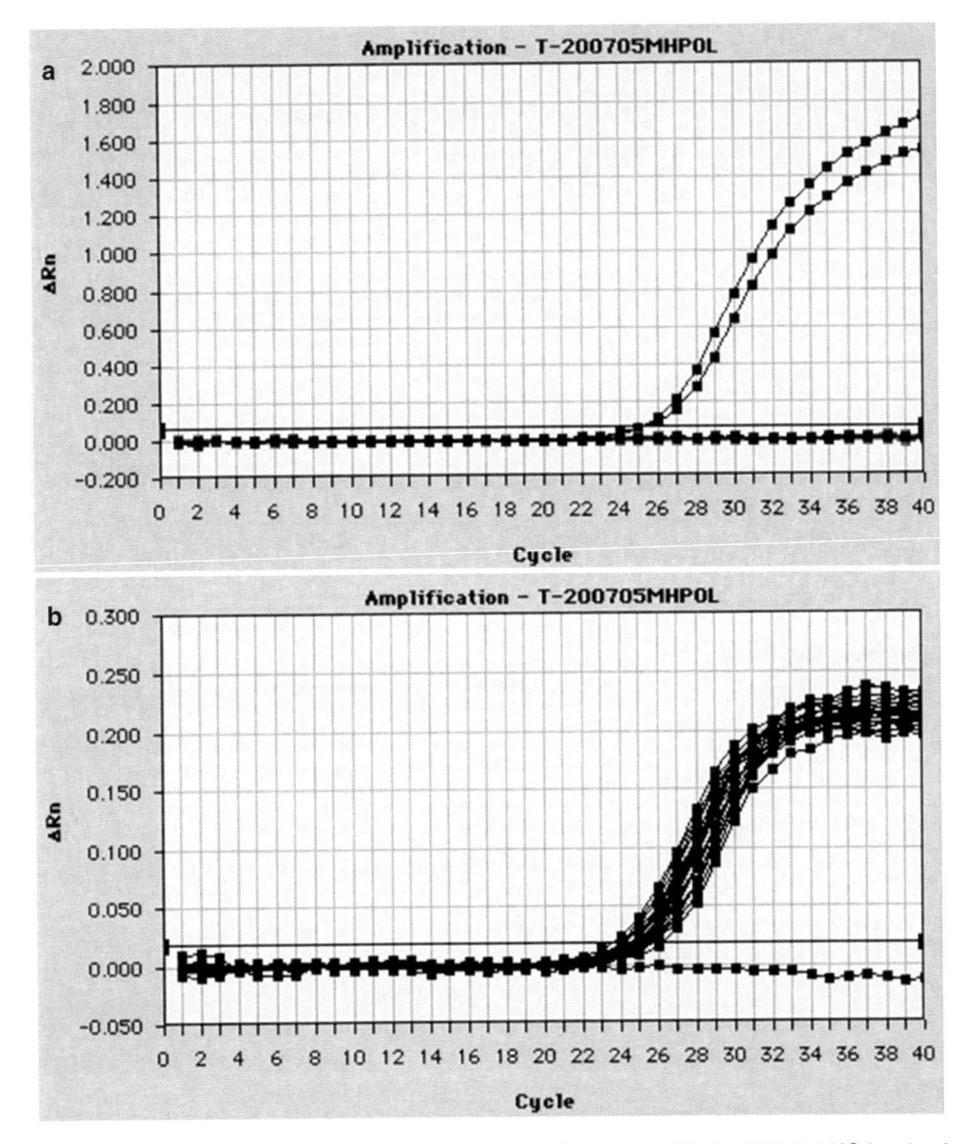

Fig. 1. Representative amplification plots for HIV-1 RNA (*a*, FAM-layer) and GAPDH mRNA (*b*, VIC®-layer) using the described Multiplex Real-time RT-PCR. During the 40 observations (one observation point per cycle) of relative change in fluorescence intensity, the typical sigmoid trajectories of the amplification reactions from an exponential phase to a late plateau phase are observed in reactive samples. The point at which amplification-derived fluorescence generation becomes significantly detectable and passes the calculated threshold (shown as the horizontal line) is called the threshold cycle (Ct). While higher concentrations of starting templates result in a lower Ct value, lower concentrations produce higher Ct values. No detectable amplification is represented by a Ct of 40. Results are shown for a test series of 7 single samples, 17 pooled samples, and 2 positive run-controls (1,000 IU/ml HIV-1-RNA) that were included in the run. One of the pool-samples showed no amplification on the VIC®-layer (GAPDH) and was therefore defined as inhibited. All other samples were found to be non-reactive for HIV-1 RNA

on the FAM-layer (Ct = 40, see Fig. 1a and legend) and inadequate signals on the VIC®-layer indicate either failure during viral RNA extraction or inhibition of the RT-PCR reaction. Both, inhibited and initial reactive samples must be retested (see Note 16).

13. Export data to Excel® and create an appropriate report of results. The report should at least include the original sample IDs, determined Ct-values on the FAM- and VIC® dye-layers, and the interpretation of the results (e.g. reactive/nonreactive/inhibited).

3.4. Assay Validation

3.4.1. Determination of Assay Specificity

1. Introduce at least 100 HIV-1-negative plasma (pool) samples to the established HIV-1 Real-time RT-PCR assay.
2. Assess data and calculate assay specificity by calculating the percentage of negative test results (see Note 17).

3.4.2. Determination of Assay Sensitivity on HIV-1 Group M Subtype B Genomes

1. Use the WHO international standard for HIV-1 RNA.
2. Serially dilute reconstituted standard material (100,000 IU/ml) with negative plasma to achieve final concentrations ranging from 4,000 down to 31.25 IU/ml (see Note 18).
3. Prepare three independent sets of eight aliquots (140 µl) for each dilution and store at less than –70°C until use.
4. Perform Viral RNA isolation and subsequent real-time detection by three different operators on different days with different batches of materials and reagents. Each operator should process eight aliquots of each input concentration.
5. Assess data and calculate the detection limit (95% cut-off value) of the assay by probit analysis (see Note 19 and Fig. 2).

3.4.3. Determination of Assay Performance on HIV-1 Non-B Subtypes

1. Use e.g. the HIV-1 RNA genotype panel for NAT assays. This panel consists of 11 quantified samples (cell culture supernatants diluted in negative plasma, 1.1 ml per sample) including eight representatives of HIV-1 group M, 1 of group O, 1 of group N, and 1 negative control (20).
2. Dilute the samples with negative plasma to reach a maximum concentration of three times the 95% cut-off value (sensitivity limit of the assay) as determined by probit analysis (see Note 18).
3. Introduce at least triplicates of each sample to the described procedures of viral RNA isolation and real-time RT-PCR amplification. Repeat at least two times on different days to allow evaluation of both, within- and between-runs variation patterns.

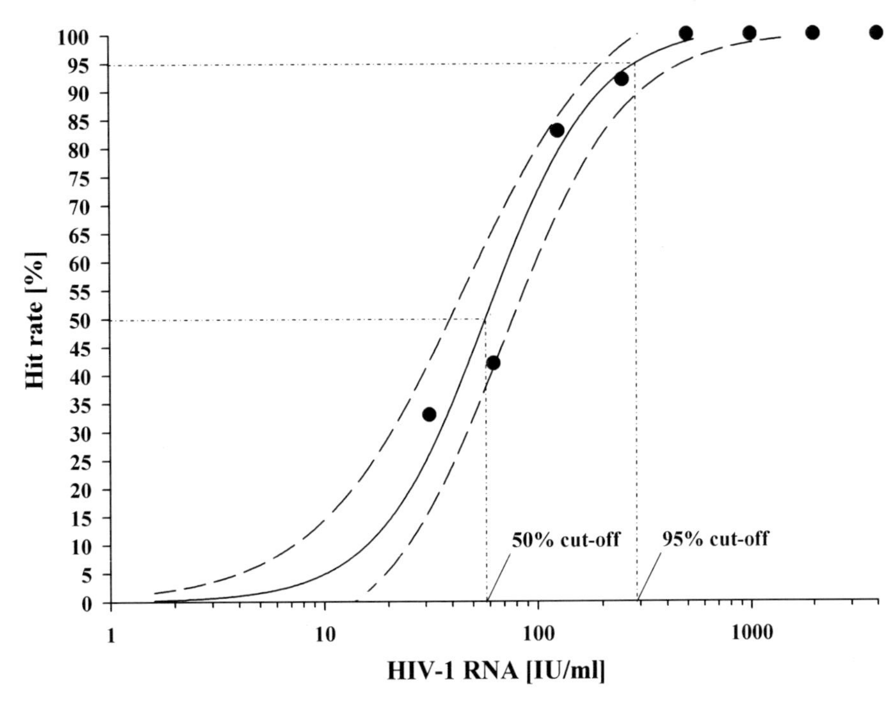

Fig. 2. Probit analysis of limiting dilution series of the WHO international standard for HIV-1 RNA. The figure shows the percentage of reactive results (*hit rate*) found at each input concentration (●) and the calculated probit regression line (*solid line*) with 95% confidence limits (*dashed lines*). The determined 95% cut-off value represents the detection limit of the assay (see Notes 18 and 19)

4. Assess obtained FAM amplification signals to evaluate assay performance with respect to the detection of different HIV-1 subtypes at low doses (see Note 20).

3.4.4. Determination of the Robustness of the Assay

1. Determine within- and between-runs variation patterns of viral RNA isolation and amplification using low-positive samples (maximum of three times the 95% cut-off value) and different batches of reagents. However, data obtained from the validation experiments mentioned above may be used for calculation of assay variation.
2. Validate the pooling process using a positive control sample according to the pre-defined sensitivity limit of 10,000 IU/ml with respect to single samples. Introduce the control together with a series of confirmed HIV-1 RNA negative samples to the pooling procedure and generate one positive pool of maximum size (e.g. the positive control together with nine negative samples [see Note 7]) along with several negative ones. Introduce all pools to viral RNA isolation/amplification and confirm expected results. At least perform triplicates of the

positive pool during viral RNA isolation and amplification. Assess HIV-1 amplification signals stemming from the positive pool in comparison to that of parallel processed run-controls (e.g. 1,000 IU/ml [see Note 8], $n \geq 3$). Repeat at least two times on different days to allow evaluation of both, within- and between-run variation.

3. Repeat the above described procedure using a highly concentrated HIV-1-RNA positive control sample to test for cross-contamination risk. However, both experiments may be performed in parallel.

4. Notes

1. SuperScript™ III RT is a mutated Moloney Murine Leukemia Virus (MMLV) enzyme with reduced RNAse H activity and increased thermal stability. The reduced RNAse H activity allows the use of the described 5′ nuclease probes in antisense orientation in a one-step RT-PCR setup (21). Even at low concentrations, the enzyme works well in various buffer systems and is therefore convenient for the development of one-step RT-PCR applications. The half-life time of the SuperScript™ III RT enzyme at elevated temperatures is 24 min at 55°C and 2.4 min at 60°C as specified by the manufacturer. Nevertheless, we observed the most sensitive detection of HIV-1-RNA when the RT-step was performed at 60°C for 20 min. However, optimal RT-temperatures might vary between cyclers (see Note 13).
2. Owing to the initially high concentration of the provided SuperScript™ III RT enzyme (200 U/µl), the solution should be prediluted to a working concentration of 20 U/µl using the described storage buffer. Store the prepared working solutions at –20°C. We observed no loss of RT-activity in diluted material over a period of several weeks.
3. ROX(6-carboxy-X-rhodamine)iscommonlyusedwithABIreal-time cyclers as a passive reference dye for normalization of reporter signals. The use of ROX improves the results by compensating for small fluorescent fluctuations that may occur in the reaction tubes. We observed unspecific rises of plotted reporter signals with real-time cyclers that do not allow normalization, leading to doubtful false positive results.
4. Primers and probes are generally delivered lyophilized. Stock solutions should be prepared by dissolving in nuclease-free water to achieve final concentrations of 100 µM and 50 µM for primer and probes, respectively. Working solutions (see Table 1) are prepared by dilution with nuclease-free water. Store all solutions at –20°C until use.

5. Primers and probes for detection of HIV-1 sequences were designed using Primer Express Software® Version 1.5 (PE Applied Biosystems, Foster City, USA) and chosen to ensure similar assay performance on both HIV-1 group M and group O sequences. The primers amplify a 160 bp fragment within the *pol* integrase (IN) gene of the HIV-1 genome. Real-time detection is performed by the use of two fluorogenic nuclease probes. The number of potential mismatches of primer and probe sequences for HIV-1 group M, O, and N templates were assessed on the basis of sequence information available at the Los Alamos HIV sequence database (17). For evaluation of HIV-1 primer and probe sequences, point your browser to http://www.hiv.lanl.gov/content/sequence/HIV/HIVTools.html and use the "Primalign" tool that allows the alignment of input nucleotide sequences to all corresponding HIV-1-RNA sequences available in the database.
6. The primer and probe sequences used for the internal control reaction are specific for GAPDH mRNA and do not detect genomic DNA sequences (22).
7. A maximum pool size of 10 single samples is used in our lab. Thus, according to the required sensitivity limit of 10,000 IU/ml of HIV-1-RNA and the determined detection limit of the described assay (281 IU/ml, see Note 18), larger pool sizes together with an appropriate run-control (see Note 8) might be used.
8. The concentration of the positive run-control must reflect the maximum pool size used. At a maximum pool size of 10 single samples, the applied positive run-control must be set to a maximum concentration of 1,000 IU/ml. Use the WHO international standard for HIV-1-RNA (or correspondingly calibrated material (11, 16)) and negative plasma for preparation of the run-controls and store appropriate aliquots (140 μl) at less than –70°C until use.
9. Creating an extraction protocol where samples and controls are assigned to ascending numbers allows more convenient labeling of columns and tubes used during extraction.
10. Isolation of viral RNA using the QIAamp® Viral RNA Mini Kit may be automated using e.g. the QIAcube sample preparation system. Alternatively, the vacuum procedure described in the handbook of the QIAamp® Viral RNA Mini Kit may be used for faster manual performance. However, the sensitivity limit of the described assay (see Note 18) was determined using the manual spin procedure.
11. A nucleic acid elution volume of 50 μl was found to be optimal for use with the QIAamp® Viral RNA Mini Kit. Besides testing for HIV-1-RNA, the eluates can also be used for parallel testing for HCV-RNA (16).

12. The real-time RT-PCR reaction mixture prepared according to Table 1 finally contains 1× PCR buffer, 4.0 mM $MgCl_2$, 200 μM of each dNTP, 200 nM of each HIV-1 forward primer, 400 nM HIV-1 reverse primer, 100 nM of each HIV-1 probe, 50 nM of each GAPDH forward and reverse primer, 25 nM of GAPDH probe, 1× ROX reference dye, 10 U of RNaseOut™, 8 U of SuperScript™ III reverse transcriptase, 1.25 U of HotStarTaq® DNA Polymerase, and 20 μl of the sample preparation in a total volume of 50 μl.
13. RT-PCR parameters were optimized for use with the ABI SDS 7700 instrument. However, thermal cycling conditions might need to be adapted if a different real-time cycler is used. Besides optimization of generated fluorescence signals, also ensure that only 2 product bands representing HIV-1- (160 bp) and GAPDH (226 bp)-derived PCR products are visible in an agarose-gel after amplification.
14. An HIV-1-RNA-reactive result must be observed for the positive run-control(s) to prove the validity of the assay. Thus, inhibition of a single introduced run-control leads to invalidity of the performed test run. This can be overcome by generally testing two run-controls in each run.
15. Detection of endogenous GAPDH mRNA that circulates in human blood and is copurified during nucleic acid isolation is used as an internal control reaction. To avoid competition with the amplification of HIV-1 sequences, limiting GAPDH primer concentrations are used. Experiments performed for validation of assay specificity (see Note 17) revealed similar levels of GAPDH transcripts in single and pooled plasma samples from healthy blood donors, yielding an overall mean Ct-value of 24.8 ± 1.0 (range: 22.1–27.2). However, the implementation of an artificial RNA target sequence that is added to the samples during the extraction step might be useful in order to completely eliminate inter-individual variations within the internal control reactions (16).
16. Retesting includes nucleic acids isolation and subsequent amplification. Retest initial reactive or inhibited samples for confirmation. Members of initial reactive pool-samples must be tested individually.
17. For determination of assay specificity, EDTA-anticoagulated plasma samples from 1,206 blood donors were available in our lab. All samples were previously demonstrated to be negative for anti-HIV-1/2 antibodies and HIV-1 RNA (17). 1095 of these samples were automatically pooled using a Genesis RSP 150/8 (Tecan, Crailsheim, Germany) as previously described (16). Altogether, 232 HIV-1-RNA-negative plasma samples (111 individual samples and 121 pool-samples)

were available. Using the described HIV-1-assay, all plasma samples tested negative for HIV-1 RNA, corresponding to an assay specificity of 100%.

18. The concentration range of HIV-1-RNA used for probit analysis might need to be adapted if assay conditions are modified, e.g. if a virus concentration step is performed. Furthermore, a half-logarithmic dilution series might be used to cover a larger concentration range. Under the described assay conditions, a detection limit (95% cut-off) of 281 IU/ml, showing a 95% confidence interval (CI) of 190–567 IU/ml was determined. The 50% cut-off was found to be 57 IU/ml (CI = 40–75 IU/ml) (see Fig. 2).

19. Probit analysis is used to calculate the probability of obtaining a positive result from a given input concentration. Conversely, the concentration corresponding to a predicted success rate can be estimated from the ascertained data set. Generally, the detection limit of an assay is defined as the concentration that results in a 95% probability of a reactive result (95% cut-off). To run probit analysis in SPSS™, enter the tested concentrations along with the corresponding numbers of both, replicates tested and observed reactive results into three separate columns of the Data Editor and choose " Analyze" > "Regression" > "Probit" from the menu bar. Set the "reactives column" as the "Response Frequency", the total number tested as the "Total Observed", and the concentrations as the "Covariates". Select "log base 10" to transform your concentrations and click "OK". SPSS™ then returns a complete list of percentage probabilities and corresponding threshold concentrations including 95% confidence intervals.

20. Alternatively, to the qualitative detection of equally diluted material for the assessment of assay performance on different HIV-1 subtype samples, quantitative results may be determined and compared to given values or values measured by a quantitative reference assay. However, this approach requires the availability of appropriate calibrators and validation of the quantitative method. We used the quantitative approach to validate the described assay with respect to the detection of various HIV-1 variants (34× group M, 8× group O and 1× group N) that were included in two commonly available and two in-house HIV-1 subtype panels (17). Calibrator dilutions were prepared from a qualified HIV-1 RNA positive control (Accurun® 315 HIV-1 RNA positive control series 500 [BBI, West Bridgewater, MA, USA]) that had been quantified by parallel testing with the WHO international standard for HIV-1 RNA. If applicable, HIV-1 subtype samples were measured using both, the described real-time RT-PCR assay and the commercially available

automated Abbott RealTime™ HIV-1 assay (9). All tested samples showed reactive results in both assays and quantitative test results significantly correlated ($R=0.86$; $p<0.001$). Based on overall test results, the mean difference in $\log_{10}$ transformed values was 0.04 (95% confidence interval [CI], 1.12–0.91). However, the agreement was lower within the group O samples when compared to group M (17).

Acknowledgments

The author would like to thank Miriam Herzig and Regina Maurer for expert technical assistance, as well as Jürgen Bach, Bernd Pötzsch, and Jutta Rox for critical reading of the manuscript.

References

1. Offergeld R, Faensen D, Ritter S, Hamouda O (2005) Human immunodeficiency virus, hepatitis C and hepatitis B infections among blood donors in Germany 2000–2002: risk of virus transmission and the impact of nucleic acid amplification testing. Euro Surveill 10:8–11
2. Paul-Ehrlich-Institut, Bundesamt für Sera und Impfstoffe (2003) Bekanntmachung über die Zulassung und Registrierung von Arzneimitteln: Verminderung des Risikos von HIV-Infektionen durch zelluläre Blutprodukte und gefrorenes Frischplasma. Anordnung der Testung auf HIV-1-RNA mit Nukleinsäure-Amplifikationstechniken. Bundesanzeiger 103:12269
3. Hourfar MK, Jork C, Schottstedt V, Weber-Schehl M, Brixner V, Busch MP, Geusendam G, Gubbe K, Mahnhardt C, Mayr-Wohlfart U, Pichl L, Roth WK, Schmidt M, Seifried E, Wright DJ, German Red Cross NAT Study Group (2008) Experience of German Red Cross blood donor services with nucleic acid testing: results of screening more than 30 million blood donations for human immunodeficiency virus-1, hepatitis C virus, and hepatitis B virus. Transfusion 48:1558–1566
4. Butler IF, Pandrea I, Marx PA, Apetrei C (2007) HIV genetic diversity: biological and public health consequences. Curr HIV Res 5:23–45
5. Hampl H, Sawitzky D, Stoffler-Meilicke M, Groh A, Schmitt M, Eberle J, Gürtler L (1995) First case of HIV-1 subtype O infection in Germany. Infection 23:369–370
6. Rox JM, Eis-Hübinger AM, Müller J, Vogel M, Kaiser R, Hanfland P, Däumer M (2004) First human immunodeficiency virus-1 group O infection in a European blood donor. Vox Sang 87:44–45
7. Barin F, Cazein F, Lot F, Pillonel J, Brunet S, Thierry D, Damond F, Brun-Vézinet F, Desenclos JC, Semaille C (2007) Prevalence of HIV-2 and HIV-1 group O infections among new HIV diagnoses in France: 2003–2006. AIDS 21:2351–2353
8. Sábato MF, Shiffman ML, Langley MR, Wilkinson DS, Ferreira-Gonzalez A (2007) Comparison of performance characteristics of three real-time reverse transcription-PCR test systems for detection and quantification of hepatitis C virus. J Clin Microbiol 45:2529–2536
9. Schutten M (2008) Comparison of the Abbott Realtime HIV-1 and HCV viral load assays with commercial competitor assays. Expert Rev Mol Diagn 8:369–377
10. Saldanha J (2001) Validation and standardisation of nucleic acid amplification technology (NAT) assays for the detection of viral contamination of blood and blood products. J Clin Virol 20:7–13
11. Nübling CM, Chudy M, Löwer J (1999) Validation of HCV-NAT assays and experience with NAT application for blood screening in Germany. Biologicals 27:291–294
12. Hourfar MK, Schmidt M, Seifried E, Roth WK (2005) Evaluation of an automated high-volume extraction method for viral nucleic acids in comparison to a manual procedure with preceding enrichment. Vox Sang 89:71–76

13. Mackay IM, Arden KE, Nitsche A (2002) Real-time PCR in virology. Nucleic Acids Res 30:1292–1305
14. Espy MJ, Uhl JR, Sloan LM, Buckwalter SP, Jones MF, Vetter EA, Yao JD, Wengenack NL, Rosenblatt JE, Cockerill FR 3rd, Smith TF (2006) Real-time PCR in clinical microbiology: applications for routine laboratory testing. Clin Microbiol Rev 19:165–256
15. Cone RW, Hobson AC, Huang ML (1992) Coamplified positive control detects inhibition of polymerase chain reactions. J Clin Microbiol 30:3185–3189
16. Mueller J, Gessner M, Remberg A, Hoch J, Zerlauth G, Hanfland P (2005) Development, validation and evaluation of a homogenous one-step reverse transcriptase-initiated PCR assay with competitive internal control for the detection of hepatitis C virus RNA. Clin Chem Lab Med 43:827–833
17. Müller J, Eis-Hübinger AM, Däumer M, Kaiser R, Rox JM, Gürtler L, Hanfland P, Pötzsch B (2007) A novel internally controlled real-time reverse transcription-PCR assay for HIV-1 RNA targeting the pol integrase genomic region. J Virol Methods 142:127–135
18. Heid CA, Stevens J, Livak KJ, Williams PM (1996) Real time quantitative PCR. Genome Res 6:986–994
19. Chassaigne M, Vassort-Bruneau C, Allouch P, Audurier A, Boulard G, Grosdhomme F, Noel L, Gulian C, Janus G, Perez P (2001) Reduction of bacterial load by predonation sampling. Transfus Apher Sci 24:253
20. Holmes H, Davis C, Heath A (2008) Development of the 1st International Reference Panel for HIV-1 RNA genotypes for use in nucleic acid-based techniques. J Virol Methods 154: 86–91
21. Boiziau C, Thuong NT, Toulmé JJ (1992) Mechanisms of the inhibition of reverse transcription by antisense oligonucleotides. Proc Natl Acad Sci USA 89:768–772
22. Mueller J, Rox JM, Madlener K, Poetzsch B (2004) Quantitative tissue factor gene expression analysis in whole blood: development and evaluation of a real-time PCR platform. Clin Chem 50:245–247

INDEX

Nicola King (ed.), *RT-PCR Protocols: Second Edition*, Methods in Molecular Biology, vol. 630, DOI 10.1007/978-1-60761-629-0, © Springer Science+Business Media, LLC 2010